Transplanting the Metaphysical Organ

Stefanos Geroulanos and Todd Meyers, *series editors*

Transplanting the Metaphysical Organ

German Romanticism between Leibniz and Marx

Leif Weatherby

FORDHAM UNIVERSITY PRESS
NEW YORK 2016

THIS BOOK IS MADE POSSIBLE BY A COLLABORATIVE GRANT FROM THE ANDREW W. MELLON FOUNDATION.

Copyright © 2016 Fordham University Press

All rights reserved. No part of this publication may be reproduced, stored in a retrieval system, or transmitted in any form or by any means—electronic, mechanical, photocopy, recording, or any other—except for brief quotations in printed reviews, without the prior permission of the publisher.

Fordham University Press has no responsibility for the persistence or accuracy of URLs for external or third-party Internet websites referred to in this publication and does not guarantee that any content on such websites is, or will remain, accurate or appropriate.

Portions of the article "The Romantic Circumstance: Novalis between Kittler and Luhmann," copyright 2014 by the Board of Regents of the University of Wisconsin, which appeared in *SubStance*, vol. 43, no. 3, 46–66, are reprinted with permission. Portions of the article, "Das Innere der Natur und ihr Organ: Von Albrecht von Haller zu Goethe," *Goethe-Yearbook* 21 (2014): 191–217, are translated and reprinted with permission.

Fordham University Press also publishes its books in a variety of electronic formats. Some content that appears in print may not be available in electronic books.

Visit us online at www.fordhampress.com.

Library of Congress Cataloging-in-Publication Data

Weatherby, Leif.
  Transplanting the metaphysical organ : German Romanticism between Leibniz and Marx / Leif Weatherby. — First edition.
    pages cm. — (Forms of living)
  Includes bibliographical references and index.
  ISBN 978-0-8232-6940-2 (cloth : alk. paper)
  ISBN 978-0-8232-6941-9 (pbk. : alk. paper)
  1. Philosophy, German. 2. Romanticism—Germany. 3. Metaphysics. I. Title.
  B2521.W43 2016
  193—dc23

                                         2015030356

Printed in the United States of America

18 17 16   5 4 3 2 1

First edition

CONTENTS

Introduction: Romantic Organology:
Terminology and Metaphysics ... 1

Part I. Toward Organology ... 47

1. Metaphysical Organs and the Emergence of Life:
   From Leibniz to Blumenbach ... 51
2. The Epigenesis of Reason: Force and Organ
   in Kant and Herder ... 72
3. The Organ of the Soul: Vitalist Metaphysics
   and the Literalization of the Organ ... 108

Part II. Romantic Organology: Toward a Technological Metaphysics of Judgment ... 123

4. The Tragic Task: Dialectical Organs and the Metaphysics
   of Judgment (Hölderlin) ... 131
5. Electric and Ideal Organs: Schelling and the Program
   of Organology ... 171
6. Universal Organs: Novalis's Romantic Organology ... 206
7. Between Myth and Science: *Naturphilosophie* and the Ends
   of Organology ... 251

Part III.   After Organology   277

8. Technologies of Nature: Goethe's Hegelian Transformations   279
9. Instead of an Epilogue: Communist Organs,
    or Technology and Organology   316

   Acknowledgments   353
   Notes   355
   Bibliography   425
   Index   449

INTRODUCTION

Romantic Organology: Terminology and Metaphysics

René Descartes's (1596–1650) *Treatise on Man* begins with the curious sentence, "These men will be composed, as we are, of body and soul."[1] Descartes is famous for leaving aesthetic speculation out of his refounding of the philosophical enterprise, but his intended treatise on the most intimate part of the world, the body, is written as fiction—as he puts it in his *Discourse on Method*, a "historical record" or "fable."[0] Descartes's hesitation to state his claims directly came from rumors of what had happened to Galileo, who had been condemned just as Descartes was composing the treatise in 1633—claims for the reality of the earth's rotation were punishable, and Descartes worried about his own doctrines in this light. Even fictional devices might not be spared. The resulting fable had an extreme realist pretense: to derive a world similar to ours from mechanical interactions alone, and then to prove that this other world is identical to ours.[3] The human body would be treated as though it did not possess a soul, as though God had directly fashioned a "statue or machine made of earth . . . with the explicit intention of making it as much as possible like us."[4] These divine machines, the subject of what Claude Bernard would call "a phantasy physiology, almost entirely invented,"[5] would be able to "imitate all those functions we have which can be imagined to proceed from matter and to depend solely on the disposition of our organs.[6] Because Descartes's metaphysics famously divides the world into thinking things (minds) and extended things (all bodies), physiology is necessarily thorny. The bodies under consideration in human physiology, after all, belong to minds, and this belonging seems obscure. Whatever might connect the two would have to *be* one of the two, and would serve little purpose in elucidating why this body is *mine*, that *hers*, and so on. Descartes's answer is as complex as it is baroque, involving the pineal gland (gland "H") and a series of nerve networks that extend out from the brain, allowing for material imprints ("traces") to become ever finer, until in their pineal imprint they are somehow read off by the thinking thing. The mind is like a pilot in his ship, writes Descartes, borrowing a metaphor from Aristotle that would retain its force in explaining the mind-body problem for some centuries.[7] The pilot must know and be able to decipher the atmosphere around the ship, the water, the sky, and to intuit through the ship's movements the course of action to be taken. The physiological pilot is not exactly a human mind—it is a fictional proxy for that mind in bodies of fabled men. And the physiology exploring this apparatus is concerned with just that: arrangements of organs, and that which these arrangements make possible.

Functions and their physical basis would remain the topic of nonfictional physiology, of course. Implicitly or explicitly, organs would take over the function of mediators between body and soul—and in fact, as this book will show, would come in Romantic discourse to define being as such. For Descartes, however, they were merely physical arrangements—points of articulation, to be sure, of specific realms of phenomena (sound, smell, and so on), but inside the fiction, merely the gathering mechanisms of a world that would then be transmitted with the aid of "animal spirits" to gland H. In fact, only the mind could glean this information; organs gathered it automatically, but were literally dumb. Descartes wrote:

> At this point I had dwelt on this issue to show that if there were such machines having the organs and outward shape of a monkey or any other irrational animal, we would have no means of knowing that they were not of exactly the

same nature as these animals, whereas, if any such machines resembled us in body and imitated our actions insofar as this was practically possible, we should still have two very certain means of recognizing that they were not, for all that, real human beings.[8]

Animals would not pass the Turing test—we would not know they were machines—because they have no language, a capacity not located in an organ. Organs would also not supply the sheer range of rational human action:

> Although such machines might do many things as well or even better than any of us, they would inevitably fail to do some others, by which we would discover that they did not act consciously, but only because their organs were disposed in a certain way. For, whereas reason is a universal instrument which can operate in all sorts of situations, their organs have to have a particular disposition for each particular action, from which it follows that it is practically impossible for there to be enough different organs in a machine to cause it to act in all of life's occurrences in the same way that our reason causes us to act.[9]

A number of oddities might strike today's reader about this passage, not least the fact that machines have "organs." We are accustomed to thinking of organs as belonging not to machines but to animals, devices as belonging to machines and not to bodies. Descartes, of course, rejected just this point—animals could have mechanisms so long as they were not possessed of the "universal instrument" reason, and even then, particular organs were merely automatic conveyors of information to reason's seat.

This sense pervades Descartes's writings: "I desire, I say, that you should consider that these functions follow in this machine simply from the disposition of the organs [*de la seule disposition de ses organes*] as wholly naturally as the movements of a clock or other automaton follow from the disposition of its counterweights and wheels."[10] This emphasis, as we shall see, is aimed directly at Galenic physiology, which includes purpose in its schema. For Descartes, "purpose" can refer to nothing except God and reason, and both seem far removed enough from the particular disposition of mundane things to ward off any false inferences from the apparently useful functions at hand. This is not only Descartes's explicit position, however. It goes to the point where he often inserts an indifferent "or" between "organs" and "devices": "Now before I pass to the description of the rational soul, I want you once again to reflect a little on all that I have just said about this machine; and to consider, first, that I have postulated in it only such organs and working parts [*organes et ressorts*] as can readily persuade you that they are the same as those in us, as well as in various animals lacking reason."[11] This conflation is not merely semantic polemics. The term "organ" simply did not carry the fullness of its modern meaning in the seventeenth century. As we shall see, it meant something more like "tool" or "device." This suited Descartes's system well: organs were just complex mechanisms, not comparable to that "universal instrument" that only humans possessed. If the universe functioned like a clock—a metaphor of great currency in the seventeenth century—the body was also a device, one with pulleys and springs. Its parts were "organs," but this merely meant tools functioning for a machinic purpose. And in that machine, reason itself was a "universal instrument"[12]—Descartes thinned and confused his metaphor, underscoring its centrality. But in effect, he did not change the sense of this term. The soul used its machine just as it had once, in Galen's system, used its instruments.

Galen's doctrines, both physiological and therapeutic, dominated European science and medicine for more than a millennium, and then crumbled only slowly from the sixteenth to the nineteenth century.[13] Descartes opposed the Aristotelian doctrine of the causes that underlay Galen's work. Where four "humors" (black bile, yellow bile, phlegm, and blood) seeped in and out of the body, justifying a therapeutics primarily of letting and balancing fluids with enemas, bloodlettings, and cupping, Descartes conceived of "animal spirits" carrying not only fluids but images through the nerves onto the sense-organs.[14] For Galen, causes fell out into material, formal, efficient, and final causes, following Aristotle. What Descartes reduced to efficient cause, the direct result of an action from another, had been conceived in richer terms, with the substance, form, and purpose of the individuals involved bearing some conceptual responsibility for the way events played out. Galen needed more causes, however, because he was interested in the etiology of disease, in which the circumstance of the body, its balance of fluids—influenced by heat, cold, sex, fatigue[15]—clearly establishes another kind of causality, an "antecedent" causality, the condition of the event of becoming ill. This combines with Galen's strong sense of teleology in the body—the final cause not only guides, as it does for Aristotle as well, but can be found operative in the very relationship of parts to whole, in what would later be called the "organs" of the body.[16] But Galen adds another cause: the "instrumental" cause, the form and shape of the tool as it contributes to the outcome. This cause is "that with which" the construction, in his example, of a bed occurs—the tool (*organum*) must be tailored to the outcome, and it contributes genuinely to the material outcome, the formed object.[17] Galen extends this argument to the parts of the body: the pulmonary arteries are thin *because* this helps them to execute their function.[18] Thus function subordinates material; the locale of significance is fixed. And even though Descartes was disputing this very "significance" in the final cause—for the world of extended things has no purpose other than God's inscrutable will, no internally significant order—the problem of the site of function remains, and this is clear in his infamous solution to the issue of bodily cognition, in the problem of sense-organs. Without an instrumental cause, Descartes cannot explain how the "spirits" convey an image that minded things can "read off" the gland H. His physiology lacks mediation, and physiology has the privilege of having organic mediation as the object of its study. This is why he calls functioning parts of the body "organs/devices" or "springs"—although they function, it is not clear what *for*.

"Organ," for Galenic theoreticians as for Descartes, never really meant what it does now, never designated the independent unit of material function in a body. "Organs of sense" was somewhat more common because, in their case, instrumentality is defined by conceptuality, thus we "make use of" what we perceive, and the metaphor can become literal, the senses tools of that perception.[19] The Galenic inclusion of the fifth or sixth cause, the "instrumental" cause—the shape of the *organon*—might have played an important role in the rise of physiology, since ascribing functions means investigating the material constitution of the parts which bear those functions. That may seem like a baroque way to say "observe organs," but historically, this is not so. "Organ," both in the designation of what counts as a function-bearer, and in the realm in which such function counts as needing an instrument, was deeply uncertain until about the 1790s. It appears to me that this is so in a tempered way in the British and French traditions, both of which adopted the term in the life sciences long before German did. The deeper problem persisted in the peculiar Enlightenment that the German-speaking lands underwent, one which was more religiously inflected, marked by the development from theism to pantheism (rather than to deism). In that context, the question of the instrumentality of the body could be raised in a way that left open the question of both the body's *telos* and whatever relation such might bear to God. Organs might well be the instruments of some larger order, but that order was itself a problem, not to be rejected (as by the mechanists in France) or to be questioned insofar as our knowledge of it is concerned (as the Scottish empiricists had it), but with the sense that an answer might indeed emerge from the interstices between scientific investigation and philosophy. And initially, that is what happened: a new notion of life emerged. That notion included, however, the underdefined semantic *novum* "organ," and the Romantics exploited the physiological and metaphysical, the epistemological and methodological senses that suddenly accrued to the word. To be sure, it seemed forced to claim that "organ" might mean one thing in all these fields. But at that time, that is precisely what the word, as a metaphor, meant: simply use, simply the bearer of a function, without any known designation of purpose. Instrumentality without subordination to a totality: this is what the Romantics discovered at the end of a long conceptual and short semantic history.

By around 1900, it had become common—as it remains—to use "organ" as a metaphor for specific functions of that universal reason once kept so separate from its physical counterpart. So the neo-Kantian Hans Vaihinger, as influential in the first part of the twentieth century as he is forgotten today, wrote in his *The Philosophy of "As If"* (1911) that

> The psyche is *an organic formative force*, which independently changes what has been appropriated, and can adapt foreign elements to its own requirements as easily as it adapts itself to what is new. The mind is not merely appropriative, it is also assimilative and constructive. In the course of its growth, it *creates its organs [Organe]* of its own accord in virtue of its adaptable constitution, but only when stimulated from without, and *adapts them to external circumstances.* Such organs [*solche Organe*], created by the psyche for itself in response to external stimuli, are, for example, forms of perception and thought, and certain concepts and other logical constructs.[20]

Vaihinger's metaphor is the opposite of Descartes's—where what might be taken to be living parts of bodies had once been demoted to mechanisms, here the "universal instrument of reason" is analogized to the animal body. Of course, Sir Francis Bacon had inaugurated observational methods with an analogy that looks, at first blush, like this one: his *Novum organum scientiarum* (1620). But *organum* means "tool"—the new tool of the sciences— just as *organe* did for Descartes. The "tools" of the mind are older than Bacon, and have outlived Vaihinger's mixture of mechanics and organics, his animal mind. The analogue is not to just any animal, but one struggling for survival, assimilating and adapting, in short, a Darwinian animal surviving only by the continual creation of new organs. Vaihinger's treatise is about fictions of the type we cannot do without, about the role of fiction in legal, logical, and practically every other context. His metaphors are the result of overcoming the difference Descartes had marked out between reason and bodies, and in particular between reason and its source of information about any body at all: organs. Just like Descartes, however, Vaihinger slides easily between "organ" and "instrument" and even "tool":

> Thought is bent on continually perfecting itself and thus becomes a more and more serviceable tool [*Werkzeug*]. For this purpose it expands its province by inventing instruments [*durch Erfindung von Instrumenten*], like other natural activities. The arm and the hand do the same, and most ordinary instruments [*Instrumente*] are to be regarded as elongations and extensions of these. The natural function of thought, which we spoke of above as a tool [*Werkzeug*], also expands its instrumentality by the invention of tools, means of thought, instruments of thought, one of which is to form the subject of our enquiry [*erweitert seine Instrumentation durch Erfindung von Werkzeugen, von Denkmitteln, Denkinstrumenten, deren Eines den Gegenstand dieser Untersuchung bilden soll*].[21]

Organs of motion—the hand, the arm—and their extensions in technology as tools are like the internal extension of thought to new domains, perhaps new fictions, for use and ultimately survival. The concatenation of instrumental metaphors at the end of this passage points up the weight of the two metaphors: the mind is an animal body; the mind is a tool. Vaihinger does not seem to care which of these analogies dominates—it is enough that the mind does not hover above its uses, but is extended into them, identical to them, forms the source of new methods, new fictions.

What, then, is an organ? The answer is not simple, and never has been. Even as internal organs form the basis of potential free markets and are given for transplant only after death by relatives, the successful face transplants that have occurred since 2010 have resulted in the addition to the legal definition of "vascularized composite allografts," such as faces and hands.[22] The limit below the organ's function—tissue, vital matter, the cell—has shifted even as the definition has over centuries. By the time microscopes began to help in identification of organs in the seventeenth century, some basics of human physiology were worked out—but the word "organ" still hovered around its Greek root, *organon*, or "tool." The term was never neutral, but attained its literal meaning only in the eighteenth century—in the German language, only late in that century. In fact, Descartes and Vaihinger are probably more typical than those who oppose organs to tools, the living to the merely physical. "Organ" is just that location which bears a function, that place that performs work, the unity of use and material, and also the only source of cognition. The organ is that which unites and divides, in the body as in the mind as in the universe; its problematic is that of the meaning and location of function as such. It became, in the works of Friedrich Hölderlin, Friedrich Schelling, and Friedrich von Hardenberg (Novalis), the tool for a new metaphysics. I call their project Romantic organology, and it is the subject of this book.

My use of the term "organology" differs from the occasional sense of "study of organics," usually used to mean philosophical uses of organic models for other purposes.[23] This sense is perhaps most firmly associated with the work of Georges Canguilhem, who uses the term in precisely this sense.[24] My use is closer to that of Bernard Stiegler, who proposes that we study function across the boundaries between bodies, tools, and social entities, and that we approach the politics of the technicized world from this perspective.[25] Social, technical, and bodily organs are all just that: organs, and this means for Stiegler that they are manipulable.

He diagnoses contemporary capitalism as an "entropic" tendency in which these three systems lose kilter and reduce humans to work in the service of a finance economy that renders us stupid as producers and consumers. Stiegler calls for resistance to this tendency, for the reversal of these tendencies. Somewhat less clear is how that reversal might occur.

Stiegler's term comes not from Canguilhem, but from the work of Gilbert Simondon, the philosopher of technology, who refers to the possibility of a "general organology":

> Infra-individual technical objects can be called technical elements; they
> distinguish themselves from true individuals in the sense that they do not
> possess an associated milieu; they can integrate themselves in an individual; a
> hot-cathode lamp is a technical element rather than a complete technical
> individual; one can compare it to an organ in a living body [*un organe dans un
> corps vivant*]. It would be possible in this sense to define a general organology
> [*une organologie générale*], studying technical objects at the level of the element,
> and which would constitute part of technology, along with mechanology,
> which would study complete technical individuals.[26]

Organology is the study of elements with variable functions. Those elements might make up part of a larger technological object or a body—or, by Stiegler's lights, a social entity. Organology is the study of instruments as technical elements, the cells of the world of making. They might be material artifacts or partial processes, physical or organizational in nature. Organology is a study of infra-objects, of usable parts that might be repurposed depending on the body or process at hand. For Simondon, it is a new analytical perspective; for Stiegler, it suggests norms for political action at every technical level. The units "object," "body," and even "process" are discarded in favor of the study of multifunctional instruments. We do not live in the Classical world of subject, object, and society, but in the weird world of symbiotic and antagonistic organs.

Romantic organology is the origin of this way of looking at the world. Although Simondon dismantled a Classical world to find its technical infra-objects underneath, this view goes back to the Romantics.[27] I think it is one of the large-scale tendencies of modern speculative thought, a way of coming to terms with the complexity of development of modern science and society. If the Enlightenment developed a philosophical empiricism that placed emphasis on "matters of fact," alongside the logical efforts to make our view of the world internally consistent in Rationalism, there emerged at the end of the eighteenth century that other dominant philosophical paradigm, transcendentalism, which takes the question of where the world is in our judgments, and what limitations our faculties might have, as the object of its study. Immanuel Kant (1724–1804) originated this view, asking how our empirical efforts and our rational consistency can coexist. Organology is a response to this question. It is not Kant's response—it does not fix a transcendental unity of consciousness above the play of historical phenomena—but a specifically Romantic integration of his question into the emergent problems of the collapse of monarchy and cosmology alike. It sees every form, process, object, as a combination of the empirical and the rational, construes it as its own capacity: every historical situation and scientific discovery becomes part of the project of retooling the cosmos. It maintains a skepticism about finality while rejecting any sense of ultimate limitation on human capacity. It sees every particular as a synthesis of rational process and brute object, and every such synthesis as a tool. It is this last point that pushes the Romantic sense of organology in a different direction from that defined by Simondon and Stiegler. Organs are the objects of analysis, to be sure, and they even provide a kind of source of practical norms, a direction, for the organologist. But they are also fragile syntheses, the location of both knowledge and potential action. They cannot prescribe that action any more than motor nerves can control the muscle. They are the location of whatever freedom might be possible in a world populated by emergent technical necessities of our own making. Because they can be repurposed even though their purpose is always in a state of expression, they are the physical bearers of both the developing real and the actual state of possibility. Organology has something akin to metaphysical *bricolage*, using historically prepared tools and developmentally formed beings to form new unities with new generalities. The Romantics I discuss below invented this way of looking at the world. Speculative and actual at once, Romantic organology was meant to be a powerful intellectual tool for the refashioning of the natural and social worlds.

Romantic Organology

Friedrich Schlegel's version of the Romantic demand for a "new mythology," presented as a task for the Jena circle by the discussant Ludoviko in his *Dialogue on Poesie*, calls for poetry to function as the instrument of an "ideal realism": "I too have been carrying the ideal of such a realism in myself for a long time, and if it has not yet come to be communicated, that is only because I am still searching for the organ for it [*weil ich das Organ noch dazu suche*]. Yet I know that I can find it only in poetry [*Poesie*], for realism will never be able to present itself [*nie wieder auftreten können*] in the form of a philosophy, not to speak of a system."[28] This philosophy remains, in this text, largely a task, but one with specifications. Both "ideal" and "real," the philosophical or rational mythology must present a cohering organism of sentences and verses, yet not appear in the form of a "system."[29] Thus Spinoza, whose philosophical style is perhaps as far away from Schlegel's notion of *Poesie* as any, and indeed was understood by contemporaries as purely systematic—of course, the subtitle of the *Ethics* does read: "demonstrated according to the manner of geometry"—is nevertheless the heart of the canon for any would-be poetic genius. Schlegel's proposal is, indeed, an impossible task, self-consciously contradictory in historical appeal and philosophical determination. Worse, its impossibility is meant nevertheless to become reality by means of a metaphor: an organ.

If Ludoviko claimed in 1800 that he was "still searching for an organ" for poetry, Schlegel himself had been more confident in 1796: "Poetry is a *universal* art: for its organ [*Organ*],[30] the *imagination*, is already incomparably more intimately related to freedom and more independent from external influence."[31] Schlegel identifies the imagination, which hovers between the determination of the senses and the determining force of the understanding, as the simultaneously lawful and local faculty of the poet. That the poet's task is mythological is also comprised in this organ: "The imagination [*die Fantasie*] is the human organ for the divinity."[32] Poetry requires a sense that is both concrete and general, both determinate and determining, and only the organ of such a sense fulfills both sides of this task. The imagination is between two other faculties, as both passages make clear; poetry thus captures the middling status of the human, the admixture of physical and spiritual central to Romantic anthropology.

This book does not devote a chapter to Schlegel, although his mythology is treated as part of chapter 7, where his focus on aesthetics and politics is highlighted. That subset of the more general project is not without its importance, but it would be a mistake to confuse the two, as I shall be arguing. Just as the organ binds and points up the difference between the organic and the mechanical (rather than merely being a "part" of the organic), so it unites and demonstrates why aesthetics and metaphysics are related, being proper to neither discipline alone.

On the Study of Greek Poetry (1796), from which the first quotation above is taken, is a response to Friedrich Schiller's sustained effort to write a new aesthetics in the middle years of the 1790s. Schiller's first (of two) failed dissertations ("The Philosophy of Physiology," 1779) at the Hohe Karlsschule in Stuttgart seeks to solve the mind-body problem through the introduction of a "*Nervengeist.*" The notion of a mediating fluid or gas in the nerves was a refrain of the anthropological movement, which insisted on the model of "physical influence" (*influxus physicus*) to unify the Cartesian human. Schiller, however, introduced his version of this claim with what this study shows to be an anomalously early systematic sense of the term "organ." There must be a "middling power" that can mediate between body and spirit, and it must be distinct from both. The collection of "lower forces" that make up our mechanical nature as a complex (the body) should be called, according to the student Schiller, the "structure" (*der Bau*), and "structure and middle-force in connection are what we dub organ. It will thus naturally come to light that the difference of the organs does not lie in the force but instead in the structure."[33] The usage is extraordinary, the work never published.[34] Structure and mediation together equal the "organ," which is functionally determined not by the force but by the structure itself. In other words, the *arrangement of organs determines their function*. Schiller anticipates here both Herder's cosmology, in which organs express laws *in concreto*, and the biologists of the 1790s who explored the relationship between physical structure and organic function.

And yet the organ of the middling force remained eclectic. Not only was the dissertation rejected, but Schiller himself appears to have given up the term. With the exception of some very late considerations of the tragic chorus (see chapter 4), Schiller does not make use of the term again, even as it experiences its heyday in his own city, Jena.

Perhaps, then, we could draw a dividing line between Romanticism and Classicism precisely in this term. (Goethe's use of the term started in the 1790s in conjunction with his biological work and his fascination with Idealism, but this did not coincide with his Classicism until well after Schiller's death [see chapter 8].) And if organology had an origin in aesthetics, it was precisely in a fundamental disagreement with the Classical Schiller.

Take this central passage from the penultimate "aesthetic letter":

> Appearance is only aesthetic insofar as it is *earnest* (takes leave of all demands on reality), and only so far as it is *independent* (foregoes any aid from reality). As soon

as it is false and simulates reality, and as soon as it is impure and needs reality for its effect, then it is *no more than a low tool* [*nichts als ein niedriges Werkzeug*] for material purposes and can prove nothing for the freedom of the spirit.[35]

In his Classical aesthetics, Schiller adopts the language of means and tools, abandoning his earlier foray into semantic innovation. The existence of "mere appearance" is central to human existence—indeed, Schiller has derived it in this passage from the way the senses extract formal information from material presentation, as Aristotle had in the *De anima*. But the senses are not given "organs" in the text, and their functioning is left an assumption. The same assumption, one might think, holds for art here. "Aesthetic appearance" is cordoned off from its correlate, reality. Schiller is following Kant, for whom aesthetic judgment seeks a unity its native capacity for concepts cannot provide. This form of judgment is not unrelated to reality, but determines nothing about it: "It is also entirely unnecessary that the object in which we find the beautiful appearance is without reality, if only our judgment of this object takes no note of reality; for as long as it takes this note, it is no aesthetic judgment . . . but to be sure it requires an incomparably higher degree of beautiful culture even simply to feel beautiful appearance in the living than to do without life in the appearance [*das Leben am Schein zu entbehren*]."[36] Aesthetic judgment becomes the sign of not only humanity but of culture—this argument will be extended in the final letter to argue for an "aesthetic state." The human and his culture have the ability to experience "mere" appearance as their essence, a point Hegel would exaggerate into the dictum "the appearance is essential to essence." Rooted in a cognitively constituted reality it can treat as less than real, the human is defined as aesthetic, his education meant to resolve the tension between the tendencies to unconsidered selfish enjoyment and unsympathetic application of universal law.

For Schlegel, however, poetry is not *merely a means* when it encounters reality. Indeed, it measures its successes in precisely that reality, where its organ is both constructed and found. The imagination must make itself concrete on either end of its spectrum, while maintaining the irreality of fiction as the form of its activity. It nevertheless has reality effects at its lower margin, where it interacts with its influences, thematizing, in Schlegel's famous phrase, the Revolution, Goethe's novel *Wilhelm Meister*, and

Fichtean philosophy. At its higher margin, it seeks to systematize reality and force it to transcend itself socially, thus rendering culture mythologically transcendent (more on this in chapter 7 below). This entire program is captured, as I will argue throughout, in the slip from Schiller's negated "common tool" (*niedriges Werkzeug*) to Schlegel's "organ." The organ names the complex relation of reality to appearance as an aesthetics that is also a metaphysics. That metaphysics does not treat of "being as being," nor does it seek out universal laws. It treats, cognitively and practically simultaneously—that is, as a *techne*—the coincidences of law and case as encounters between the real and the possible. And by insisting on that combination as the history not merely told but made by the writer, Schlegel and his friends invented an interventionist metaphysics for a world in which Newton, Kant, and the Revolution were already *problems*.

The terminological difference also serves to clarify a key difference between this form of Schiller's Classicism and Romanticism. Classicism contains aesthetics in the realm of appearance[37] and allows its effects to flow back into social reality only pedagogically. This has the apparent consequence of eternalizing the human essence captured in the aesthetic condition: Schiller writes of "a comprehensive intuition of his humanity [eine vollständige Anschauung seiner Menschheit]" as "a symbol of his *essence carried to completion [ein Symbol seiner ausgeführten Bestimmung]*."[38] The order of these appearances is permanently separated by the human from the historical and natural worlds, set across from these. For the Romantics this division remains, its form in judgment, its heritage Kantian. Yet both factors are alterable, changing in the very natural history in which they are embedded. And the function that separates and binds reality and appearance—the *organ*—is itself subject to this change. In other words, the point of interaction between reality and appearance is a shifting function. Schiller's determination is metaphysical in the old sense: it gives the formal law of being, albeit in anthropological guise. That would be the metaphysics of the final chapter of the quarrel between the ancients and the moderns. It is not hard to see that the Romantics moved beyond that aesthetic debate by proposing a three-part structure in which Romanticism was the progressive solution to precisely that debate, a resolution and a horizon beyond its terms.[39] That resolution is both theoretically and practically historical: Romanticism is the self-conceived *historical answer* to the debate.[40] This self-conception is captured in the slip from the Classical tool to the Romantic organ, from a metaphysics about being to a metaphysics woven into being's historical state. From the Greek *organon* and the life-scientific *Organ* came the efficacy of a tool to work on a world constituted as malleable. The Greek is in turn derived from an Indo-European root ("*uerg") for *work, cause, effect* (German *wirken*). Jörg Henning Wolf has shown that its modern sense did not emerge until the eighteenth century[41]—we can already see that its etymology continued to confuse and fascinate in two editions of Johann Heinrich Campe's *Dictionary for the Explanation and Germanification of Those Foreign Expressions That Have Been Forced upon Our Language*.[42] The entry covers the sense-organ meaning of the term, going on to its etymological use specifically in language-functions in the body (the "liebliches Organ," or lovely organ, as the voice of the actor or singer), and then uses the term to separate the organic from the inorganic (the latter does not have organs by definition). Campe goes on: "In the following passage of one of our authors: "this aether is the means = *organ*," one could say intermediate layer [*Zwischenmittel*]: this aether is the intermediate layer." Organ might be the sediment that grows between two rock-formations, holding them apart. In the earlier edition,[43] Campe had written: "This aether is the means = *organ*," one could say means of execution [*Wirkmittel*]: this aether is the means of execution etc." As we shall see, this terminological/etymological richness, even vagueness, is essential to the term's creative use. Both the separation of organic from inorganic and the wavering sense of the organ's "effectiveness" (*wirk-*) contribute to the semantic field of *Organ* around 1800. The Jena Romantics used the concreteness and functionality of the medical concept to make an analogy to the normativity and desired concreteness of a set of ideal or social circumstances.

This terminological conflation produced an impossible term, a logico-aesthetic concept of a passive yet formal function: the Romantic organ. "Organ" was meant to unite form with content, the general with the particular. As such, it was the central term of a new metaphysics, one open to real development and responsive to the historical conditions of knowledge and political life. The organ thus made ontological innovation in the historical world a possibility, bringing system and antisystem (Cassirer's "two tendencies" of the Enlightenment)[44] into an intentionally impossible identity.

The literal organ is both a physical location and a manner of operating, a set of rules: the location or part of the body performs a function with respect to the whole. By analogy, the "organ" is a set of rules for thinking and the concrete application of those ideal rules—the ideal "organ" thus makes thought real and makes thinking efficacious. The medical concept was intentionally conflated with the philosophical concept of an *organon*, the tools for philosophy itself. *Organon* was the name given in the tradition to Aristotle's logical corpus, and important echoes in the Early Modern period were to be found in Francis Bacon and Johann Heinrich Lambert— as we shall see.

Since real and ideal, for Schlegel and his compatriots, are meant to be complements in an admixture of organic, developing reason, the concept "organ" operates on a continuum of materiality and ideality, and its metaphoric force attains its value along this continuum. Its distance from the one or the other pole makes it relatively figurative or literal, but its figurating activity is not primarily or finally at a (representative) "distance" from those poles. Rather, it is itself an agent of metaphysical change. The sense in which we mean "organ" as an operator with a determinate range of effects in a given system both comes from and is here applied to the traditional problems of metaphysics. As the active principle in a developmental monistic metaphysics, "organ" is both absolutely general and entirely particular. All possible rules must be real within an organ of their application, yet the real must be organological as much as the ideal. When we say that a publication is the "organ" of an ideology within party politics, we are only shifting this specific usage from metaphysics to sociology. But in doing so, we also deprive it of its organological pretension, the ability of the term (for and in Early German Romanticism) to undercut dualisms like "metaphoric/literal" and "real/ideal" by operating as the generator of such necessary antinomies. When we demote organs from their metaphysical status after organology, we remetaphorize a term that once served to open out onto and bind speculation and pragmatics.

That binding served the purpose of achieving metaphysical cognition: the metaphorical but "real" ability to range over the sliding scale between real and ideal also is meant to afford us insight into and power over the reasons for our cognition of being and beings, being in beings.[45] Yet the link to the specificity of disciplinary knowledge, and even to the possibility

Introduction   17

of ethical action, is also retained. Organology develops an instrument for the mixing of speculation and observation even as it also crosses the divide into action, allowing for a systemic (but not deductive) relationship between metaphysics and politics.

I will be arguing here that this metaphysics, standing in the tradition of those systems, from Leibniz forward, that think of the scientific and democratic revolutions as the occasion for a new determination of the "queen of the disciplines" (rather than signs of its irrelevance), neither necessarily produces regressive social viewpoints nor determines in advance what sort of an empirical world we live in. My investigation treats the concept of the organ for metaphysical cognition and action as the foundation of an open system.[46]

Aristotelian Terminological Problems

Romantic organology is the center of Romantic metaphysics,[47] and in no way reducible to what we might later call "organics." It is not an analogy based on a stable concept from an established discipline (biology), but a literal doctrine based on a neologism that exploits various ambient senses to determine the relationship between literature and metaphysics. In demanding an ideal yet concrete organ, Schlegel was drawing on and innovating in a terminological history that goes back, as I have briefly indicated above, to Aristotle. Indeed, both *organon* (the term used to classify the logical works in the Aristotelian corpus) and *organ* are ultimately of Aristotelian descent. The German *das Organ* (unlike its English and French counterparts) did not come to have its present meaning—"functional part" of a living being: internal or sense-organ—until the late eighteenth century.[48] We can mark out three distinct but interlocking semantic fields of the Greek *organon*'s heritage in Aristotle himself, in order then to see what the Romantics were doing by conflating the modern meanings of organ and *organon*.

The most general definition[49] given to the term reveals the *organon* as that which is potential with respect to a field of actuality on which it is concentrated.[50] This technical definition fits well with the sense of a "tool": the flint houses a possibility, the reality of which we call fire. We can note that this example wavers: flint is only an *organon* when it is used to make fire.

And this is precisely the framework in which Aristotle develops his term *organon*. The comparison of nature (*physis*) to artifice (*techne*)—artifice is, in his terms, an "imitation of nature" or *mimesis*—provides the conceptual background on which to develop the notion of function, both for nature's *teloi* and for human uses:

> Whatever is formed either by Nature or my human Art, say X, is formed by something which is X *in actuality* out of something which is X *potentially*.[51]

This analogy—which is sometimes called "technomorphism" or the "*techne-physis* analogy"—allows for the passage, whether natural or technological, from potential to actual. And it does this by means of the *organon*—indeed, this is the latter's most fundamental meaning. The "organ," we can say, is that functional part—in any order, natural or human—which is so organized as to bring about a specific effect within a field of possibility its own specificity circumscribes. Human purposes mimic—indeed, are a mimesis of—cosmological *teloi*, and the concrete actualization-apparatus is called, in both cases, *organon*.

Because Aristotle develops the concept in parallel to a notion of common undifferentiated material and "differing" structures (*anhomoiomere*) in animals, he is near to the concept of the organ in applying the term. *Organon* is used precisely where the *anhomoiomere* take on functional or "practical" characteristics.[52] And where differentiation occurs, it does so for a natural purpose. The organs of nature that Descartes rejected in Galen are exempla of the *techne-physis* analogy, without which they could not exist. Their functions would fall apart with that analogy, and the Aristotelian cosmos would become a general organology.

The concreteness of the functional part, however, does not predetermine it to physical existence (except in the sense of *physis*, which corresponds to "nature"). So, in a first—and determinative—metaphorical application of the term, Aristotle defines the senses as the instruments of perception. The "sensor" (*aistheterion*) is "that part which is potentially such as its object is actually."[53] The concrete sense-"organ" (we can say, with terminological anachronism) is an *organon*, a functional part covering a field of potentiality—in his example, the tactile—and making perception possible through the characteristic transfer from possible to actual. Epistemologically, the point is that the senses cannot transfer the material they interact with to the mind, but instead only the formal elements of that field. We do not get an eyeful of wood when we look at a tree, but a representation of that tree. Aristotle continues:

> By a "sense" is meant what has the power of receiving into itself the sensible forms of things without the matter. This must be conceived of as taking place in the way in which a piece of wax takes on the impress of signet-ring without the iron or gold . . . but it is indifferent what in each case the *substance* is; what alone matters is what *quality* it has, i.e. in what *ratio* its constituents are combined. . . . By an *aistheterion* is meant that in which ultimately such a power is seated.[54]

The *aistheterion* is thus a dynamic formation in the natural world, and its position in the passage from potential to actual in representation singles it out—once it gains its metaphorical usage as "sense-organ"—as a concrete version of the definitional problem presented by the term *organon* itself. To speak of the "instruments of perception," as authors of the eighteenth century so often did, was to invoke the very problem of the connection of mind and body, and in turn, the ontological problem of the structure of the universe. As the *techne-physis* analogy came into doubt with the crisis in metaphysics (and the end of the Aristotelian "schools") at the end of the seventeenth century, the term "organ" was released into a metaphorical field where it eventually found its literal home in medicine. But there were some detours along the way.

Although it was not Aristotle himself who gave the name *organon* to the logical part of his works,[55] it is possible to see, in a third semantic field opened up for the term by the Philosopher, an overlap between the logical *organon* and the cosmological *organon*. This overlap would be exploited after the word emerged in German at the end of the eighteenth century.

As we shall see in chapter 2, the problem of an *organon* for metaphysics in particular would exercise the young Kant. He rejected what he saw as the Rationalists' continued adherence to a key Aristotelian dogma—that judgments could be unproblematically formal and material at the same time, that they could refer without further consideration to the world. He connected the problem of the instruments of perception to the grander problem of logic itself. Aristotle's own repetition of the categories in the *organon*'s treatise of that name—*The Categories*—and in the *Metaphysics* (albeit in different form) was the paradigmatic error of this kind.

This type of error is perhaps clearest in the Pseudo-Aristotelian treatise on *Problems*, where, in the section on thought, intelligence, and wisdom, the author repeats the doctrine that knowledge is the tool of the mind, just as the hand is the tool of the body. Pseudo-Aristotle writes: "For the mind exists within us among our natural functions as an instrument [*organon*]; other branches of knowledge and the crafts are among the things created by us, but the mind is one of the gifts of nature."[56] The organ of knowledge, it would come to seem in the eighteenth century, was a natural point at which to doubt that metaphysical and epistemological categories overlapped neatly with each other. If not Aristotle, then at least the author of a treatise that went under his name for millennia saw the instrumental function of the human animal at physical and cognitive levels as of a piece. He confirmed: "But nature itself is prior to knowledge, and so also are the things that are produced by it."[57]

What seems a confirmation of Aristotle's commitment to the empirical is much more. Because *episteme* is here the tool of *nous*, nature's priority means that there can be no fundamental break between the order of knowledge and the order of things. For Kant, that would be a—perhaps *the*—methodological error.

And yet, it was an error that Aristotle had commented on even while he committed (or inspired various Pseudo-Aristotles to commit) it. To be ensouled, the treatise On the Soul (*De anima*) tells us, is to have instruments at one's disposal (412b). The soul itself is in the way of a tool—a potentiality in which the body "is" (412b–413b). The organon is associated with any effecting of the passage from potential to real, and the form that this passage takes. Which is why, Aristotle argues, we cannot apply the term to the mind. Among the views of his predecessors which come up for consideration and rejection in that work, Aristotle singles out Anaxagoras's assertion that mind (*nous*) must be completely distinct from that which it cognizes, its material. The Philosopher affirms this point alone among the earlier views: the mind must be pure "in order, as Anaxagoras says, to dominate, that is, to know."[58] For Aristotle, the mind is in fact not real until it knows: cognition is the actualization (passage from potential) of *nous* itself. He concludes:

> Therefore, since everything is a possible object of thought, mind in order, as Anaxagoras says, to dominate, that is, to know, must be pure from all admix-

ture; for the co-presence of what is alien to its nature is a hindrance and a block: it follows that it too, like the sensitive part, can have no nature of its own, other than that of having a certain capacity. Thus that in the soul which is called mind (by mind I mean that whereby the soul thinks and judges) is, before it thinks, not actually any real thing. For this reason it cannot reasonably be regarded as blended with the body: if so, it would acquire some quality, e.g. warmth or cold, or even have an tool [*organon*] like the sensitive faculty: as it is, it has none.[59]

Nous must be general, or its goal of general and certain knowledge cannot be secured. Any "admixture" of specificity cannot be *nous* but that on which *nous* works. And so there can be no organ of reason, no circumscribed field of application of the mind. It must operate definitionally the way that the *organon* does—passing from potential to actual, and causing this passage— but it cannot be merely a tool. The paradox is given most succinctly at *De anima* 431b: "In every case the mind which is actively thinking is the objects which it thinks." The mind must have the organological function—it must perform the teleological "fitting" of things to representations. And yet it cannot be specific, since its field of function is purely general. This anticipates the problem of an *organon* for metaphysics in the eighteenth century, a problem that could only arise when this epistemological passage from potential to actual was no longer hidden within the technomorphic image of the cosmos. As long as the mind was that analogy to nature's purposes, its security rested in its imitation of nature. It could not have[60] or be an instrument, but its analogical resonance, even for these impossible areas, was clear. Thus Aristotle sums up his use of *organon* across all three problems with: "It follows that the soul is analogous to the hand; for as the hand is a tool of tools, so the mind is the form of forms and sense the form of sensible things [*kai gar he cheir organon estiv organon, kai ho nous eidos eidon kai he aisthesis eidos aestheton*]."[61] The hand—a key figure in organology, as we shall see in chapter 8—characterizes the human as producer. His means are *organa*, and his hand is the condition of their use. This gesture toward a transcendental technology is then made analogous not only to the instruments of perception and their formal uptake of sensible things, but also to the mind: *nous* is almost an organ in the Romantic sense, a truly transcendental tool for the grasping of all things.[62] From animal-generation to metaphysical methodology, Aristotle laid the groundwork for the palette of problems the Romantics inherited. And yet, in order for them to do so, Aristotelianism would first have to perish on the emergent world-picture of the experimental natural sciences.

Organology sought precisely the means of actualization of cognition that Aristotle rejected. Following the late-seventeenth-century crisis in Aristotelianism, the dominance of Leibnizian metaphysics in Germany from the 1720s to at least the 1760s, and Kant's critique of all previous metaphysics, the Romantics sought to refound the discipline, using the concept "organ" to move from rationalist metaphysics to organicist rationalist metaphysics. As we shall see, this move entailed rejecting a narrowly defined representationalism about thought. Aristotle's problem is binding so long as we remain in that model: cognition itself must be constituted as too general to have any particularizing means determining its object, as each sense does. Without relinquishing the notion of representation (*Vorstellung*) at large, the Romantics looked to the activity of cognition in general, and the result of this search was a model that included efficacy and development as features of thought. Organology was a "realism" in Schlegel's sense: a metaphorically grounded metaphysics bordering directly on both politics and history, derived from old problems and confronting a new world.

Transplanting the Organ

The German term *das Organ* indeed comes, as I have pointed out, from the Greek *organon* ("tool"), but it did not emerge in German-speaking Europe—to mean "functional part of a living being"—until the 1780s, in the wake of Kant's public polemics with his former student, Johann Gottfried Herder. The term remained caught between its ancient instrumental and modern biological senses well into the nineteenth century. Early German Romanticism (1796–1800) thus occurred in a small window of semantic confusion just before the naming of the discipline "biology" in 1800. Organs were not merely parts of animals, but also faculties of spirit and instruments of literary and literal production. I argue that by repurposing or "transplanting" the nascent term "organ" into a central functional term for a new metaphysics, certain Romantics attempted to reinvigorate speculation after Kant and provide theoretical justification for human intervention in natural and historical processes. While it has become a historiographical refrain that the Romantics aestheticized metaphysics on the basis of an organic model of the universe, I argue that the rare intimacy of metaphysics and aesthetics in their thought is instead based on the combination of the functional, instrumental, and physical senses of "organ," making their literary metaphysics what I call Romantic organology.

The first part of the present book, "Toward Organology," examines the development of these various senses in the sciences and philosophy in the German Enlightenment. In chapter 1, "Metaphysical Organs and the Emergence of Life: From Leibniz to Blumenbach," I examine how Leibniz used the term in a metaphysical sense, but almost exclusively in French. The meaning that emerged in the Rationalism in which Kant was trained was something like "expressive nexus," the connection between individual example and general law. Meanwhile, the modern biological meaning of the term emerged as a *desideratum* in the debates around the emergence of life. The extraordinary fact is that figures like Albrecht von Haller used the term *organe* in French, translating it eclectically and leaving a semantic void in German.

The second chapter, "The Epigenesis of Reason: Force and Organ in Kant and Herder," treats the adoption of *both* of these meanings by Kant and Herder. Before and after their acrimonious split, these philosophers used "organ" in its biological sense (indeed, they may have been first to do so) as an analogy for the critical capacities of the human mind. The literal meaning of "organ" thus emerged *as a metaphor*, projected in its creation in a literal sense that did not yet exist, and applied to the mind in a semantic transfer with no starting point. For Herder especially, the term retained the sense of "expressive nexus." Indeed, Herder even exaggerated this sense, rooting a cosmology in a series of organs. And Kant came very close to combining the new biological sense with the traditional logical term *organon*, a "tool for metaphysics" that he finally rejected, supplying—as I show—the outline of a metaphysics in keeping with the larger precepts of his notion of Critique. The collective analogy of animal body to human reason left out the "organ"—the Romantics would complete the association and exploit it to simultaneously literary and metaphysical ends.

The third chapter, "The Organ of the Soul: Vitalist Metaphysics and the Literalization of the Organ," is an interlude, describing the rise of "Vitalist Metaphysics" in the 1790s in the debates around *Lebenskraft* among such figures as Kielmeyer and Reil. This interlude tracks the first literalization of the term "organ" in German, and shows that the Romantics had access to a newly branded term, which was already both metaphysical and biological. The surprising remetaphysiciziation of biological discourse just before the name "biology" first occurs shows that the complex semantic history here had not yet settled into a stable and ignorable concept (in Hans Blumenberg's phrase, was not yet "terminologized").

Romantic organology—the topic and title of the second part of the book—had a short but intense career. This distinctive use of the term "organ" started with Hölderlin (as discussed in chapter 4, "The Tragic Task: Dialectical Organs and the Metaphysics of Judgment"), who borrowed it from Samuel Thomas Soemmerring's metaphysics of the brain. The poet turned it into a term for finite syntheses of knowledge, an alternate for "judgment," and made it dialectical by insisting that there be an "organ" of (infinite) "intellectual intuition." He made three contradictions as extreme as possible, in order to make the infinite distance between them organologically bridgeable. These were thought/being (resulting in what I call the "metaphysics of judgment"); mechanics/organics (implicitly transcendentalizing the "organ" above these metaphorical registers); and *physis/techne* (articulating the task of creating a radically new world).

Hölderlin's use of "organ" is an extension and literalization of the analogy of animal body to human reason proposed by Kant and Herder. The species of that body—German *Art*—becomes the *Art* or *genre* of a literary metaphysics. The tragedy became an organ of cognition in Hölderlin's work, the organ's concreteness used to denote both nonabsolute knowledge and its relation to a putative absolute knowledge presented in the tragic form itself. Empedocles, whose self-proclaimed apotheosis as nature's god led to his political ostracism and eventual suicide by volcano, was the figure of that cognitive tension. The narrative of his death presents us with a devolution into the *aorgic*—Hölderin's term for the unorganized, which corresponds to "intellectual intuition" or cognition of the absolute. Yet when read against the genre-theoretical writings, the tragedy itself becomes the figure of nonabsolute knowledge: the *organ*. The contradictory notion of an organ of intellectual intuition—a concrete cognition of the absolute—thus comes to light between philosophy and literature, in genre theory. The organ perches lightly between its general, philosophical sense and its actuality as tragic writing. Ultimately, the word helped Hölderlin to articulate the task of producing a systemically new world—it would remain for Schelling and Novalis to describe the means of such production.

The ubiquitous claim that aestheticization was Hölderlin's primary move—and that of the Romantic movement at large, especially in Jena—can here be refined. Literature was instrumentalized, but not in the senses given to us by the Enlightenment or its photo-negative in Critical Theory. Instead, the reality of literature was claimed—as Novalis put it, *poetry is the authentically absolutely real*—with a subtle etymology of the "instrument" itself. This did not only aestheticize reality—it also made the literary real, ontologically relevant, concrete. Writing could be speculation, which was in turn restrained by writing. Organs are the interface of ideal function and physical constraint, of metaphysics and literature.

Schelling's version of organology (chapter 5, "Electric and Ideal Organs: Schelling and the Program of Organology") articulated the program, combining physiological and instrumental senses of the term to shift philosophy away from the antinomies freedom/necessity and subject/object and toward the picture of a world filled with organs containing various syntheses of these contradictions. The focus on the description of being disappeared from this metaphysics, which instead engaged a world already essentially constructed, already aesthetic or even artificial. Schelling's organology is a metaphysics for dealing with the emergent properties of human creation(s).

In his various attempts at a natural system, Schelling uses the term "organ" as a connecting element within and between orders. He begins with a classical physiological conception of the organ as bearing a "force" that circumscribes its function, but later alters the conception so that the already-dual organ also forms a "third" in relations of force, connecting orders of Nature to each other. The highest of these orders—sensibility—is singled out, and I read its primacy on the altered Kantian method Schelling uses throughout these essays.

Take the famous statement by Schelling of the primacy of the aesthetic:

26   Introduction

> If aesthetic intuition [*Anschauung*] is just transcendental intuition become objective, then it should be understood that art is the only true and eternal organon just as it is the document [*Organon zugleich und Dokument*] of philosophy, which always and continuously authenticates [*beurkundet*] what philosophy cannot present outwardly, namely the non-conscious [*das Bewußtlose*] in acting and producing and its original identity with the conscious part.[63]

There is no doubt that this statement runs counter to the Kantian division of judgmental cognition into intuition and concept, and thus no doubt that epistemology finds a significant aid in the aesthetic. But Schelling's statement—which perhaps most clearly represents his break with Fichte, who thought that intellectual intuition must remain entirely unconscious—also suggests that there is a source for truly metaphysical cognition we already possess. Because that possession is both cognitive and "non-conscious" (which is identified with the "object" by Schelling), its source—its *organon*—suggests the possibility of an intervention in the "deed" or production of cognition. This possibility was adumbrated in the methodologically uncertain drafts of the *Naturphilosophie* in the 1790s, and then first suggested in *System of Transcendental Idealism* in 1800. There, the "organ" of philosophy was a self-intuiting *I*, and the *organon* of that philosophy was the experience of art in "aesthetic intuition." The productive capacity of the latter was the "poetry of the world," a world which fit imperfectly into the greater world it might mimetically reproduce. Schelling articulated a conceptual determination of organology—no longer merely a task, in his work the organ came to signify both the interface between different orders of being and the tool for knowing them. We live, for the young Schelling, in a world entirely composed of such organs. If the world is filled with aesthetic syntheses of freedom and necessity, then the metaphysics meant to alter the direction of their totality (which is called mythology) anticipates the problem of machinofacture as depicted by Marx (chapter 9). Schelling determined what Romantic metaphysics had to look like; Novalis would make this organology universal.

From Novalis's *Notes for a Romantic Encyclopedia*: "Under 'phil[osophy],' higher physics, or higher mathematics or a mixture of both has hitherto always been understood. One always sought through phil[osophy] to make something workable [*werckstellig*]—one sought an organ that could do anything [*ein allvermögendes Organ*] in phil[osophy]."[64] The task assigned to the organ makes clear that both the Aristotelian sense of a logical instrument and the medical sense of a concrete function are in play. Novalis comments on the history of philosophy's search for general knowledge (metaphysics) even as he insists in his own construction on the practical focus of philosophy itself. Indeed, we can see his text replacing the notion of an intellectual faculty—Kant's *Vermögen*—with that of an organ.

Novalis describes his unfortunately-named doctrine of "magical idealism" in nearly identical terms: "The active use of organs is nothing other than magic, thaumaturgic thinking, or the *willful* use of the world of the body—for will is nothing but magical, *forceful* thought-capac[ity]."[65] Joining the chorus of critical continuation of the Kantian and Fichtean projects, Novalis wrote the most robust version of organology (chapter 6, "Universal Organs: Novalis's Romantic Organology"). By including a real contradiction in the literalized neologism "organ," he universalized the possibility of principled but effective intervention into both nature and history. His *Notes* sketch the enormity of organology's task. So far from the irrational optimism sometimes caricaturized as Romantic attempts to idealize the natural and historical objects around them, the task Novalis sets himself is the unification of the speculative and the literally disciplinary. Disciplinary knowledge is put into dialogue with the consideration of being itself, and of cognition's relation to will. It is precisely this relation that Kant had attacked in 1798 in *The Conflict of the Faculties*—a book that attempted to bridge this gap was, for Kant, an *organon* where a *canon* was needed (more on this shortly). Novalis agreed, but saw every reason to create that tool. His eventual doctrine was based in a cosmology that constituted the universe as lacking or possessing a necessary hole. That doctrine was filled in by a revision of Kantian Criticism that made faculties into historical organs, and sought the active transformation of everything into organs and organs in turn into the vehicles of a better world. The transition to moral and political organs is clearest in Novalis and makes his version of the doctrine the most complete.

Novalis wrote the most concentrated version of organology. Drawing on his friend Friedrich Schleiermacher's radical theology, he developed a notion of the organ as a kind of hole in the universe, a constitutive incompleteness made up by human cognition and injecting negativity into the spatial cosmologies of the eighteenth century, especially Herder's. These "cosmological organs" made up the distance between the possible and the actual, constantly recasting that difference but never sublating the organ-like nature of the ordered universe. In his encyclopedia and Bible projects, Novalis sharpened the philosophical notion of this organ with explicit reference to Kant, placing the organ where Kant had placed the faculty. More than just a condition of possibility, though, this Critical organ was also the possibility of possibility, or possibility raised to the second power (*in zweiter Potenz*). The mutable nature of organs—their possibilities of development—meant that the categorical system could not be fixed, and that syntheses of cognition were historical. Novalis called this pluralizing of the transcendental argument "organibility," or the possibility of making anything into an organ (an interface between possibility and actuality). Novalis's central example was the notion that a sense could be a tool. Because a "sense" is already the field of an organ's activity, to instrumentalize the sense is to raise it to a higher power, to make its specific ability to actualize a field of the possible itself an organ. This strategy—making an existing organ a tool, or making an involuntary organ voluntary—is the letter of the doctrine Novalis called "magical idealism." This idealism was based on the negativity of the organ at the cosmological level, and resulted in the new genre of the Romantic encyclopedia, meant to stand between philosophy and literature and use the constraint of its genre (the formal conditions of science as such) to allow the organ to write itself as the container for speculation. Metaphysics became concrete through generic constraint, as it had in Hölderlin. It responded to Schlegel's demand that it be both real and ideal, in the very writing of the encyclopedia. Its concreteness allowed for application, as Novalis elaborated: adopting the term "moral organ" from the Dutch philosopher Franz Hemsterhuis, he claimed that the critical organ's procedure could be used to make perception itself—indeed, the world itself—moral. This operation occurred in three steps. First, the moral organ had to be established and rigorously conceptualized for use. Then, a social medium of its use needed to be conceived—Novalis borrowed terms ("higher" and "lower" organs) from Fichte's *Foundations of Natural Law* for this purpose. Finally, an organological politics would have to be written. Novalis claimed the need for an "intellectual intuition of the political," a mysterious phrase that, in my interpretation, points to the methodological—rather than metaphorical—use of "organ." Political organs were the potentiated loci of collective human activity. They shared the synthetic histories and capacities of sense-organs, but were also the composite objects of a kind of historical idealism seeking to combine systems and histories (orders of things) with a view to changing them. Statecraft was political organology.

Thus "organ" was made literary—or more strictly, genre-theoretical—in Hölderlin; metaphysical and moral in Schelling; and universal but also concrete in Novalis. Where Hölderlin saw the task of producing the new, Schelling determined a program for organology, and Novalis universalized that program. Organology introduced the encounter of the absolute and the historical, and it held on that basis the possibility of systemic intervention in history in reserve. Reason's transition to the absolute—and back—was meant to have a means. The philosophical discourse of metaphysical means around 1800 was rooted in the etymological field of the *organ(on)*.

Romantic organology went in widely different directions after Novalis's death in 1801 (chapter 7, "Between Myth and Science: *Naturphilosophie* and the Ends of Organology"). Friedrich Schlegel, Novalis's philosophical intimate, used a version of organology based in transcendental imagination to form what I call the "Jeffersonian strain" of Romantic organology. Schelling's call for a "new mythology" paralleled that of Schlegel—but the Schellingians Gotthilf Heinrich Schubert and Joseph von Görres, while they produced historical justifications for radical governments and cosmologies, ignored the methodological rigors that had marked the early program. Their doctrines were fantastic projections onto a transcendent screen—shot through with mentions of "organs," but ultimately wilder, uncritical versions of Herder's cosmology. By the time Schelling moved to Munich in 1806, Franz Joseph Gall's phrenological system had gained popularity. Schelling and Hegel both rejected this organ-based conception of the commerce of body and soul. Schelling's rejection was a waving of hands, but Hegel's went far deeper, presenting us—in the *Phenomenology of Spirit*—with his deepest flirtation with and ultimate rejection of Romantic organology. Schelling's only published response to Hegel's dismissal of their earlier shared program came in what I take to be the last gasp of organology, a "Romantic metaphysics of morals" (my phrase) in his 1809 *Essay on Human Freedom*. There, he articulated human being (as human will) itself as constituting a nature it transcendentally observed in its own actions.

This human was God's organ, and Schelling's recuperation of traditional theological terms under the mark of the methodology that was Romantic metaphysics was the final moment of the doctrine's proper expression.

In the third part of this volume, "After Organology," I locate the aftermath of organology in an oblique dialogue between Goethe and Hegel in the 1820s. Goethe and Hegel clashed in the 1820s over this philosophical legacy (chapter 8, "Technologies of Nature: Goethe's Hegelian Transformations"). Goethe had witnessed the rise of the term "organ"—and indeed had contributed to it—but only came to test its etymological capacities after Hegel had accused him of operating without means in science. Goethe combined the classificatory drive of his Classicism with the Idealism he had gathered over decades in Weimar and Jena to produce a late response to the Romantics, a revised organology. He thought, on my reconstruction, that science itself should cut a middle path between the emergent positivism of Paris and the waning *Naturphilosophie* of Schellingian and Hegelian stripes. The social task of science would be to impress norms upon the world, to alter its constitution categorically but tenderly. This program reflects Goethe's long struggle to come to terms with the political upheavals of his time.

I conclude the book with a consideration of the organological project's legacy in the writings of Karl Marx (chapter 9, "Instead of an Epilogue: Communist Organs, or Technology and Organology"). From his early debates with Feuerbach to his self-construal as analogous to Darwin to his long and influential analysis of machines in the factory, Marx returned to the term "organ" throughout his career. The term points up the crucial conflict in his writings between genetic method and radical politics—"organs" are, for Marx, the coincidence of method and real force. Thus a crucial impulse of Romantic organology was included in the philosophical radicalism that became Marxism.

This book pays more attention to Kant, and especially the *Critique of Pure Reason*, than is usual for studies in Romanticism. This is because, as will become obvious, I tend to think that Fichte faded from centrality in Jena Romanticism's philosophical project as the century drew to a close. In fact, two main trends in Romanticism historiography have left this lacuna because of a temporal problem. Attempts to include Romanticism (especially Hölderlin, Schlegel, and Novalis, but also Schleiermacher) in the history of Classical German philosophy and especially Idealism (Dieter Henrich, Manfred Frank) have focused on the years 1794/95, when the Romantics independently yet collectively discovered Fichte, then at the height of his lecturing fame in Jena. (An exception to this trend is Frederick Beiser, whose work I engage throughout.) The second large trend of interpretation—which focuses on Romanticism as a practice of writing, and takes up impulses from deconstruction (Maurice Blanchot, Jean-Luc Nancy/Philippe Lacoue-Labarthe, Ernst Behler, Winfried Menninghaus)—focuses its analytical attentions on the years of the production of the periodical *Athenäum*—that is, 1798–1800. There are, then, two missing years from Romanticism's developmental history. And it is hardly coincidental that it was in 1796/97 that Hölderlin, Schelling, and Novalis all turned back to the *Critique of Pure Reason*. Kant had provided the larger methodological framework in which the Romantics took themselves to be working, and there was, in this sense, a return to the issues raised in Kant's first Copernican book. Framing the project of Romantic organology this way allows me to focus on the centrality of judgment for Romanticism. Judgment seemed to have both the flexibility and the productive contradictory essence that also characterize the organ.

What follows is a revision of the theoretical conjuncture of Early German Romanticism. That movement found a point of indifference between writing and thinking, a point that has historically been construed as upending dichotomies given by Kant, Fichte, Spinoza, or else the very dichotomies of Western metaphysics as such. On Paul de Man's reading, this has its home in the practice of irony, which produces a "permanent parabasis," an aside or exit from the philosophical tradition, which is already an "allegory of tropes."[66] Similarly, the order of signs and their origins and destinations in consciousness is "reserved" in Early German Romanticism for Manfred Frank, and consciousness is therefore confronted with an infinite task that becomes ontological—an existentialist crisis. For Winfried Menninghaus, too, what Kant produces as an impossible relation between sign and being is maintained as the trace of some relation in Romanticism—this is called "ontosemiology." But how can we square this circle? On the one hand, there is this permanent undermining of all posited knowledge and all positing (which is surely true)—and on which reflective consciousness is based—and on the other, positive Romantic engagements with vast realms of real knowledge, especially in Hölderlin, Schelling, and Novalis, where we see the adoption and adaptation of figures of scientific knowledge, philosophemes, and sociohistorical complexes? How can there be both the abyss and its determinate constituents? I think that this is the question the Romantics asked *themselves*, especially after and probably because of their clustered enthusiasm about Fichte's writings from 1794 to 1796. And I think that the answer was expressed in this deceptively simple term "organ," the place where the infinite and the finite take synthetic form, the place where Romantic writing is indifferently philosophical and literary, the place where, in an unpredictable manner, over and over again, being, sign and metaphorical content are joined, constituting and proliferating patterns of historical necessity in all its continuing contingency. The impulse to intervene in those sites of production is what I hope to point to in these central Romantic texts. The discourse that was built around that impulse is what I call "organology."

The book tries to demonstrate that philosophical and literary Romanticism are rooted in a common methodological and metaphysical project. The readings I give of texts by Hölderlin, Novalis, and Goethe are not merely consistent but coterminous with the methodological concerns I am highlighting. I think that the organological project partly explains the mysterious sense of reserve in Romantic literature, the sense that natural laws are both undiscovered and possibly fungible. Still, the book is not primarily an analysis of literary style. In a sense, it is a book about literary theory—but then, it claims that this theory is also a rewriting of metaphysics.

I put two common notions about German Romanticism into question in the pages that follow. On the one hand, the ubiquitous thesis that the Jena Romantics promoted aesthetics above all other disciplines deserves deepening. As I argue in chapter 3 and after, the intimate association of metaphysics and aesthetics in Romantic thought replaces the discursive pair "aesthetics/anthropology," which Carsten Zelle has proposed as one dominant note in the Enlightenment. The Romantics did not merely aestheticize metaphysics, nor did they simply grant aesthetics metaphysical power. The "organ" complicates either picture, raising questions about what metaphysics is, and what literary and technical production might have to do with it. That "organ," a stubborn contemporary intuition tells us, is bodily, is "organic," is part of the order of the living. I will ask the reader to suspend this sense of the term, which did not take on this intuitive meaning until well into the nineteenth century. Suspension of terminological disbelief is one thing; another is the argument I will make throughout that *organology is not biological*.[67] The ambitions and products of the Romantic program exceed any given disciplinary structure intentionally, yet we have awarded the "living" or the "biological" with some pride of place in the interpretation of Romanticism. Far more difficult than suspension of disbelief about a historical term will be the argument that the Romantic universe is not *alive*, not *organic*, although to be sure it uses this model among others. What the Early German Romantics wanted from the term "organ" was much more than a heuristics of mutual causality, a metaphor for form, or a model of the cosmos as a living animal (hylozoism)—as we shall see.

I frame Romanticism here in a long history. This history is intellectual and literary, scientific and philosophical, because Romanticism, too, is all those things. This is justified, in my view, because the larger-scale histories of the period in which Romanticism falls have tended to excise the Romantic episode, or to see it as eclectic within a greater narrative arc, contributing in turn to the tendency to look at most forward to aesthetic experimentation inspired by the so-called "first *avant-garde*." I hope that this book goes some way to correcting those historiographical gestures.

The reader will notice that the story told below is predominately German, involving German authors, scientists, and cultures. This focus is certainly a matter of expertise, but it is not entirely cynical.[68] The Romanticism that developed organology was, to the best of my knowledge, German, and the engagement with science skewed German as well, although significant exceptions punctuate the chapters below. The development of a doctrine around a word necessarily concentrates this history in its linguistic character, although it by no means restricts it to one language. Nevertheless, I am not aware that the word enjoyed a comparably grandiose conceptual development in other European languages. Organology also emerged in a period of punctuated rather than smooth royal-road science history, between the developments in physiology in the eighteenth century and the "taking form" (Lynn Nyhart) of modern biology in the discovery of the mammalian egg, the articulation of cell theory, and the hypothesis of natural selection, from Albrecht von

Haller's articulation of the function-problem—immediately a problematic of force and its location, both spiritual and physical force—to the microscopy that allowed the cell to be seen.

Albrecht von Haller (1708–1777) towers in the history of physiology. The Swiss polymath wrote magisterial works on the functioning of mammalian bodies, incorporating his Newtonian scientific views into his didactic poetry. His experiments on chicken embryos led him to the view that the ability of the muscle to be excited, to react to stimulus by shortening—its "irritability"—was at the basis of the development of the organism. The heart would start only through "irritation," and the subsequent formation of the embryo would develop units of corporeal function only on this condition. Haller's wide-ranging writings on bodies consistently call functional units in those bodies "parts," as in the definition of irritability itself: "I say that part of the human body is irritable which become shorter through contact with something external . . . and I call that part of the human body sensible contact with which is represented in the soul."[69] "Irritable parts" of the human body—distinguished from the "sensible parts" in which stimulus causes representation "in the soul"—would become the focus of physiological and biological research. The question of how irritation could occur, what the bodily conditions would allow for this singular interaction, dominated the life sciences in one way or another for two generations after Haller's discovery. As we shall see, this problematic played a direct role in the so-called "Brownian revolution" in medicine and in the debates around animal electricity or Galvanism at the close of the eighteenth century. Physiology was a matter of the interaction of functional units and stimuli. If the heart could *respond* to the new arrangement of fluids in the ovum, the specificity of that response needed clarification.

It seems from a contemporary perspective obvious that the responsive bodily unit should be called an "organ." But while Haller, as we shall see, used the Latin variant *organum* sometimes, and regularly made use of the French *organe*, in German the term "organ" was not widely used until the 1780s. Even when Haller used the term, he clearly meant it in the sense of a "tool"—the *auricle*, for example, is the "instrument" of hearing.[70] It is easy to recognize the degeneration of the Galenic system as leading to this lacuna in vocabulary—while Haller can retain the traditional phrasing, "organs of motion and of sense" (*nervos organa esse diximus sensus motusque*),[71] it seems likely that the lack of systematic usage of the term "organ" stems from physiology's lack of confidence in the *purpose* for which corporeal tools are used. The body is the instrument of the soul when intentions are involved—we use the eye to see, the leg to walk, etc.—but when they are not, functions are without overarching reference. "Organ" was not yet a neutral term—it would have to obtain its current literal sense, or in other words lose its etymological association with the "tool," to emerge. But as soon as it did, it became a candidate for Romantic system building.

By the time Theodor Schwann began to isolate cells under his microscope and construct the cell theory in the late 1830s, organs had become normal for science. Indeed, the experiment with the conceptual border of the term was largely over, and the speculative biologies that proliferated around 1800 in Germany had been forcefully rejected by a cohort of physiologists who would define modern experimental methods.[72] In 1827, Karl Ernst von Baer had demonstrated the existence of the mammalian ovum, an entity that had long been assumed to exist but had never been seen. Schwann's work (and Rudolf Virchow's) on the cell, however, changed the scale at which the study of life was conducted, on which the question of function could be posed. There had always been the question of the differentiation of organic matter into functional units, going back to Aristotle and including Haller's division of bodily materials into "fibrous" and "fluid." The cell, however, was a structured unit that could vary in function, the "common principle of development" that united plants and animals, and thus went some way to explaining the physical basis of life—as Virchow would generalize, *omnis cellula e cellula*.[73] It was a building block for organs, as Schwann wrote:

> The same process of formation and transformation of cells inside a structureless substance repeats itself in the formation of all organs of an organism, just as it does in the formation of new organisms, and the basic phenomenon through which productive force expresses itself in nature is thus the following: first there is a structureless substance there, which lies either inside of or between cells that are already present. In this substance, cells form themselves according to specific laws, and these cells thus develop into the basic parts of organisms in manifolds ways.[74]

Organs are no longer the lowest functional unit. Cells mediate between structureless substance and the "elementary parts of organisms." The very phrasing shows the shift in focus: what is "elementary" is no longer the object of investigation: organs have functions for a reason that does not lie above, with God, with the essence of the animal, or the like, but below, in the variable functions of cells.[75] Timothy Lenoir has described von Baer's, Schwann's, and Virchow's work as the pinnacle of a research program of "teleomechanism," in which research into the mechanisms of functions are animated by the regulative belief in purpose.[76] The delicate balance between these mechanisms and purpose is clearly visible, but the theory of the cell reopened the question of function's purpose at a micrological level. This question was answered only with the discovery of the double helix. Denise Gigante writes that cell theory undercut the Romantic analogy between biological life and poetic form: scaling the unit of analysis down allowed for the exclusion of integral selfhood and unknown forces from the emergent studies of biophysics and biochemistry.[77] The shift was a slightly smaller one for the German Romantics, from the problem of functional units to the emergence of variable biological functions as a scientific problem. Cell theory shifts the paradigm to a smaller and nonspecified functional unit, meaning that biology is no longer potentially organized around organs—this makes the inclusion of biology in the positivistic paradigm possible, since it fully precludes the possibility of organicist attitudes dominating experimental method. It leaves open the question of the use of organs—which is then studied by physiology—and the history of the development of individual organ-patterns (species), which is taken over by evolutionary theory and emergent ecological thought[78]. But as far as the *concept of life* is concerned, the rise of cell theory leaves the organ behind. Biology is no longer—if it ever was—a question of organs. Romantic speculation was excluded from scientific development, its language henceforth construed as metaphorical.[79]

During the period between Haller and Schwann, a functional physical unit—an organ—was a candidate for the organizing principle of the life sciences. Indeed, transcendental morphology and *Naturphilosophie* coquetted with the possibility of an organological biology, although I do not think they ever carried out that potential program. Function remained the focus of physiology after the methodological polemics that resulted in what we still largely regard as normal scientific method after the 1840s; biology broke into subdisciplines operating at different scales (the ecological, the micrological, and so on), the unity of its object—life—projected into a future that would rest with the discovery of DNA and remain contentious beyond that watershed. The two doctrines that might have retained the analytical prominence of the organ were those of two embittered rivals, Cuvier and Lamarck. The layout of animal organs was the empirical escape from essentialist notions of "type" or species for Cuvier, while Lamarck defended the obvious plasticity of organ development. As we shall see, when Darwin combined these two lines of thought, interjecting an almost infinitely long graded process of development through adaptation (of organs), he included the problematic of the organ within a revisionist notion of life. In the distance between the unnamed location of function in Haller's theory of excitability and the cooption of the organ in modern biology—this is where the discourse of Romantic organology played more than a metaphorical role, where it contributed directly to the development of the discursive matrix on the basis of which the new biology came into being. That new biology, like its sister sciences physics and physiology, based its efforts on a rejection of Romantic science. That is why this is not a book about science, but instead about the mutual discursive constitution of philosophy and science in a period when the two were inextricable. It is as though Romantic speculation was deflected from a science just beginning to normalize itself—perhaps, then, Romanticism played a minor part in that normalization, as Thomas Kuhn suggested long ago.[80] But that is not the emphasis of the book that follows. It is not a book about a failed scientific movement, but instead about a discourse that sought to reconceive the dynamics of scientific knowledge—indeed, of knowledge about the world as such—into the very basis of its cultural imagination. "Science" in its various polemical guises around 1800 plays a major role in Romantic thought, and the Romantics may have contributed ideational materials of import especially to the field of biology. But that is not the point. Romanticism in this vein, Romantic organology, was an attempt to conjoin the *possibilities* of science in its development with the possibilities of an unstable and apparently open-ended historical and social horizon. It was an attempt to refashion speculation not simply to make it adequate to the historical and scientific situation, but to think functional systems in development for the refashioning of that situation itself. "Organ" was the term that represented the norm within the representation, the acknowledgment that in the continuum of human science and the life of cultures, no element was without possible function. The aspiration of organology thus differs sharply from the emergent scientific discourses from which it drew. Romanticism's home in presentist discourse can hardly be in direct conversation with today's sciences, but might instead be in the need for new tools for social critique, new resources for discovering the critical intersections between large-scale historical processes and the constitution of our knowledge of ourselves and the world. Sometimes Romanticism is characterized as having a "holistic" worldview. I will be disputing that tag, but organology is meant to be general in the sense that the problem of the organ transcends the empirical limits of any slice of reality or any discipline constituted to address only that slice. Indeed, one crucial element of the narrative that follows is that organology was articulated just before the institutional crystallization of the disciplines, especially the university's organization of the sciences, with which it might be taken to be in competition. This is both why it is of special relevance to the humanities today, and why its history has largely remained unearthed up to the present. Because its relevance and the cause of its neglect are overlapping, I will dwell on them briefly before going over to the main investigation of the book.

The Missing Organs of Historiography

When Ernst Cassirer came to write the fourth volume of his gargantuan *The Problem of Knowledge*, he noted that the titular problem had shifted from philosophy, and the conviction that it could be "the perfect and wholly adequate *expression* of the total motif of knowledge," to the sciences. The problem of knowledge had drifted from philosophical speculation to the experimental and disciplinary settings of physics and biology (and especially physiology), and history:

> There are now as many single forms of theory of knowledge as there are different scientific disciplines and interests . . . accordingly we choose for the present treatise a method different from that of the preceding inquiry. We shall not follow out the findings attained in the theory of knowledge by reference to the philosophical classics of the period under discussion, but shall try to penetrate the motives that led to their discovery. . . . [T]he problem of knowledge is clearly seen to have taken a new turn and acquired a much more

complex structure than in the earlier centuries. . . . The era of the great constructive programs, in which philosophy might hope to systematize and organize all knowledge, is past and gone. But the demand for synthesis and synopsis, for survey and comprehensive review, continues as before, and only by this sort of systematic review can a true historical understanding of the individual developments of knowledge be obtained.[81]

The protocols of specific disciplines became, according to this thesis, the locale of developments in epistemology. Both Hans Blumenberg's sense that metaphor won out against metaphysics and Michel Foucault's notion of the disciplinarization of knowledge play out against the background of Cassirer's larger claim. If we accept that epistemology—not to speak of metaphysics—became a scientific and disciplinary matter in the period after Hegel, we might see Romanticism as the last gasp of a universalizing program for knowledge. I argue instead below that the perceived contradictions in Cassirer's approach—between disciplines and universality, pure problems and historical and institutional approaches to those problems—are the very motor of Romantic thought. For the Romantics, it makes little sense to think of the particular complexes of laws and their applications as isolated from general questions of philosophical method. Perhaps this is because they lived during the transformation that Cassirer, Blumenberg, and Foucault analyze. In that case, their thought on this issue is precisely the "expression of the total motif of knowledge" that would, according to Cassirer, later seem impossible. But that expression was of the very transformation in question, and so retains relevance as the process of disciplinarization develops from then to now. In what follows, I want to argue that Romanticism retains precisely this metaphysical relevance down to the present, because its history in this regard has not been told. Nor could it have been. I devote the rest of this introduction to examining why, in the cases of Blumenberg and Foucault, it was not.

METAPHOR AND METAPHYSICS: BLUMENBERG'S METAPHORICAL ORGAN

Hans Blumenberg's notion that we might approach intellectual and textual history metaphorologically partly inspires my approach here. And yet, I will offer an alternative to his—and to Foucault's—account of the period around 1800 because the historical picture seems to me in both cases incomplete. The emergence of modern antinomies—as real, discursively but also pragmatically contradictory powers—like those between the "visible" and the "invisible" or the "metaphorical" and the "metaphysical" was already a part of the reflexive legacy of Romanticism as it emerged. Its answer to some of these antinomies was contained in its organology, as we shall see.

Blumenberg's "metaphorology" amounts to the tracking of subterranean shifts underneath conceptuality, the functionalization of conceptual unities for differing human purposes over time.[82] It is based on Blumenberg's overall sense of European intellectual history, which centers on a putative major turn with the destruction of Aristotelian ontology. The broadly painted Aristotelian worldview holds that Being is complete, and that, therefore, all human making is of the order of imitation.[83] Not restricted to aesthetics, this distinction is meant to apply all the way down—indeed, the first innovations with respect to it are noted by Blumenberg in the works of Nicholas of Cusa, and are references to craftsmen, not to "artists." As this ordering system came to an end, according to Blumenberg, the emergence of the theoretically creative human became possible: the possibility of the radically new came into view.[84] The end of the *techne-physis* analogy meant the possibility of ontological innovation, but this innovation was unmoored from the dock of Being, and nested ultimately in antinomical orientation. With the end of Aristotle's order of being, the metaphor itself could emerge.

Blumenberg himself is a practitioner of this metaphorization. In *Shipwreck with Spectator*, Blumenberg returned to Husserl, expanding the metaphor's range to include an interaction with the *Lebenswelt*, understood as "the universe of the self-evident"[85] He writes: "To adopt Husserl's terminology, metaphor is, first of all, 'resistance to harmony.' This would be fatal for the consciousness whose existence depends on its concern for identity; it must be the constantly successful organ of self-restitution."[86] This line of thinking was developed in notes posthumously published as *Zu den Sachen und Zurück* (*To the things and back*).[87] Blumenberg is concerned to identify the "analyst" in both phenomenology and psychoanalysis, and to develop some point of resistance that these schools find in their attempts—practical and theoretical—to determine the nature of consciousness. Collecting his luminaries—Kant, Cassirer, and Husserl—Blumenberg starts from the notion that consciousness is both "self-constituting and self-restitutive."[88]

In its self-constitution, consciousness opens its own contents to itself. Blumenberg notes that Kant had already established this when he tied the table of categories (synthetic unity-producing forms of judgment) to the table of judgments (logical—mind-given—possibilities of that judgment). This makes—for all involved—consciousness *genetically available to itself*. Blumenberg strenuously objects to a possible charge of "idealism" here, since consciousness is taken by none of these figures to be an ontological parthenogenesis. It merely gives itself form. Phenomenology makes experience the source of this form instead of its product, and thus establishes a "world" in which self-correction, -repair, or -*restitution* is possible. This world is characterized by the dynamics of remembrance, formation, and reference to other humans and other things—all in the tension between *Anschauung* and judgment. What, then, *is* consciousness?

> Consiousness is a purposeful "organ" [*zweckmäßiges "Organ"*] with respect to the essential typology of its contents, which—if they are not this already—as "objects" exhaust the complete measure of their manner of offering themselves [*das volle Maß ihrer Selbstdarbietung ausschöpfen*] [to that consciousness].[89]

Resistance to the "impurities" of the empirical ego (Kant) or the psychological approach to philosophy (Husserl) or transference (Freud) is necessarily performed by consciousness. Consciousness is an instrument of that resistance, and it forms and repairs itself with that goal. Its objects appear in organological form—this is *transcendental "I," intentionality,* and *Freudian ego* in one. Without hypostasizing itself, it serves the function of preservation—constitution and restitution—as bulwark against the interruptions proper to its dynamic openness. It is the "organ" of that very preservation.[90] The approach taken by Blumenberg's treated authors—and by Blumenberg—is not "idealistic," but rather insists that consciousness itself idealizes life: "As its last product and 'organ,' elementary achievments of life take it to its extreme, to its pure presentation, 'exaggerates' ['*ubertreibt*"] in the double sense of this word that must be sustained. Thus the judgment is . . . pure 'exaggeration.'"[91] Blumenberg here integrates transcendental philosophy—of Kantian and Husserlian stripes—with Freud's metapsychological dynamism. The permanent frustration of the transcendental philosopher, his openness to the empiria without an explanation for that openness, is made the content of transcendental consciousness itself.

Construed as a legitimate problem, this duality only resolves, and then tentatively, with the notion of intermittence.[92] Blumenberg contributes to our terminological history even as he provides some orientation in the stakes of doing terminological history.

Blumenberg's overall history is not only methodological, but narrates an emergent domain of the metaphor, placing the end of meta*physics* and the critical emergence of the autonomous metaphor in the eighteenth century: "Metaphysics proved itself to us often as metaphorics taken literally; the disappearance of metaphysics calls metaphorics back to its place."[93] This is to say that conceptual determinations lie at the basis of metaphysics, and that doubt about that determination dominated eighteenth-century philosophy, culminating in Kant's critical position.

This is why, for Blumenberg, Romanticism often seems to be the semantic grasping after a lost stability. It is positioned between the fall of the *ancien régime* and Aristotelianism and the rise of the consciously modern metaphorical play of a Paul Valéry: "The Romantic fragment accepts both functions: the awaiting of infinite perfection and remembrance of lost totality."[94] Romanticism does not grasp the nature of the metaphorological possibility in which it stands. Instead, it "aestheticizes the world," which in turn "makes its reality paradoxically superfluous, because it would always have been more beautiful to have only imagined it."[95] The philosophical pretension of Romantic writing appears almost laughable, but only almost. Blumenberg diagnoses Romanticism with a pre-Hegelian hope in totality rooted in nothing, misunderstanding the disappearance of substance, taking its metaphors for realities. The judgment could easily enough (and did) come from Hegel, in the name of Spirit rather than the para-reality of metaphorics. The Romantic "organ," however, was not exactly a metaphor, and its use was not a fall back into a substance-metaphysics. Indeed, as we shall see, it was the semantic lynchpin on which a functional metaphysics would be based.

Blumenberg's own use of "organ," however, rejects the Critical legacy[96] of metaphysics in dialogue with metaphorology. Where his organ reifies the Aristotelian problems noted above, the Romantic counterpart was supposed to engage those problems directly. Indeed, while Blumenberg's opening of phenomenology to psychology—without being psychologistic—shares the practical and concrete orientation of Romantic organology, the absence of this chapter[97] in his historical itinerary allows his methodological contribution to maintain the antinomy metaphor/metaphysics uncritically.

I maintain that the loosening of the completeness of Being and the end of the *techne-physis* analogy also gave way to a new metaphysics—that of the organ, a "technological" metaphysics paradigmatic of modern concerns with the point at which speculation and politics, theory and praxis, can communicate. Romantic organology, as I will show, reveals an alternative to Blumenberg's narrative: the relation between metaphor and concept was indeed reevaluated after Kant, but not to the final detriment of the concept—rather to its instrumentalization, quite literally. The orientation provided by metaphorology is indispensible, but Romantic metaphysics already includes a kind of metaphorological awareness. Blumenberg's "organ of consciousness" is a metaphor for an all-too-real metaphysics. His concept is suspended between methodology and historiography, and we need a fuller sense of that history. The investigation which follows owes a great deal to Blumenberg, but seeks to replace the Romantic part of his narrative; this replacement entails a partial revision of the larger thesis about metaphor and metaphysics.

A last note on Blumenberg: his notion of an "intellectual history of technology" (*Geistesgeschichte der Technik*)[98] provides a key intervention in the debate about technological thinking at the end of technomorphism. This suggestion returns to earlier engagements with the late Husserl,[99] whose attempt to "re-start" within Reason a line of metaphysical thinking free of the conflation of method and being[100] proves essential for this new history. Blumenberg writes of twin dead-ends for the intellectual history of technology, one "chronicling" events (but ignoring the circumstances which gave rise to them), and another (Marxian) explaining those circumstances, but powerless to connect them to the dispositions to which they putatively gave rise. The alternative, which draws on both Husserl and Heidegger[101] but seeks to refashion their "new starts" into responsible historiography, requires that we track the intellectual emergence of technology as a problem, not merely the history of invention or of artifacts. This study will similarly ask how a *technological imagination*[102] stepped out of history, and how it asked concretely what technology in the broadest sense had to offer speculation. If metaphysical and metaphorical thought combined could respond concretely to the "technical will" of European modernity, it will have been in Romantic organology.

WHO'S AFRAID OF REPRESENTATION? FOUCAULT'S TRANSCENDENTAL-EMPIRICAL ORGANS

If metaphor and metaphysics share a home in Romantic organology, and if this home is in critical dialogue with especially the cultural effects of technologization, then the story about Romanticism's relationship to representation will have to be revised. Michel's Foucault's engagement with both of these themes has proven determinative for the field. Indeed, he characterizes his investigation of the "microphysics of power" in just these terms: "Rather than seeing this soul as the reactivated remnants of an ideology, one would see it as the present correlative of a certain technology of power over the body.... The soul is the effect and instrument of a political anatomy; the soul is the prison of the body."[103] In searching for the roots of "technologies" of power, Foucault introduced the technological problematic into historiography. Discursive and—as he came to call them—disciplinary means were investigated in their very genesis. And yet, this project had arisen from an earlier one, that of an intellectual history of the disciplines themselves. *The Order of Things* studies this problematic precisely in terms of representation. The breakdown of the mode of knowledge production (*episteme*) called "Classical"—the early Enlightenment, characterized by *mathesis universalis*, or the spatialization of beings in a transparently classifying order—entailed a crisis of representation. Representation, a sufficient instrument for analysis in the Classical episteme, was destabilized around 1800. This took place in three parallel forms, each providing a principle for an emergent science. The principle was withdrawn into the order of the invisible, but organized the visible itself. Thus: "labor" in economics; "organic structure" in the life sciences; and "inflection" in linguistics. These principles were "alien" to the parts of the domains (representations) that they organized.[104] The emergence of intellectual disciplines was based on the withdrawal of their object of investigation from the representative order—organic structure can be investigated, but life, no longer a subset of being, is instead an isolated ontological area, forming the unrepresented basis of a new discipline: biology.[105]

Foucault interprets Kantian criticism as paradigmatic for this move. Kant invented the "transcendental field," dividing it from the "empirical"—Foucault refers to this as the *transcendental-empirical couplet*.[106] Although he recognizes that Kant intended the two to be united in consciousness, Foucault maintains that the division of the disciplines is underpinned by this philosophical intervention because it allows for empirical fields to be divided from each other by local transcendental elements, like "life." These "transcendentals" function beyond the realm of representation but are the object of representational investigation. Kant's questioning of the legitimacy of representation forces the "the withdrawal of knowledge and thought outside the space of representation."[107] The new metaphysics will not be about representation, but about the "source and origin of representation."[108] Thus two types of conditions of possibility are identified: first, the conditions of experience itself, and second, the conditions of objects. While Foucault sees that the second governs the first for Kant,[109] he proposes that the emergence of *disciplinary transcendentals* was made possible by the misrecognition of or disagreement about that identity after Kant. Foucault concludes that "the criticism-positivism-metaphysics triangle of the object was constitutive of European thought from the beginning of the nineteenth century to Bergson."[110] Synthesis is split off from the field of representations, and transcendental subjectivity is divided from the mode of being of objects.[111]

This thesis has received some support from Azade Seyhan, whose *Representation and Its Discontents*[112] examines the critical attitude of Jena Romanticism with respect to problems of representation. She writes that "the journal [*Athenaeum*] envisioned its intellectual task to be re-presenting representation, in other words, recasting narrative accounts of philosophy, history, literature, and art in terms of their present or modern configuration."[113] This focus on something more than representation, and indeed the notion that the re-presenting of representation gives way to "critical praxis," is salutary.[114] As we shall see, not only textually but also theoretically, the Jena Romantics sought literally to instrumentalize representation, to employ it for the speculative purposes of organology. And their recasting of earlier metaphysics in this mold indeed engaged Kant's critical legacy directly.

And yet, in Foucault's description of the development of biology, which focuses on Cuvier's functional anatomy, the Romantic legacy is missing.

The "discontinuity" of the representational and ontological "spaces" that Foucault sketches with Cuvier is one in which organs disappear into functionality. They form series, which are hierarchized according to an unavailable (withdrawn) notion of life. A local metaphysics—that of the life sciences—opens up as a disciplinary space.[115] And indeed, Cuvier did break with an analogical tradition (still defended, in the famous *Academy Debate*, by Geoffroy de St.-Hilaire—more on this in chapter 8) in favor of comparison by function. The *orga*nization of beings was no longer the methodological point: their comparative functions dictated classification.

What if, however, there had been a discourse about organs that saw them neither in terms of pure functionality nor in terms of analogy? And indeed, what if this discourse had been engaged—precisely through the lens of Kantian Criticism—with the splitting of the disciplines and the philosophical stakes of that division of knowledge? What if that discourse had, in fact, been interested in the emergence of technicity and the disciplines at just that moment when the Classical episteme fell? I will defend in this study the notion that just that discourse is at the heart of Early German Romantic metaphysics, and that this metaphysics, for that reason, meant to provide something like a technology of orientation, a technological metaphysics, concretely engaged with institutional and technological developments as they emerged, and attempting, using the organ of reason, to idealize them and with them life (cultural and biological). That legacy is the object of the following study.

PART ONE

Toward Organology

> And so must the Uniformity in the Bodies of Animals, they having generally a right and a left side shaped alike, and on either side of their Bodies two Legs behind, and either two Arms, or two Legs, or two Wings before upon their Shoulders, and between their Shoulders a Neck running down into a Back-bone, and a Head upon it; and in the Head two Ears, two Eyes, a Nose, a Mouth, and a Tongue, alike situated. Also the first Contrivance of those very artificial Parts of Animals, the Eyes, Ears, Brain, Muscles, Heart, Lungs, Midriff, Glands, Larynx, Hands, Wings, swimming Bladders, natural Spectacles, and other Organs of Sense and Motion; and the Instinct of Brutes and Insects, can be the effect of nothing else than the Wisdom and Skill of a powerful ever-living Agent, who being in all Places, is more able by his Will to move the Bodies within his boundless uniform Sensorium, and thereby to form and reform the Parts of our own Bodies. And yet we are not to consider the World as the Body of God, or the several Parts thereof, as the Parts of God. He is an uniform Being, void of Organs, Members or Parts, and they are his Creatures subordinate to him, and subservient to his Will; and he is no more of the Species of Things carried through the Organs of Sense into the place of its Sensation, where it perceives them by means of its immediate Presence, without the Intervention of any third thing. The Organs of Sense are not for enabling the Soul to perceive the Species of Things in its Sensorium, but only for conveying them thither; and God has no need of such Organs, he being every where present to the Things themselves.
>
> –NEWTON, *Opticks*

Query 31 of Isaac Newton's (1643–1727) *Opticks* embeds the term "organ" in methodological problems spanning cosmology, epistemology, and theology. The passage is a terminological paradigm of the century that followed it. Organs are the biological marker of humanity and the object of that humanity's scrutiny of its own limitations. They are also the stamp of the divinely determined profane, the bearers of information and impulse, the animal part of the human that separates him from God. The shape of the complex of organs determines the organism, and this determination reveals the "intelligent Agent" who has built and might alter the construction. That agent, though he has a Sensorium extended throughout the world, needs no mediators, no "third thing." Organs are the essence of the mundane, the mark of its "very artificial" construction and of its limitation—they separate us from the "Things themselves."

Although Newton is still well-known for his dictum, "I do not feign hypotheses," the Queries he appended to his *Principles* project a theological cosmology in which the dividing and providing work of "organs" figures centrally. Organs, it will turn out, are of both cosmological and technical significance, since they are the letters in which the naturalist must read for syntax (regardless of who "invented" the language). They also, however, characterize the finitude of human cognition and volition—they separate us from the things they also allow us to represent to ourselves, to judge, to consume, to destroy. This double meaning of "organ"—metaphysical and epistemological—is here couched in terms that have remained determinative from Spinoza to cybernetics. "Organ" designates a cosmological function that demands explanation, whether theological or experimental. But "having organs"—the original meaning of the German *organisch*—also means being separated from precisely the knowledge of the order that bestows specific organs and arrangements of organs onto individual beings. From Spinoza's "aspects" to Kant's faculties, from the physiological differentiation of functions from Albrecht von Haller to the quantitative determination of proportional nerve impulse speeds in the Fechner/Weber law, the composite structures bearing functions are the object of speculative and experimental inquiry into the epistemic limitation of the inquirer. Thus God's lack of organs, which stands in this passage from Newton for the parallel tradition of unmediated representational knowledge. Epistemologically, the tradition stretches from Spinoza's *scientia intuitiva*, to Leibniz's "intuitive adequate knowledge" and Kant's and German Idealism's "intellectual intuition"; cosmologically, Laplace's Demon and to a lesser extent Maxwell's present us with organless knowers who, in the immediacy of their presence to the world, *produce* or alter that world according to the clarity of their insight.

Thus the organ ties the question of being to the question of knowledge, and as we shall see, the cosmological problem of force to the physiological problem of impulse. The organ unites and divides; its very semantics determine a duality that the term reconciles as a single condition—being "organic," or living. These overlapping problems played out in naturalist inquiry and philosophical speculation during the eighteenth century, not least because the unified physics of Newton spurred debate about the dual and apparently multiple conditions of life even as confidence in Aristotelian metaphysics disappeared. In each of these discourses, "organ" presented the problem of method. The "Organs of Sense and Motion" were the expressive node of whatever content the relevant science projected: the mark of the design of the "intelligent Agent"; the expression of the universal law in the individual phenomenon; the impulse-receiving and volition-bearing functions of the human; and the "tool" of a new kind of philosophy, the *organon* of metaphysics in an unstably self-designated modernity. They were methodological problem and object simultaneously. And yet, their very name—*das Organ*—was born only tortuously over the course of the Enlightenment.

ONE

Metaphysical Organs and the Emergence of Life:
From Leibniz to Blumenbach

No World: Metaphysics and Physiology

At the beginning of the eighteenth century, the world disappeared. It would be another century before Friedrich Heinrich Jacobi would coin the term "nihilism," heightening a charge—originally made against Spinoza—and a philosophical worry about acosmism, or lack of world. By that time, however, as Kant had put it in 1781, metaphysics had lost its status as "queen of the disciplines" and had become a mere "arena" for speculative polemics.[1] The early-eighteenth-century disappearance was of another sort: it was not the cosmos itself that was gone, but its contents. The richness of that world—its interlinking, lived reality, its creation, the knowledge of it—was reduced in the European theater to something else: regularities, laws. Isaac Newton's claims to nonknowledge about the true nature of forces (in spite of the cosmology of his *Queries*), complemented on the Continent by Christian Wolff's (1679–1754) claim to generate the world's content from the principle of noncontradiction, are only the antagonistic symptoms of the same felt disease, a simultaneously liberating and crippling doubt about the nature and the knowability of the world. Metaphysics had been legislated against in favor of lawfulness pure and simple: the world had disappeared into its own order.[2] The crisis of metaphysics—or the sense that the long eighteenth century was a series of unfolding metaphysical crises—was complemented by the political crises that erupted in 1776 and 1789.

The Romantics inherited these problems together, and their attempt to suture them by combining and redressing them with a metaphysical sense of the term "organ" is the object of this study. First, however, we need a sense of where the term came from—in other words, we need a discursive story about what made Romantic organology possible.

That story goes by way of the disciplinary boundary between philosophy and the life sciences. The literal sense of the term "organ"—functional part, whether internal or sensory—emerged only slowly in Germany. The conceptual locus of its emergence was the debate about the development of life, the question of "preformation" (development from a tiny, preorganized model) or "epigenesis" (gradual formation of the organism from unorganized matter). This debate was marked, however, by an absence of the term (at least in German). It was Immanuel Kant, in his 1790 *Critique of the Power of Judgment*, who gave epigenesis the term "organ," naming the proper object of a very old research agenda.

That debate, however, was also metaphorized by philosophers throughout the eighteenth century. Leibniz—who subscribed to preformationism—articulated a metaphysics of the organ, a world which was filled with organs, all the way down. Writing in French and Latin, he defended this vision not only metaphysically but also for the human intellect. With that shift to epistemology, the first glimmer of a metaphorology of the organ came into existence.

The Divine Preformation of the Mind: Leibniz's Metaphysical Organ

Sometime in the early 1690s Leibniz published a short treatise called *Plan for a German-Loving Fraternity*. Building on his plans to create learned and poetical societies that would extend the brief of organizations like the first

German-language academy, the Fruitbearing Society (1617–80), he argued for the Republic of Letters to be also poetic, poetic circles to be also scientific. Anticipating the cultural ambitions and anxieties of the German-speaking Enlightenment, he pointed to the instrumental superiority of the Germans: "We humans now have arrived, through the grace of God, at such great means to investigate the secrets of nature."[3] The "Telescopia" and "microscopia" "armed" human eyes—a metaphor that Novalis would take up more than a century later. These weapons could help humans to "a great insight into the deepest inner side of nature [*das innerste der natur*]"[4] were they only used correctly. For Leibniz, the vanity of German nobility stands in the way of his project, and the missing realization is that the German language, above other European languages, is a "mirror of the understanding" (a point he makes at length in his *Unanticipated Thoughts Relating to the Use and Embetterment of the German Lanuage*, ca. 1697). The apparatuses of science require a culture of science and inquiry, one linked to poetry by the national language. The analogy between these weapons finds it literal referent, however, in logic: "And beyond all this we also have improved the *organum organorum*, namely correct logic and the art of invention [*erfindungskunst*], far better than the ancients, both through the *analysin speciosam* as well as other means, such that one can say, our understanding is not less armed hereby, as our eyes are by the telescope."[5] Leibniz's combinatorial program, the dream of a logic that applies fully to the phenomenal world, has also "armed" the understanding.[6] But the "instrumente" of science and their implied military counterparts are German. "True logic," as much as it is German in its linguistic essence, is named in Latin: the "tool of tools" is an *organum*. If Leibniz's metaphysics stands at the beginning of a crescendo of calls in Germany for a philosophically and scientifically conscious culture and literature, it supplies the soon-to-be deafening roar of that cultural program with an essential yet ambivalent metaphor: the "organ." Leibniz never uses the term in its biological sense[7]—indeed, this sense developed only late in the eighteenth century. And yet, in giving the term a metaphysical meaning which could be extended to language, scientific apparatuses,[8] and indeed the operations of the understanding, Leibniz laid the semantic groundwork for the next century.

The beginnings of the metaphorization of the "organ" for metaphysics are in Leibniz. Indeed, he runs the categories of force and organ—in the larger debate about generation—through each other in a manner not meant to be metaphorical, but instead metaphysical. Adopting the discourse of preformation, he contains it in a larger static order, a world in which simple substances are in a harmony pre-*established* by God. Nevertheless, he uses the term in a quasi-metaphorical way to describe the mind, and it is this use we can track forward to Kant, and to Herder.

The *New System of the Communication of Substances, and of the Union of Body and Soul* of 1695 was the first public statement of a new emphasis in Leibniz's thought.[9] This was the doctrine of "pre-established harmony," meant to respond to a simultaneously scholastic and Cartesian problem about the interaction of substances.[10] The larger problem was that of the interaction of substances in general: given that substances are simple and have accidents, how are they to interact? Descartes tied this question to the sciences, defending a strict mechanicism that allowed for efficient causality (billiard ball on billiard ball) to be the exclusive model of physical interactions. This generated, however, another problem, that of *commercium mentis et corporis*: how could the two heterogeneous substances, mind and matter, interact? The answer developed in its classical form by Malebranche was that of "occasionalism": God's touch intervenes at each moment of such interaction, ensuring that representations are correct and that physical actions correspond to them.

Leibniz agreed that substances do not interact. In fact, the final statement of his metaphysics, the monadology, is absolutely isolationist in this respect: even physical substances do not actually interact. What establishes the appearance of interaction—and the correctness of common language about it—is a harmony preestablished by God.[11] There is no need, as he would emphasize against the Newtonian Clarke, for God to intervene again and again: God has not made the motion which decays over time (empirical interactions) but the force which is conserved.[12] The laws of the universe are overseen by this harmony between substances, including the body and the soul, mind and matter. Substances reflect each other harmonically, preordained by God—no interaction is necessary. At one important place, Leibniz describes this doctrine as "an agreement produced by divine pre-formation,"[13] putting a point on the analogy.

In an important passage in the *Theodicy*, Leibniz echoes Aristotle's rejection of organs for the mind, taking issue with Cartesian attempts to show that the mind can "guide" the body, a bit like a knight, though he gives no force to the horse he mounts, nevertheless controls it in directing that force as it seems good to him. But as this is done by means of the bridle, the bit, the spurs, and other material aides, one can easily see how this is possible; but there are no instruments of which the soul can avail itself for this effect, in the end nothing in either the soul nor the body—that is to say, neither in thought nor in mass—which could be used to explain this change the one brings about in the other. In a word, two things are equally inexplicable: that the soul change the quantity of force, and that it could change the direction of its movement [*la ligne de la direction*].

Johann Christoph Gottsched (1700–1766), the Rationalist philosopher and playwright, translates this passage into German with *Werkzeug* for "instrument"—no material tools mediate between body and soul.[14] And although Gottsched dissented from his teacher, suggesting that the soul was like a bow drawn taut, his suggestion seemed no better than the fineries of Descartes's pineal gland. The mind can have no organ, and where the body serves as its instrument, the materiality of that service is a mystery.

The conservation of force leads, in this case, to the preestablished harmony: no change in the quantity of force is possible, and so humans are restricted to the rearrangement of forces already present in the natural order. Further, the interaction of mind and body cannot be clarified by removing this element of putative spontaneous generation of force: any possible instrument for such an interaction would have to be *either* material or ideal, precluding the very possibility of interaction. And no minimalization of that interaction makes it more probable: "guiding" the force would require some means of guidance, some point of interactive possibility, some instrument. The metaphorical use of "instrument" is here precisely used as a rejection of an "organ" of material knowledge or action. It is God's world we know, by God's harmony of bodies and souls. The metaphor of the instrument is put under erasure: no philosophy of technology in the reflective, organic order of forces.[15]

Leibniz's increasing emphasis on the metaphysical nature of force can help us to see the connection of his system to the metaphorology he instigated. The category "substance" is both that which corresponds to the subject of a judgment[16] and that which is held together by a force. He writes in the *New System* that, considering the nature of a "true unity," he was led back to the substantial forms[17] of the scholastics: "I found then that their nature consists in force."[18] Indeed, an "original *activity*"[19] must be postulated to explain real unity at large. Force, it seems, is that which binds actively into a unity (we shall see Kant exploiting this notion in the next section). While the instruments of force can arise and decay, however, original activity cannot begin or end (as Herder would also maintain). The substances bound by force cannot, therefore, originate or be annihilated. Leibniz reaches to the model of preformation, expanding its sense from a narrow, biological principle to a metaphysical doctrine, indifferently referring to bodies and souls.

The *New System* holds that true substances have always existed, a doctrine that forced Leibniz to confront the problem of the transmigration of souls. If the soul always exists, could it be that it is attached to different bodies at different times? Rejecting this possibility, Leibniz develops the notion of a substantial unity he calls an "organic machine": the preformed and always-existing animal. This machine is testament to God's infinite power, compared to human *techne*, which operates or seems to operate in an entropic world.[20]

Rather than transmigration, so Leibniz, "transformation" is the appropriate term.[21] Referencing the experiments of Swammerdam and Malpighi (seventeenth-century preformationists armed with early microscopes), Leibniz offers a reason for the doctrine of preformation: since simple substances cannot be generated—since their force is original—they must instead be preformed. Rejecting "atoms of matter," he suggests that there are "atoms of substance" or metaphysical "points," held together by this original activity. This force-bound unification extends—but only by God's power—to the attachment of one soul to one body. Indeed, the body and the soul are always attached—as the *New Essays* read: "Death can only be a sleep, and not a lasting one at that."[22] In the *New System*, Leibniz expands: "Consequently, instead of the *transmigration* of souls, there is only a *transformation* of the same animal, according to whether its organs are differently enfolded and more or less developed."[23]

The organs in this passage are simply the organic parts, their distribution changing in transformation, their existence contemporaneous with the whole. *Organ* is, then, a medical concept. The metaphorical sheen of the passage is merely an empirical prejudice we (or perhaps Hume) might hold: *organ* seems like a metaphor because the metaphysics in which it occurs is deeply unfamiliar—for Leibniz, literal organs occur in all substance, corporeal or rational.

Indeed, the nature generated by God is full of organs, is an infinite intension of organs: "Every natural machine (and this is a true but rarely recognized *distinction between nature and art*) is made up of an infinite number of other organs."[24] Leibniz here anticipates the arguments he would make against the Newtonians in his famous correspondence with Clarke: arguing against occasionalism broadly and mere mechanism locally, he there would state that God's creation is not a "watch" which loses time—if God is a craftsman, his machines have *perpetuum mobile*. In the *New System* the "moderns" come up for criticism on precisely this point: they lack a distinction between natural and artificial things, a difference "not simply . . . of degree, but a difference of kind."[25] The natural machines "have a truly infinite number of organs . . . [and] remains a machine in its least parts . . . being merely transformed through the different enfolding it undergoes."[26] The functionality of the organic is machinic all the way down: each part down to infinity is a preformed recapitulation of the whole, bound by an original activity supplied ultimately by God. What in the late work would be called "monads" are here imagined—at least for nature's beings—in infinite unities with each other, not substantially interacting but reflecting each other.

Human making, on the other hand, merely unites aggregates that can decay. Its organs are merely instruments, expressive of perfections of the whole when enough artistry is involved, but nevertheless parts capable of disintegrating from the unity established. The difference between the organic and the mechanical is the *degree of artifice*, infinitely higher in the organic, but differing only in degree rather than kind from its machinic model.[27] Leibniz's assumptions here are nevertheless not purely Aristotelian, and he has indeed established a more general positive sense of the term "organ" by using it indifferently for bodies and for souls (for substances in general, that is). Yet the technologies of God and of humans differ essentially—any possible philosophy of *techne* is suppressed in the fullness of being preestablished and preformed. The quasi-metaphorical organ is the sign of perfection, the expression of force's regular unity, but in the hands of the human, a mere point of possible decay. The "fullness of nature" denies the possibility of innovation. In Blumenberg's terms, metaphorics has not yet escaped metaphysics. As we shall see, the problem is that an organological metaphysics has not yet appropriated the efficacy of metaphorics for itself.

This indifference with respect to bodies and souls (in favor of a substantial metaphysics), combined with the static system in which force is active, is reflected in Leibniz's epistemology—and this model survived into rationalism at large.

Where nothing radically new could be created by humans, Leibniz's epistemology and its formalization in the various *Logics* in the Rationalist school look very much like a toolbox, a set of instruments for knowing that larger Being in all its forced-based dynamics—and this despite the refrain that the soul can have no instruments.

Leibniz's epistemology is a matter of how clearly and distinctly we represent: revising the criteria for Descartes's "clear and distinct ideas,"[28] he established a framework for making distinct that which is already clear. The Leibnizian cognition-tree runs as follows: what is confused is representation without objects. The ability to identify an object makes knowledge clear. Clear knowledge can become distinct when we can identify the properties that allow for recognition of the object over multiple instances. (Leibniz's favored example here is that of an assayer recognizing gold by its yellowness combined with a certain density, etc.) Distinct knowledge can become adequate when the primitive qualities of the object are included in the representation (so, for example, the atomic makeup of gold). Even adequate knowledge, however, can be bettered, if the elements of that primitive makeup can be included in a single representation of the object: Leibniz calls this type of presentational knowledge intuitive as opposed to symbolic. The latter uses signs as markers or reminders of elements not presented; the combination of synthesis (into a single representation) and analysis (from constituted object all the way to primitive property) is called "intuitive adequate knowledge."[29] The logical treatises of Wolff, Baumgarten, and Meier all take up this model, offering refinements and explanations of these means of arriving at true knowledge. Kant was weaned on them, and taught them to Herder. This search for a means of betterment of knowledge is where the creative historical energy of the rationalist movement is to be found—and it is the field that Kant intentionally entered with his early methodological considerations, as we shall see.

What is established by this system is both the notion of epistemological means (not yet metaphorically "organs") and a static order known by them. David Wellbery has labeled this model "progressive semiosis":

> Sign-use is situated, then, between two types of intuition, the sensate intuition [the basis of confused "knowledge] of our perceptual experience and the intellectual intuition of divine knowledge. The Enlightenment myth of the sign localizes sign use between two experiences of plenitude and presence—an original perceptual experience in which the world reveals itself directly, the *arché* which grounds man's subsequent symbolic representations, and the *telos* of divine cognition in which the world is again experienced intuitively, but with total distinctness. The movement from one pole to the other, the advancement of culture itself, is a process of progressive semiosis.[30]

Wellbery focuses on the representationalism of this system but points to its cultural ramifications: critical development of that system meant "advancement" in the direction of critical reflection, although the represented system remained the unalterable substratum.[31]

There is a deep structural analogy between the force-model of substances in their preestablished harmony and our cognition of beings or progressive semiosis: each of these models posits a fundamental dynamism contained in a static order. And indeed, they touch at the point of God's knowledge. This analogical presentation makes the organ—whether of an organic machine or the ladder of knowledge—a candidate for reflection. Methodology—*how we know*—comes into view, albeit encapsulated, preformed, in the order of things. As the *Discourse on Metaphysics* assures us: "It is evident that all true predication has some basis in the nature of things."[32] The exploration of means has not yet undermined faith in the representational principle—by a literal analogical structure, we can know the enfolded and unfolding, preestablished world.

It is this substance-based structural analogy—the world of Leibnizian metaphysics—that provided the basis for Leibniz's initiation of the metaphorology of the organ. In that metaphysical world, with those methodological means, the question of mind had to be treated, too.

Leibniz's doctrine of the intellect was articulated *in nuce* in a single comment. Responding to the Aristotelian claim that "nothing is in the intellect which was not first in the senses" (*nihil est in intellectu quod non prius in sense fuerit*), Leibniz appended: *nisi ipse intellectus* (except the intellect itself).[33] With this statement, Leibniz entered a controversy about whether there were "innate" ideas. Locke, in particular, defended an empiricist sensationalism in which the mind was *tabula rasa*, and all representation arose from physiological interaction. Leibniz thus took direct issue with Locke in order to elaborate his statement into a doctrine of the mind.

The *New Essays on Human Understanding* were written in 1704, withheld because of Locke's death, then published only in 1765, when they were quickly translated into German. They take the form of a dialogue between a defender of Locke and a mouthpiece for Leibniz. The dialogue is a running commentary on the chapter of Locke's *Essay Concerning Human Understanding* (1690). The preface immediately invokes the central difference: the constitution of the human intellect. Reviewing several doctrines of "innate" ideas (the *prolepses* of the Stoics, for example, and the *zopyra* or "little sparks" of Julius Scaliger), Leibniz announces that he will oppose Locke's view that the mind is *tabula rasa*. The establishment of the reasons for things, and the universal necessities observed in mathematics, metaphysics, and ethics, Leibniz tells us, may well only occur to us when the senses awaken our attention—but they cannot come from the senses. Instead, Leibniz asks, "why could we not provide ourselves with objects of thought from our own depths, if we take the trouble to dig there?"[34] Ultimately, the entire universe is reflected in each monad, and Leibniz here expresses this in terms of the doctrine of unconscious perceptions, *petites perceptions*, which go unnoticed as individual waves are not heard in the roar of the ocean, though we know that they constitute the latter. This indifferent ocean, perceptive and perceived, *organ*ic to infinite smallness, establishes the appearance of equilibrium but is in constant activity, productive of our knowledge and even our actions by pushing us in certain, preformed directions, by producing tendencies.[35]

Book 1 then begins with an account of the "new system,"[36] repeating the arguments about preformation and metempsychosis we saw in the *New System*. Responding to Philalethes's (Locke's) assertion that there are "no innate ideas," Theophilus (Leibniz) states that "the new system takes me even further and . . . I believe indeed that all the thoughts and actions of our soul come from its own depths and could not be given to it by the senses."[37] For Locke has failed to distinguish "the origin of necessary truths, whose source is in the understanding, from truths of fact, which are drawn from sense-experience and even from confused perceptions within us."[38] Philalethes objects that for there to be innate thoughts, there must be a (necessarily obscure) faculty for such thoughts. Theophilus rejects two premises in this argument: on the one hand, no such faculty is needed, only rather a potential; on the other, there are no innate thoughts, rather only innate general principles, rules which we do not always consider but function like muscles or tendons for walking.[39] Certain truths—those of necessity—form an affinity with the human mind: "that is what makes us call them innate. So it is not a bare faculty, consisting in a mere possibility of understanding those truths: it is rather a disposition, an aptitude, a *preformation*, which determines our soul and brings it about that they are derivable from it."[40] The disposition of the mind, from which it internally draws necessary truths—those of metaphysics, arithmetic, geometry, and ethics—is thus cast metaphorically in the terminology of the biological debate. The mind is imagined as a slab of marble, which, far from being indifferent to the sculptor, is instead veined. The sculptor has sought *this* piece of marble—as God has preformed *this* type of mind—because he sees that its form is amenable to its intended function: to represent Hercules, or to represent and act upon the enfolded but unfolding truth. Unlike "vulgar" forms of innatism, Leibniz's doctrine is founded in a sort of preformation of the mind,[41] which has metaphorical veins as principles of application, an established field for the investigation of truth. The mind is not generated but created, given in shape but appearing only slowly, as our attention to its given affinities is applied. The means for knowledge might remain obscure to us, but our process of self-discovery is just that: the exploration of an order formed in advance, dynamic in its confused and even apparently chaotic development, but complete and finally known in the intuition of God. Organs are implied in this preformationist metaphor—its use in this context would not emerge until Kant, and then Herder, entered this discourse given by Leibniz.

Life's Origin: Preformation, Epigenesis, and the Question of Force

Leibniz characterizes the comos as "full of organism" (*pleine d'organisme*) in the singular. An "organism" is the set of relations that reflect the functional plan of an individual; the term came to mean "living being" only in the 1830s.[42] As we have seen, Leibniz's organs are metaphysical, demonstrating divine preformation and presenting, much as Newton's organs had, the dividing line between God's *techne* and ours. One might reasonably expect that the life sciences—which developed the term "organism"—also contributed the central elements to the emergent term "organ." And this is indeed the case, but the story is long and twisted.

Adelung's *Grammatical-Critical Dictionary of High-German Speech* (*Grammatisch-kritisches Wörterbuch der hochdeutschen Mundart*) contains a short, early entry for *Organ* in its 1777 edition. This entry defines the term as stemming from Greek and Latin, and having the meaning "especially the tools of the external senses [*die Werkzeuge der äußern Sinne*], of sensation, though in a wider sense also of changes on and in bodies."[43] What the soul can make use of is slowly transferred, at first metaphorically, to the noninstrumental body. This seems to be the "high German" use that the *Adelung* is registering. Note the difference to Gehler's *Physical Dictionary* [*Physikalisches Wörterbuch*] just over a decade later in 1790: "We call organs or tools in general bodies that are built such that certain purposes or effects can be achieved, as e.g. the eye, ear, and the remaining tolls of the senses of the animal body. The vessels in which juices circulate, which serve the nourishment of animals and plants, are thus also organs, and one attributes an organic structure [Bau], or an organization, to those natural bodies in which such a circulation of juices through vessels occurs."[44] The semantic extension of "organ" to cover the various function-bearing complexes of "animal bodies"—and the *Organisation* that was then slowly associated with the phenomena of the living—presents a sea change in the term's history. Gehler's definition is the paradigm of the modern meaning, achieved through the attribution to circulation systems. Organs function to carry blood, electricity, and other "fluids" through the body, and this function has no assigned purpose beyond itself. But this meaning was registered only slowly, through the absence of the very term "organ" in life-scientific discourse. If Leibniz had given the metaphysical and order-based concept to the Germans in French and Latin, the biological concept was supplied by scientists who also did not translate their term. The century-long standoff between Newtonian physics and the science of embryology resulted in a lexical lacuna filled only later, when the entire debate was reiterated in new philosophical terms after Kant.

Aristotle writes, in *De generatione animalium*: "Nature acting in the male of semen-emitting animals uses the semen as a tool [*hos organoi*], as something that has movement in actuality; just as when objects are being produced by any art the tools [*ta organa*] are in movement, because the movement which belongs to the art is, in a way, situated in them."[45] The *physis-techne* analogy is here applied to the moment of animal generation itself, and Aristotle thereby left a terminological legacy within this problematic that was taken up again in the eighteenth century.[46] Aristotle's own application of hylomorphism to the problem of generation (sperm is active form, ovary passive matter) formed the basis for a fierce ideological debate—again, precisely as scholastic Aristotelianism came to an end—about the origins of life. And in fact, the debate about life's origins—precipitating into preformationist and an epigenesist positions—was marked by the singular absence of the *term* "organ," even as it prepared the systematic background for the metaphorical emergence of the term.

The debate about generation was, far from academic, a deep-structure confrontation between physics in its post-Newtonian ascendance and the emergent life sciences. The formation of organic beings from seemingly inorganic matter took on urgency with the rise of the microscope[47] and the rise of Newton's mechanical model. The microscope allowed more precise observation of the fluids which went into reproduction—yet the debate transcended observable formations.[48] Newton's system had two relevant sides: on the one hand, it allowed no incalculable forces to be postulated, and this, on the other, resulted in a catholic mechanicism about nature. For Newtonians, there was simply no way to explain the development of organic bodies—no force or principle could be feigned, since it could not be formalized mathematically. This led to the position called "preformationism," essentially the doctrine that a small version of the body preexisted either in the sperm (animalculism) or in the egg (ovism). (The most celebrated authors holding this view were Charles Bonnet (1720–1793) and Haller, who wavered on the issue.) As Peter Reill has pointed out, this position dovetailed nicely with theological presumptions (God created all organic bodies in a fixed number at the Creation), and supported absolutistic politics.[49] Attempts to find these preexisting *Keime* (seeds) failed, but the microscope's limited capacity left their possibility open.

The limitation of hypotheses brought the question of the nature of force into focus. Without turning their back on Newton's methodological modesty, it was possible to maintain that force was not merely quantitative, but instead also qualitative. Blumenbach understood himself explicitly in precisely this way:

> I hope that it is superfluous for most readers to remember that the term *formative drive* [*Bildungstrieb*], just as the words attraction, gravity, etc. should serve to signify nothing more or less than a force, the constant effect of which has been recognized in experience, the *cause* of which, however, just as the cause of the aforementioned forces of nature that are so widely acknowledged, is for us a *qualitas occulta*. What Ovid says is valid for all these causes:—*caussa latet, vis est notissima* [the cause is hidden, the force is most in evidence]. The gain in studying these forces is simply this, to determine their effects more closely and to reduce them to more general laws.[50]

Appended to the obvious reference to Newton is a footnoted quote from the *Opticks*, including, "I use that word [attraction] here to signify only in general any *force* by which bodies tend towards each other, whatsoever be the *cause*."[51] Blumenbach's points of comparison are contractibility, irritability, and "sensility."[52] These forces were not physical, formalizable in the manner of dynamics. Yet their postulation was a matter of method, indeed, the modest method of Newton. The cause was not speculated upon, while the expression of the force became the object of investigation.[53]

This view thus developed into the competing notion, *epi-genesis* (generation "on" or "above") to articulate a nonmiraculous, qualitative, force-based generation. Various versions of the force molding the produced being were articulated and defended in the eighteenth century. Georges-Louis Leclerc, Comte de Buffon (1707–1788) proposed an "interior mould" (*moule intérieure*); Caspar Friedrich Wolff (1735–1794) an "essential force" (*vis essentialis*); Johann Friedrich Blumenbach (1752–1840) a "formative drive" (*nisus formativus; Bildungstrieb*). Each of these attempted to infer from the forms of organic beings form-producing dispositions or *habitus*. The mechanism of development was missing, and repeated attempts at naming it failed to specify its work. This problem, of course, remained virulent until the 1950s. But in searching for a "device" that guided generation, Enlightenment physiologists might have reached into our metaphorical field.

Yet the discourse is marked by the absence of a significant use of the term "organ."[54] Wolff, who claimed that an epigenetic force (the "essential force") guided growth, sometimes use the term "organic" to indicate those beings that included nonorganismal parts as functions in their bodies—the term "organ" was needed, but occupied no systemic place in either conception.[55]

The debate was largely settled by Blumenbach in 1781, with his "On the Formative Drive" (*Über den Bildungstrieb*). His description of the positions runs as follows:

> Namely, one either assumes that the mature but otherwise raw and unformed generative material of the parents, when it reaches its determined place at the right time and under the required circumstances, then is gradually formed [*ausgebildet*] into the new creature. This is what epigenesis teaches.
>
> Or on the other hand one rejects all generation [*Zeugung*] in the world, and believes on the contrary that for all humans and animals and plants that have ever lived and ever will live, *the seeds* were created immediately at the Creation, such that now one generation after another merely needs to *develop* [*entwickeln*]. This is why this doctrine is called Evolution.[56]

Blumenbach points to the similar semantic content of German *entwickeln* and Latin *evolutio*: to "unwrap," "to fold out," not to change in form but merely to grow. The debate centered on the problem of the nature of matter, or rather matters. Indeed, the eighteenth century saw a fierce debate in the British, French, and German worlds (and between them) on the nature of force, a concept Newton had made urgent with his investigations into gravity. The question of how force could be properly treated in science and in philosophy caused a debate about the nature of matter that dovetailed with the more intimate knowledge of that matter afforded by the microscope.

If the epigenesists adopted Newton's conception of force, however, they did not adopt his suggestive use of the term "organ." In Query 31 of the *Opticks* (the epigraph to this chapter), Newton had written (in keeping with British physiological usage) of the "organs of sense and motion" (eyes, ears, but also arms, bladders, etc.) in humans. The passage is famous both for its defense of the metaphysics of force and for its presentation of the "method of analysis." God formed the organs of animals, but had none. The origins of our mediators were to be found, the passage suggests, in the unmediated but universal mediator (more on Newton's conception of God's *sensorium* in the section on Herder in the current chapter below).[57]

Albrecht von Haller, in the author-overseen 1770 translation of his physiology textbook, wrote: "Thus, we want to go through the senses piecemeal, and then observe what is common to all of them and what in the soul follows on the changes of the *tools of sensation* [*Werkzeuge der Empfindung*]."[58] The orginal, however, reads *mutationes sensorium*, the changes of the sensing apparatus, translating *aistheterion*.[59] Haller, whose doctrines of irritability and embryogenesis veritably cried out for the term "organ," wrote in prose that avoided the term. In fact, the conceptual-semantic nexus was missing—Haller articulated the conceptual problem of the organ most precisely but did not change the terminology or the conception.[60]

Haller use the term (for the heart) when writing in *French*, as does Leibniz.[61] But here it seems to mean "device" or "mechanism." In the case of sense-organs, some terminological indeterminacy was productive: one could say a "tool" of sense since an intelligence uses that tool for a purpose—knowing.[62] The term remained attached to the problem of purpose, which followed physiology, medicine, and biology well into the nineteenth century. And yet its lack of systemic importance meant that it still seemed a foreign word in German until about 1790.[63]

The problem of the emergence of organic beings from seemingly "dead" matter thus drove scientific debate squarely into a philosophical register, one in which Leibniz participated from the 1690s onward, arguing that vital force was a metaphysical principle that had to be included, although only partially, in physics for its results to sync up with that of philosophical investigation. Kant's first publication was a consideration of this question, too. And when Herder entered the public sphere in the 1770s, the word *Kraft* was never lacking—nor was *Organ*. Thus, the confrontation of the life sciences and physics—and the emergence of biology as a discipline around 1800—involved both substantive and methodological debate offering central material for argumentation in an uncertain period for first philosophy. The borrowings of the terms "preformation," "epigenesis," and "organ" in Leibniz, Kant, and Herder all center around questions of force and our knowledge of it. Indeed, in each case, what is at stake is the mirror-question to that posed by the biological debate: not "what is the nature of matter?" but "what is the nature of reason?" Or, in other words, with uncertainty reigning about the nature of the known, the nature of knowing became a methodological problem, indeed, a metaphysical problem.

It is not, however, a merely analogical process by which the metaphorization of the epigenesis debate played out. That debate was always a question, at heart, of organicity. In §64 and §65 of the *Critique of the Power of Judgment*, Kant separates between mechanical and organic causalities. The first is a necessary link between two phenomena that is unidirectional: cause results necessarily in effect. This had been a generative explicandum of the *Critique of Pure Reason*: when the concept "cause" is judgmentally linked to a concrete set of phenomena, its necessity cannot be derived from those phenomena, *pace* Hume. This problematic led Kant to the invention of the "understanding" (*Verstand*) as that faculty which establishes and bears the weight of "objective validity," or "legislates nature"—a faculty supplying the necessity lacking in raw sensation, constituting nature by contributing elements like necessity to our phenomenal experience of it. For various reasons (to which we will return), Kant's "understanding" only functions according to this unidirectional necessity in the establishment of causes. In the *Critique of Pure Reason* (§65), he calls the causal link *nexus effectivus*—the effective link, with obvious reference to the Aristotelian concept of the "efficient cause," in Early Modern terms. Here, however, Kant allows for another type of causality thinkable by reason: mutual causality, where effect is cause and vice versa. This type of causality can be thought but makes no contribution to the understanding's legislation of nature. It is a candidate (a good one, it turns out) for regulative judgments, the types of judgments we must make to function practically (and, in this case, scientifically). The causal link here is called *nexus finalis*—Aristotle's "final cause," or *telos*. Indeed, Kant thinks we must use the concept of an "end" to interpret organic beings. And we encounter the type of judgment described here daily, for example, whenever we make judgments about human artifacts. Kant's example is that of a house which is both cause of the income that comes to its renter, while possibility of that income was also the cause of the construction of the house. Here the "end" is easily placed in the reason of the maker. The concept (house or income) has to determine the mutually causal relations *a priori*: the one is for the other in the reasoned representation (*Vorstellung*) of the renter/builder.

This example applies safely to all artificial objects—indeed, this is why the *Critique of the Power of Judgment* juxtaposes *aesthetic* and organic objects, under the heading "technical," because they are apparently built according to a plan or idea that precedes their objecthood. The final cause is in the mind of the maker. In the case of the organic object, however, the appearance of organization cannot be explained by immediate reference to a concept determining the object (that is, without introducing a *deus ex machina*). Ultimately, Kant argues for a regulative conclusion to God's moral universe on the basis of our encounters with organic beings. Here, however, we only need see that his model of the organic itself picks out the problem at the heart of eighteenth-century debate about life.

Kant defines the organic object as follows:

> In such a product of nature each part is conceived as if it exists only through all the others, thus as if existing for the sake of the others and on account of the whole, i.e., as an tool (organ), which is, however, not sufficient (for it could also be an tool of art, and thus represented as possible at all only as an end); rather it must be thought of as an organ that produces the other part (consequently each produces the others reciprocally), which cannot be the case in any instrument of art, but only of nature, which provides all the matter for instruments (even those of art): only then and on that account can such a product, as an organized and self-organizing being, be called a natural end.[64]

Even as Kant reaches for the Greek origin to determine the concept of the "part" in a nonmechanical sense (he goes on to point out that a watch can have no true organs), he makes a structural analogy (and this is all we are, in fact, supposed to be able to make, in the life sciences) from the judgmental *nexus* (that which relates part to whole and vice versa in the reasoned judgment) to organic "part": that which establishes the mutual mereological relation is, at its root, simply a tool. But this tool cannot be one of mere "art"; instead, it must belong to that nature which "delivers all material up to tools." Organization occurs by making organs that contain the traces of the force that builds them: nature itself.

The question of which force is using the tool was always at the heart of the whole debate. Indeed, the organ's correlative force is what Blumenbach sought in explicitly Newtonian (albeit qualitative) terms. The literal use of the term in biological debate, Kant shows us here, is already a matter of how we judge, and the ontological problematic is analogically mirrored by differing faculties (the understanding; reason). This became a central problematic for the next generation.

Here, however, we need merely see that the debate was always about this problem: the functionality of the literal organ (say, the heart) is exactly what Newtonians lacked an explanation for, what had to be pushed back into the mystery of God's creation. Microscopic explorations of the part-whole relation, as they increasingly tended to be the investigations of unobservable but organic forces (eventually called *formative drive* by Blumenbach), became investigations of the formations of organs as the immediate bearers of those forces. Looking to the Haller-Wolff debate, for example, we find that the bulk of their correspondence is taken up with questions of the formation of organs from blood flow. As Wolff struggles to make his epigenetic point to the physiological master, he focuses his microscope on pulsations in the first hours of gestation, hoping that the formation of the functional animal will become not only visible, but also communicable.[65] His hesitating tone does not prevent him from insinuating that a guiding force is forming the organs in the earliest stages of life—his *vis essentialis* accruing matter to the form of the animal.

As Helmut Müller-Sievers has argued, Blumenbach's intervention did not come as the "hard" result of an experiment (he calls it a "purely textual event").[66] Rather than seeing this development as merely "ideological" (as subject to a possible revisionist "critique"), I think this ideational shift should also be seen as productive, specifically of a terminological field in which the borrowing of the term "organ" became increasingly likely. And indeed, in this sense, we can see Kant less as a philosophical commentator on natural science and more as a producer of importance discourse about the organic. He set the terms of the debate going forward, if not the experimental program.[67]

But while the term came late, the philosophy did not. The major fault line between Wolff and Blumenbach is in the nature of the force inferred from the observation of formation: as Blumenbach points out, the *vis essentialis* is an almost mechanical force, merely "gathering" the necessary material for formation. The *Bildungstrieb*, on the other hand, actually does the organizing itself—it *is* the form of that organization. Blumenbach's *aperçu* came from the regeneration of parts in polyps. As he noted, that regeneration occurred in a miniaturized form: the new limb was formally similar but smaller than the original. Rather than merely collecting matter for an already (but inexplicably) organic being, the *formative drive* actively formed whatever matter was available to it, and this matter, while not itself organic, had "traces" of that forming activity in it:

> One cannot be more deeply convinced than I am of the powerful chasm that nature has fastened between enlivened and unenlivened creation; and I cannot ignore, in spite of all my respect for the sharp sense with which the proponents of the gradual sequence or continuity of nature have put up their ladders, how they want to get through, at the transition from the organized spheres to the anorganic, without a really somewhat provocative leap. But this doesn't disallow one from using appearances in the one of these two main parts of creation for the explanation of appearances in the other: and thus I regard it as not one of the lesser arguments that proves the existence of the formative drive, that the traces of formative forces are so unmistakable and general even in the anorganic.[68]

The question of organicity is here brought to its literal apex: although different types of matter—with differently investigateable forces—exist, they interact, and the traces of that interaction are the very moments of unification establishing the organic order. Blumenbach makes no special use of the term: Kant combined its literal meaning with its etymology to give conceptual articulation to the problem.[69] Kant turned Blumenbach's objection to Wolff on Blumenbach himself, asking if the *formative drive* was not "the explanation of one unknown by another unknown"[70] a charge he would also level against Herder. Kant calls the theory "generic preformation," a condition for talking about individual organization—as we shall see, it is this necessary and necessarily lawful, yet unknowable, basis, that provides Kant with the analogy to the cognitive apparatus.

The question of organization is, in fact, rooted in the organ, as the immediate expression of the interaction of forces. The mutual causality of "part" and whole—that is, the unified phenomenon—is rooted in the conceptual problematic of that union itself. Kant brought his perennial concerns with synthesis to bear on a problem with deep theological and scientific charge, opening up the term "organ" to philosophical and metaphorical use. The analogy between the predicative nexus in judgment and the causal nexus in scientific reasoning opened the field of natural philosophy to critical or transcendental investigation—this investigation would come to be called organology. Before that naming, however, came a metaphorization of the terms of the debate about life for the debate about cognition. Kant and Herder, as we will now see, literalized the term for the purpose of analogizing the mind to the body. Literalization's first phantom form was instrumental: the concept was first needed so it could be used metaphorically to describe Reason.

TWO

The Epigenesis of Reason: Force and Organ in Kant and Herder

> Observation and analysis of the appearances penetrate into what is inner in nature, and one cannot know how far this will go in time. Those transcendental questions, however, that go beyond nature, we will never be able to answer, even if all of nature is revealed to us, since it is never given to us to observe our own mind [*Gemüt*] with any other intuition than that of our inner sense. For in that lies mystery of origin of our sensibility. Its relation to an object, and what might be the transcendental ground of this unity, undoubtedly lie too deeply hidden for us, who know even ourselves only through inner sense, thus as appearance, to be able to use such an *unsuitable tool of investigation* [*ein so unschickliches Werkzeug unserer Nachforschung*] to find out anything except always more appearances, even though we gladly investigate their non-sensible cause.
>
> IMMANUEL KANT, *Critique of Pure Reason*

Our tool for analysis, according to Kant, is "ill fitting" or unsuitable for what might be taken to be its ultimate goal: the knowledge of the origin of our cognition. To the extent that we know our own faculties, however, we can inquire into nature with no expectation that we will find limits. Because, for Kant, we give the law to a nature that we constitute for our own experience, our knowledge of that nature is potentially infinitely deep, even as we are unable to trace that very knowledge into source. For the Kant of the *Critiques*, our means of research are tools. But this was not always so. Earlier, Kant had considered these means "organs," and had even considered his project the search for an *organon* for metaphysics. By the 1780s, he had demoted *organ* and *organon* to aids, guides, dogmatic counterparts to the analytics of Critique. But the earlier project remained in conceptual traces, and it did not take the Romantics long to unearth a layer of organs in the silt of Critique.

Kant's work is filled with the dual problematics of the *organon* (the tools for practicing a discipline) and the organ. From his earliest works, in which an *organon* for metaphysics is sought, to his critical system, in which that *organon* is replaced by a *canon* (a positive body of law) *of the understanding*, he brings the term "organ" to bear on similar problematics. By also characterizing the critical system as an "epigenesis of pure reason," Kant furthered the metaphorology in question, tying a critique of Leibniz to a shared—and borrowed—natural-scientific terminological apparatus. He thus came close to anticipating the Romantics' intentional conflation of the two terms. Ultimately, however, his contribution to the metaphorology of the organ is methodological. By thematizing the necessity of radical methodology in metaphysics as a response to its century-long crisis—and by placing the terms *organon* and "organ" front and center in that revisionary effort—he supplied a benchmark and a warning for any innovative use of the term. That warning, of course, extended to (any possible future) metaphysics itself.

Johann Gottfried Herder, who had studied with Kant in Königsberg in the 1760s, supplied the cosmology—the *Weltbild*—of organology. Herder came to reject the term "epigenesis" (generation "on top of"), favoring the notion of *genesis* pure and simple. Arguing for a cosmic plurality of fundamental, pullulating dynamic forces, Herder set the agenda in terms of content for the emergence of Romantic organology. His system, like Leibniz's, was entirely composed of corruptible organs expressing dynamic forces, but he added the possibility of the new, the emergence of genuinely new being in the order of things. He extended the metaphorology of the organ to the mind, as Leibniz and Kant had done before him, but he did so using analogy as his tool. For him, the "grand analogy of nature" gave way to the encompassing perception (and, ultimately, love) of God in a *sensorium commune* of language, time, and space. Reflecting this knowledge through the midpoint of nature *as analogy* were the human organs, on a scale from material to spiritual, and rooted finally in an "organ of language," an autonomous ability—the ability to identify one's proper species—reflecting the godly analogy but also affording freedom.

Reason's Awakening: Kant's "Epigenesis of Reason" and the Methodological Organ

Kant's contribution to the metaphorology of the organ—and indeed, his contribution to Enlightenment philosophy at large—was driven, in my reading, by a single question: How is it that we are rational in a world whose proper rationality we must remain agnostic about? From the 1760s onward, when the Academy forced the issue of metaphysical certainty with its prize competition of 1760–63, Kant pursued this question. Developing a critique of rationalist and eclectic[1] metaphysics simultaneously, Kant noted that they shared a lack of an articulated means by which cognition was to fit with being. By prying open that question—by bracketing, if not completely, the world—Kant opened a field of methodology in metaphysics of which, I shall claim here, his own articulation of the critical philosophy was but one possible result. At the opening of that field, the *organon* for metaphysics was a possibility. The critical standpoint, however, rejected such an *organon* in favor of a canon (of the understanding), and a discipline based on that canon. Our rationality, whatever the status of the world, was self-guiding and developing, its law capable of explicit statement. But the reason for its laws—what an instrument of reason was meant to provide—remained necessarily obscure. Kant cast this problematic in terms of biological debate, writing that the understanding's contribution to the world of its cognition was an "epigenesis of pure Reason." The metaphor of the rational organ emerges, then, with Kant, but remains an unexplored possibility. Leibniz had supplied the metaphorological field; Kant would offer the critical methodology.

At the methodological apex of the *Critique of Pure Reason*, Kant writes that the result of his "transcendental deduction" of the categories is a sort of "system of *epigenesis* of pure reason."[2] The Leibnizian problematic of the origin of necessary truths was always central to Kant, too. Indeed, he begins the introduction to the B edition of the *Critique of Pure Reason* with a translation of Leibniz's revision of Aristotle: "But although all our cognition commences *with* experience, yet it does not on that account all arise *from* experience."[3] The paradigmatic case is causality: the necessity inherent to the causal judgment cannot accrue slowly in experience, as Hume had had it. Experience is necessary for empirical causal judgments, of course, but the element of necessity cannot be derived from that experience: order does not come from phenomenal presentation but from intentional *re*presentation. This problematic reveals why Kant uses the metaphor from the life sciences: as he reasons in §27 of the *Critique*, either experience makes necessary ideas possible, or those ideas make experience possible. The first alternative is characterized as *gen-*

eratio aequivoca, the inexplicable generation of one kind of substance from another. The second alternative is embraced: the system of pure reason is epigenesist because there is a genuine formation process between intuitions and concepts, and this mixing process actually constitutes experience. Guiding the process is the "system" of that epigenesis, its own origins obscure but its rules organized around the categories and beyond doubt for us humans, like an "animal body, the growth of which adds no member but which, without changing its proportion, makes each member stronger and better suited to its purposes."[4] Kant embraces epigenesis as a metaphor[5]—how completely, we will see below—to characterize the constitution of experience, the "legislation of nature" which the understanding performs. Yet in the elaboration of the "architectonic" of pure reason, his metaphor ambivalently reaches to the preformationist model, which dictates the persistent proportion and preexistence of the parts. No plastic organ is entertained, then, in the final account, and Kant even speaks of the elaboration of a science of pure reason as an "idea" which "lies like a *seed* in reason, in which all the parts, still quite wrapped up and hardly recognizable under microscopic investigation, lie hidden."[6] The elaboration of the science of pure reason, the justification of which is the transcendental deduction, is the transformationist model of preformation.[7] The deduction itself defends an epigenetic model for reason, then, while its elaboration has an invisible but original model to follow. Like the epigenetic models of the late eighteenth century in general, Kant starts *medias in res*: the form of reason (or of the organic being) is the given object of investigation. Our lack of insight into its ultimate origin cannot speak against our investigation. We cannot determinately know the world to be rational; our system of judgments, however, is a rational architectonic. Insofar as those judgments make synthetic contributions to our experience of that world, metaphysical rationality is both called into skeptical question and made genuinely possible. Kant's derationalization of the world and intended full rationalization of human cognition and action proceded by way the metaphorology of the organ. It would take the disintegrative and systemic textual efforts of the Romantics to observe that metaphorological possibility after Kant nevertheless rejected its speculative possibilities.

The transcendental logic is based on principles for empirical cognition,[8] rules for judgment regardless of content. Kant describes this logic as a canon, excluding the term *organon* intentionally: "The greatest and perhaps only utility of all philosophy of pure reason is thus only negative, namely that it does not serve for expansion [*Erweiterung*], as an organon, but rather, as a discipline, serves for the determination of boundaries [*Grenzbestimmung*], and instead of discovering truth it has only the silent merit of guarding against errors."[9] Terminologically, Criticism is based on a turn from a synthesizing (actually metaphysical) *organon* to the positive law indicated by the term "canon." The canon of the understanding—as opposed to the *organon* of pure reason (see the section on Lambert *infra*)—is the "the sum total [Inbegriff] of the *a priori* principles of the correct use of certain cognitive faculties in general."[10] "Canon" serves as shorthand for the principled nonknowledge about the ultimate source of our cognition's activity. We can produce legitimate knowledge through self-generated rules of the understanding, but we cannot go further than that—for that, we have "a too unsuitable tool."[11]

The canon allows us to determine our knowledge of objects fully, while not determining those objects qualitatively at all. It thus responds to Kant's imperative of rationality in a world lacking its own indices (for us) of rationality. If an *organon*—in this conception, necessarily metaphysical and dogmatic—had once been in Kant's program, by 1781 it was gone. Now,

> an organon of pure reason would be a sum total of all those principles in accordance with which all pure a priori cognitions can be acquired and actually brought about. The exhaustive application of such an organon would create a system of pure reason. . . . On this account [pure logic] is also neither a canon of the understanding in general nor an organon of particular sciences, but merely cathartic of the common understanding. [12]

Pure logic's role is made purely negative, while transcendental logic bases itself on a syntheses fixed in the derivation of the table of categories from the table of principles. Judgment gives itself form, but this form is predetermined by the possibilities of judgment. The examination of judgment's forms reveals a complete table of categories—and indeed, justifies their use—but the transcendental cause of their unifying activity (their production of objects) lies hidden within their very genesis, invisible to us.

The systematic location of the two terms reveals that the biological metaphor complements the rejection of the *organon*: we cannot know the genesis of the forms of judgment—we must start with them as a totality (or a table), and ask only such questions as they themselves legitimate. The epigenesis of pure reason is also a kind of generic preformation. Whatever methodological restrictions are placed on investigation here, Kant's metaphors bring reason very close to the organic, indeed to possessing organs.

THE "PRE"-CRITICAL FIELD OF METHODOLOGY

Already in the 1760s, Kant had established the contour of his philosophical program: the search among cognitive and especially judgmental forms for the ultimate sources of our experience.[13] A number of documents (published texts, fragments, and also Herder's lecture notes) bear witness to this turn of mind. By examining this period (up to 1770) with respect to the term *organon*, we can better understand the double valence of Kant's metaphorology, opening up the methodological field to organicist metaphor on the one hand, and continuing in the Critical system as a version—albeit a canonical one—of that field's potential.

The question, for Kant, was whether there could be an internally justified means for carrying out the work of metaphysical speculation. He called this means an *organon*, writing in his announcement for his lectures in the winter semester of 1765/6:

> In this way I add to the end of metaphysics an observation about the peculiar method of the same, as an organon of this science. . . . The teacher must admittedly first possess the organon before he lectures on this science, so that he orients himself accordingly, but he must never present it to the audience except last. The critique and statute of all of worldly wisdom [*Weltweisheit*], as a whole, the complete logic, can thus only have its place in instruction at the end of all of philosophy, since only knowledge of this already achieved and the history of human opinions uniquely make it possible to establish observations about the origins of our insights as well as their errors, and to sketch the exact outline according to which such a building of reason enduringly and lawfully should be enacted.[14]

As we shall see, Kant would come to reject the concept of an *organon* entirely. Here, he not only entertains it, but employs it in his pedagogical vision in a democratic manner, suggesting that a new metaphysics may come into being on its basis. As Giorgio Tonelli has made exhaustively clear, this usage of the *terminus technicus* "organon" is among the first distinctive mentions in Kant's corpus.¹⁵ In the preceding passage, Kant divides between "general logic," or a *catharticon*, a "purifier" of the understanding, and local logics or *organa*—tools for the investigation of specific slices of the phenomenal (the logics of "disciplines").¹⁶ Tonelli notes that this particular *organon*—for metaphysics—can only come at the end of philosophy, but there is more to the story than this. For it belongs to the nature of this particular *organon* to be not simply a purifier of the understanding but its fulfillment. Since metaphysics is the study of the "most general predicates of being," its specific logic is not merely also the most general—it is the meeting point *of the specific and the general*. It is the "manner of application" of the two classes to each other. The desired *organon* would be what the philosophies Kant was weaned on failed to investigate: where method and being could be one. In addition to having the formal characteristics necessary for metaphysical warrant, the *organon* imagined here is also synthetic.¹⁷ This key term will play a role throughout Kant's career (and throughout our investigation here). Indeed, in this passage, the *organon* comes close to the formal definition of "organ": a special logic is meant not only to inform us of the nature of the discipline over which it has jurisdiction, but is also supposed to supply the synthetic principle by which we move from representation to action (in the sciences, the pragmatic "action" is usually judgment). Thus, Kant separates between the type of logic we mean when we say that P and Not-P cannot obtain simultaneously, and that type of logic we mean when we say that there is a "logic" to, for example, sartorial selection or "modernity." While the first type of logic serves as adjudicator of judgments and nothing more (has only a negative use), the second type, which Kant thinks is operative in each discipline, serves as the fulcrum from which we pass to reflection on the method of a discipline to its performance. Thus the conceptual proximity of the *organ* and the *organon*: each is a function, both organizing or synthesizing material and containing rules for that organization. Even the precritical Kant contributed to the metaphorology of the organ.

A NEW *ORGANON?* LAMBERT'S MATERIAL PRINCIPLES AND THE END OF THE *ORGANON* FOR METAPHYSICS

In 1764, Johann Heinrich Lambert published his *Neues Organon*, arguing that "the nature of an organon implies that it can be applied in each part of human cognition, and thus in all sciences, and that one must achieve an ability in its use, if one does not wish to lag behind. . . . [T]his premise is that much more natural because an organon, as far as one can bring it in the sciences, will always be newly applicable.[18] An *organon*—the tool for a discipline—is essentially a matter of application, indeed of invention, proceeding always synthetically to the new. It is a necessity for the practitioner because it bridges the gap between theoretical and pragmatic knowledge.

In the case of Lambert, the *organon* should help us to bridge the gap with which Kant was struggling: we should arrive at a genetic picture of the categories of human understanding, including our knowledge of the material world, and that picture should put our knowledge on sure speculative footing.[19] Indeed, in doing so, Lambert included a class of judgments that would become paramount for Kant, a class he too had been probing in the 1760s with relatively little progress: "material" judgments, or those judgments that did not stand under the principle of contradiction.

Kant had already very early separated between "logical" and "material" truths, and indeed sought a principle to unite them.[20] He thus recognized that Lambert's efforts were connected to his, for Lambert had sought a manner—or better, a tool—for understanding the application of judgments and concepts and intuitions to their putative objects. He called this tool an *organon*. Lambert began a short correspondence with Kant, one which, while not producing the intended collaboration between the authors, bears the traces of a sharpening of Kant's notion of the "material" of judgments, and shows us the historical reason for his rejection of the term *organon* in the *Critique of Pure Reason*.

On 13 November 1765, Lambert wrote to Kant, acknowledging the overlap between their methodological projects. Having read Kant's *The Only Possible Proof* (and having clearly focused on the material I have presented above), Lambert addressed the issue of the application and origin or the nonlogical elements of our judgments: we must indeed seek among the first principles of human cognition, and

> in fact not only the *principia*, which are reasons taken from the form, but also the *axiomata*, which must be taken from the material itself, and actually only arise in the case of simply concepts (as those which are in themselves not contradictory and in themselves thinkable), and the postulates, which give universal and unconditioned possibilities of composition and connection of

simple concepts. One comes from the form alone to no material whatsoever, and one remains in the ideal, stuck in mere terminologies, if one does not take cognizance of the first, in itself thinkableness [*das erste und für sich Gedenkbare*] of the matter or objective material of cognition.[21]

Lambert's attempt to agree with Kant reveals two terminological fault lines with methodological consequences, pointing up both the true openness of Kant's early systematic writings and his departure from Lambert's approach. On the one hand, the problem of the "material" in judgments is a matter of true agreement between the two thinkers: it is not obvious which principles can legitimately be used in such judgments. The obvious need to exclude the principles of logic was already in the Rationalist tradition—but what other principle (or *axioms*) could be used?[22] Lambert suggested the answer should be sought in judgmental syntax:[23] the way we produce and synthesize "simple concepts" (imagined, with Locke, as immediate *qualia*, like extension, duration, existence, motion, etc) is our true source—our true instrument—for legitimizing cognitive claims.

On the one hand, this system or *organon* addressed a problem Kant had identified as unaddressed across the spectrum of contemporary philosophy: the problem of the forms of judgment insofar as they were applied to "material" and not merely other judgments or concepts. Indeed, it seems Kant was prepared to pursue this path a great distance. Yet he perceived something else, yet another hidden assumption—in fact, the same assumption merely transferred to the area that both men thought might hold the solution. This hidden assumption was that somehow this judgmental syntax also could justify the isomorphism it claimed between the concepts it dealt in and their real-world referents. While Lambert had gone to great trouble to analyze just that piece of the rationalist puzzle Kant thought undertreated, he had simply put the assumption Kant wanted addressed into that syntax: the *organon* synthesized, producing real metaphysical knowledge, but it did so—as Kant would later put it—dogmatically.

This dogmatism is the reason for Kant's rejection of the very term *organon* in the *Critique of Pure Reason*. And indeed, by the time Kant came to name this problem left unaddressed by Lambert (the problem, that is, of the legitimate application of judgment to the material of experience), he had a new term for it, under which a new concept of the "material" was housed. The term was "transcendental logic," and the conception of the "material" had been reduced to the "material of sensation" as it was received by the cognizer and (pre)organized into *intuitions* (*Anschauungen*).

"Transcendental logic" treats the contents of cognition without respect to how objects are given to that cognition,[24] that is, it treats what minds do, including their negotiation of real-world materials, without determining anything about those materials. It is that specific logic for which the canon of the understanding (which we are about to explore) replaces the *organon* of reason (Lambert's dogmatic judgmentalism), and thus the point at which the metaphysical reduction of Criticism is performed—as Kant puts its, where "the proud name of ontology gives way to a mere analytic of the understanding."[25] But this moment of grandiose theoretical creation—a confluence of efforts with effects almost literally unparalleled in the history of philosophy—really included two innovations which, for the next generation, did not seem necessarily linked. On the one hand, the efforts to produce Criticism had shown that attention to the dynamic interactions between concepts and judgments were a condition of serious metaphysical efforts. And indeed, as the conversation with Lambert had shown, analysis was not enough: some warrant was needed for the ontological security of these investigations. On the other hand, the Critical system included a specific source for that security: the intuition, modeled on the space and time of the geometers and astronomers, but formed in fact as a primitive element of our own cognition.[26] Thus, when Kant came to reject Lambert's *organon* of reason for his own "canon of the understanding," he was combining a generic philosophical creation (the precritical analysis of concepts and judgments combined with a sense of the need for warrant) with a determinate version of that genre—one we might call "intuitionism." Although the precritical writings were unavailable to them, it is a strange achievement of the Romantics to have separated out these two elements and to have pursued the original project on different terms. They thus expanded the genre of "Criticism" without agreeing with Kant's own version of it.

THE CANON OF THE UNDERSTANDING

According to Paul Guyer, the signal contribution of the *Critique of Pure Reason* is the "invention of the understanding."[27] What, then, *is* understanding, and why does Kant attach to it a *canon* rather than an *organon*? As we have seen, the *organon* was historically articulated as a dumb instrument, a tool to which the problems of method in metaphysics had been merely transferred. Kant's own attempt to solve this problem went, after 1769, by way of the elaboration of a fundamentally different kind of knowing he came to call intuition.

The separation of two "roots" of our cognition[28] was, in fact, a response to rationalism, for which, as we have seen, knowing exists in a continuum rising from obscurity to adequacy. Starting in 1770, however, with his *Inaugural Dissertation*, Kant disagreed: presentational knowledge of concrete phenomena was a different basic source for knowledge than the concepts and judgments on which he had been so focused in the 1760s. It is important to see that this is a specification of the search for "material principles" in judgments. Indeed, the *Dissertation* is famous for just this, the first statement of "transcendental idealism," the doctrine that the forms of intuition (space and time) are necessary but ultimately cognizer-based elements of phenomena, not of "things" as they "are," or *noumena*. In one sense, Kant is saying to Lambert (obliquely), "Yes, *basic concepts*, unanalyzables: but why are we calling these "concepts" at all?" This is a moment in which Kant departs from his rationalist tendencies and training, permanently giving up on the putative rationality of things (while not abandoning our potential rationality). As Giorgio Tonelli has put it, the realization of 1769 was that of the "separation of sensibility from the understanding."[29] As Tonelli recognizes, Kant is asking a question about the legitimacy of "non-rational" judgments, or judgments not deriving their truth from logical statements. The result is that, in addition to the "syntax" offered by Lambert, there is a "parataxis" offered by another source: the intuition. The intuition functions to offer us the singularity and particularity of the object-world, not, as Tonelli puts it, "well-founded phenomenon" (Leibniz's *phaenomena bene fundata*) but a "phenomenal *generality*."[30]

This doctrine of intuition does not solve the question Kant was asking with Lambert, but it does specify what kind of combinative effort is taking place in "material" judgments. What Kant calls "transcendental logic" or the "canon of the understanding" is the manner of syntactic judgmental application of concepts to the "material" of intuition. Thus, defining that discipline, Kant writes (much as he had in the *Announcement* of 1765) that there is pure logic—a "catharcticon" (*sic*) of the common understanding abstracting from all content in favor of the purely formal—and "applied logic," the specific logics of different disciplines. Further, there is that mixed breed we identified above as the specific logic of generality itself. That logic is called "transcendental" because it deals neither in mere empirical experiences nor in relations of ideas: it treats the possibility of the application of the latter to the former, that is, not the content of intuitions (immanent) nor the potentially transcendent "ideas" of reason, but the way in which concepts play a role in the establishment of experience itself.[31] There will ultimately be twelve functions allowing this application, and they are dubbed "the categories."

Transcendental logic is the name for the discipline of the understanding itself, and its content is the "canon" of that understanding. Kant takes the word "canon" in its traditional sense—"body of positive law"—and thereby restricts what he had found in Lambert to be a synthetic rational *organon*. This reduction is at stake in the metaphorology of epigenesis. That metaphor is, in fact, a characterization of the status of the canon: autonomously truthful, fully legitimate, and given in its form without rational insight into its origin. Judgments gain validity through the canon, the means by which categories functionally unify concepts and judgments. And this sort of activity is the only kind which counts as cognition: we are, in fact, capable of many other types of intellectual and volitional activities (indeed, we are capable in this way because our truly cognitive, or constitutive, activity is limited thus)—but they do not make up the world we experience, they do not "give nature the law," in the famous formulation. That work is done by the epigenetically metaphorized bearer of "being" itself: the understanding. As we saw above, that understanding is autonomous and self-formed, yet its form, being given, is "generically pre-formed." Where, in the *Critique of the Power of Judgment*, Kant approvingly reproaches Blumenbach for giving a name to problem without analyzing that problem—thereby shifting a charge to Blumenbach that Blumenbach himself had leveled against Wolff—Kant seems to reproduce the problem in the intertwined realm of epistemology and ontology here. The understanding's form—the categories ruled by the unity of the "I think" and in exclusive application to intuitions—is the law of cognition, but its origin remains obscure.

CANONICAL COGNITION: THE DEDUCTION BETWEEN
THE VOID AND BRUTE FACT

Classical German philosophy (from Kant to Hegel) is based on the recognition that consciousness is self-consciousness, that cognition cognizes its cognizer as itself.[32] The loop first signaled by Kant in the Deduction of the Categories (in the second edition—the "B-deduction") of the *Critique of Pure Reason* did not only, however, enjoy a rapid efflorescence in philosophy as the problem facing Reinhold, Fichte, Schelling, Novalis, Schlegel, and Hegel[33]—it also was inherited from an earlier eighteenth century. The *Port Royale Logic* separated between "conceiving" (the formation of objects in thoughts) and "ordering" as a function of judgment, the syntax or "grammar" of thought. The latter works upon the former as its exclusive sphere of reference. Kant reflected this division in the structure of the *Critique*. There, the judgments of the understanding make Nature (for us) both possible and binding—they are constitutive, legislative of whatever we might take ourselves to be cognizing. This "whatever" seemed too little to the *Critique*'s first readers, however.[34] Kant drew therefore in the second edition on a series of drafts stemming from the 1770s concerning this problem,[35] namely: how can we be sure that what we cognize really suits the way we cognize, in other words, how can we be sure that the "nature" we constitute corresponds to the inputs we surely receive and mold from the outside? Kant makes an extraordinary move: he attaches the validity of judgment in general to the human capacity to preface any possible judgment with the phrase, "I think":

> The: I think must be able to accompany all my representations; for otherwise something would be represented in me that could not be thought at all, which is as much to say that the representation would either be impossible or at least nothing at all for me.[36]

The: "I think," as a potential form for any possible cognition, is meant to demonstrate not the validity of individual judgments but the legitimacy of judgment as a cognitive tool. This is the sense in which Kant means "deduction," as he clarifies: the question is not what we do when we cognize (*quid facti*), but "what is legitimate (*quid juris*)."[37]

The point of this move, I think, is to walk a narrow path between two equally unappealing options: first, that of allowing judgments to lose contact with the world (to "spin in a void"), and second, that of making claims about human capacities not grounded in a space of arguable propositions (to state "brute facts about us").[38] To avoid floating away into the ether or staking all too rough roots in declarative anthropological statement, Kant bases the rationality of philosophy on its medium: judgment. To emphasize his middling position, Kant draws on the discourse of the life sciences: "Consequently only the second way remains (as it were a system of the *epigenesis*) namely that the categories contain the grounds of the possibility of all experience in general from the side of the understanding."[39] *Epigenesis* designates the "generation" (*genesis*) of the order of living beings "on top of" or "out of" (*epi-*) that of dead matter. It was perhaps the single most polemical term in the eighteenth-century life sciences.[40] The other options—the argumentative dilemma that allows him to select judgment as the centerpiece—are *generatio aequivoca* and "pre-stabilized harmony." The question is one of orders: is there a law that allows for the description of the organization of living beings in a physical world organized according to Newton's laws? *Generatio aequivoca*, or "spontaneous generation," maintains that organization is instantaneous and needs no separate explanation—Kant analogizes this view to that of Locke (and, by implication, Hume), who, at least according to Kant, derives the higher-order functions of reason from an origin in the senses without providing laws for that development. If this holds, Kant is suggesting, then we simply list our capacities, but learn nothing about them. The statement of these brute facts about us is the equivalent of Linnaean biology, where classification leaves open questions of ontogenetic and phylogenetic development.

The opposing view—again analogized from biology to epistemology—is that of in-born categories. "Pre-established harmony" points to the epistemological version of Leibniz's doctrine that there are no truly informative exchanges between substances, but instead only mirrored parallels prescribed from eternity by God.[41] Pre-establishment is meant to root interactions between all substances (of which cognition is only one case) in the highest possible authority. For Kant, however, this would undermine the project of critique, the apex of which is the legitimization (deduction) of the categories. Since the categories are the specifically human contribution to the cognitive process, they require a deduction that makes no arguments based on putatively judgment-external entities or processes. Pre-established harmony minus God (as an authority for philosophical argument) equals a void in which to spin. The argumentative procedure that Kant's notion of Critique requires can thus rely on no external tribunal, and must develop its legitimacy from within its proper activities. Just as a justificatory discourse about organizing forces led to the creation of the independent discipline of biology, so the Deduction separates Critique from other styles of philosophy. This argument is Kant's own reason for the final line of the *Critique of Pure Reason*: "Only the critical path is still open."

So much for the argumentative stakes of the metaphor "epigenesis." Needing an internal anchor for the argument, Kant suggests that judgment itself must produce its own legitimacy, and that this production implies that "experience" is precisely the internally valid emergence of one order (self-consciousness) from another (consciousness). The means by which this transition (constantly) occurs is the judgment, which therefore serves to unite two different orders. In doing so, it "generates" the categories (the unifying functions of judgment).[42] The metaphor points, then, to the elements which bind orders together or *organize* orders—the judgment as sole medium of legitimacy in argumentation is also the part that unifies whole and part.

Kant calls the universal "I think" as a capacity the "transcendental unity of apperception." The unity is transcendental because it forms the condition of possibility for all cognition—everything we think is premised on this unity. And yet, this point could remain trivial, if apperception remains only a container for our judgments—after all, saying that we "think" something hardly legitimizes the faculty of judgment. The contents of cognition are gathered under the marker "I think," allowing us to analyze them regardless of what they are—this is why Kant calls this first unity the "analytic" unity of apperception. For the desired justification of judgment, our judgments must have a necessary relation to those contents—that is, judgment must be capable of forming a lawful connection to any possible content. Kant calls this connection the "synthetic" unity of apperception, since the judgment in this case combines elements (concepts and intuitions) in a manifold and relates them through the forms of judgment, thus forming a unity between separate orders.[43] Judgment is the part of our cognitive apparatus that unifies its parts (intuitions, concepts, categories) with its whole (*intellectus*). As the generative factor that both separates and combines all types of cognition, judgment serves as the unity that precedes the very category of unity.

To legitimize this conception of judgment, Kant needs to show that there is a necessary relation between the analytic and synthetic unities of apperception, or, in other words, that our ability to unify what we think is necessarily related to our ability to think phenomenal contents.[44] If he can point to this connection between the two unities, then the world as we experience and constitute it will be the eminent sphere of our judgment, which will therefore be legitimate as a cognitive tool. In order to carry this out, Kant suggests that judgment must unite its action with its unifying function: "But in order to cognize something in space, e.g., I must *draw* it, and thus synthetically bring about a determinate combination of the given manifold, so that the unity of this action is at the same time the unity of consciousness (in the concept of a line), and thereby is an object (a determinate space) first cognized."[45] This second-order unification simply *is* judgment,[46] which therefore operates both as the general form and the discriminating function of cognition.[47] The repetition of the term as a container and as a set of rules is what interests me here. We could even say that, since this is the point that was so crucial for the Romantics, that judgment is the self-legitimizing distance from consciousness to self-consciousness. If this structure is epigenetically developed, then its parts—the categories—are metaphorically "organs." Perhaps this is the sense in which Kant can speak of metaphysics as a "physiology of reason"[48]—not only the method, but also the image. If the mind is generated like an animal body, it must develop organs according to its type.

The transcendental structure "I think"—the judgment as self-consciousness—becomes the canonical articulation of the problem I mentioned above, namely: if the "I" is structured as self-cognition, then what is it cognizing? This was the problem that occupied Fichte in the 1790s: the self is that structure which knows itself (self-consciousness), which leads to the very odd conclusion that the self presupposes itself, but can neither preexist "itself" (whatever that would mean) nor, strictly speaking, ever exist for itself as an object.[49] An object that is also a subject, and remains suspended in the tension between the promise of such cognition[50] and an abyss of ontological uncertainty:[51] that is the subject that was extracted from Fichte by Hölderlin and Novalis in 1795, and by Schlegel and Schelling[52] shortly thereafter, as we shall see.

The canon law of the understanding provided legitimacy but relinquished the principled search for reason's grounding at precisely that moment where it touched upon the real: neither at the pole of the "thing in itself" nor at the ultimate origin of the intellect could determinate grounds be given for or by the intellect. The human cognizer, as the anthropological discourse of the eighteenth century had it, was a "middle-being," caught between the twin unrationalizable extremes of the noumenal "world" and the noumenal "soul." This, in fact, was the message of the Transcendental Deduction itself: between the sempiternal fact of judgment (the "I think" with its capacity to accompany any cognition) and the synthetic unity of intuitions as they are received contingently by us cognizers, the understanding does its legitimate unifying work. This is the realm of the properly human judgment about the world—the answer to the problem of "material" judgments, written as a judgmental capacity, indeed, *the* judgmental human capacity. That capacity, whatever its origin, is characterized as the "epigenesis of pure reason," a designation of the fundamental spontaneity of the capacity to judge (the understanding) as the type-basis on which experience (combinations of concept and intuition) would "develop." This term thus places experience foremost in the methodological register—at least one of Kant's lines of thinking is simply devoted to the analysis of experience—and curbs investigation into the ends that form this experiential middle. In biology, the ultimate source of type is not knowable—in epistemology, the same is true of the ultimate source of our sensibility (in particular) and our categories. Critical procedure must start *medias in res*, taking the formative middle as its most basic object of investigation.

That fundamentally middle status was formed as an inheritance from the seventeenth century's logical tradition. When we attend to Kant's final table of capacities (*Vermögen*), we find that, after "intuition" (the independence of which had been championed by Baumgarten, but the specific constitution of which was original to Criticism), the "understanding" and "reason" are written in accord with the discourse stretching back to the *Port Royal Logic*. Here we find—as we do with Kant—that the "understanding" consists in simple propositions, essentially the predication of concepts one to the other, while "reason" deals in syllogisms, in the reflective action of judgments on judgments. As in the course of the *Critique of Pure Reason* we leave the realm of the simple, constitutive judgments making up the understanding, we enter a realm of complex syllogistic judgmental stylings fundamentally connected to that simpler realm but also removed from it by the absence of that element, intuition, which had conferred secure status on the former. Kant calls this unintuitable realm that of the "ideas"—a basically "dialectical" or truthless place which can only be regulated, not known. This was his name for those disciplines—rational psychology, cosmology, and rational theology—which were known, in the Wolffian system, as "special metaphysics."

But it is the transition to that realm of ideas that interests us here, for it is at that moment, where the possibility of intuition falls away and the formal similarity of judgments, whether those of the understanding or those of reason, comes to light.

The *Critique of Pure Reason* is a notoriously difficult book, but there is one confused issue that derives, I think, from the "specification" thesis I have offered above. The Transcendental Deduction is plagued by a central difficulty: it is meant to provide information about how we legitimately separate between "objectively valid" and other types of judgments. Having stated in the lead-up to the argument that the understanding is itself the "capacity for judging," and that the categories are the functions of unification within that judgmental sphere, Kant has circumscribed his argument with a focus on judgment that, as we have seen, goes back to the 1760s. The unifying functions (Kant groups twelve categories under four headings: quantity, quality, relation, and modality) must be seen, in the course of the argument, to apply to *any and all experience* we might have. Because the "I think" is attachable to any cognitive content, Kant is at pains to describe the relationship of the categories to the "I think" (which is termed the "transcendental unity of apperception"—which means, in English, the unity of consciousness which is a condition for our awareness of anything at all). As Paul Guyer has observed, the different versions of the Deduction follow different strategies. The B edition of 1787, however, attempts to prove that this link between the functions of unity and the sempiternality of the unity of apperception conditions all our experience, intuitive or not (although nonintuitive experience doesn't count as constitutive knowledge of the world).[53] Here the argument hits a kind of snag, however: if that link operates (and operates legitimately) outside the intuitive sphere, then the "objective validity" of its claims seems to derive ultimately from judgment itself, in its connection to consciousness (in its judgmental, apperceptive form). This leaves us, to use contemporary language, "spinning in a void," unsure on the question of to what our judgments might apply. Kant never goes the full way toward this argument, as, for example, the "Refutation of Idealism" shows, arguing for the necessity of the presented object-world for the very syntax of judgment we are otherwise exploring in itself.[54] On the other hand, if this presentation-in-intuition, and not the link between the categories and apperception, is the measure of validity, then the stated argument of the Deduction fails, since its conclusion will only be valid in and because of the nature of our intuitionism. Kant, then, is here forced to make a decision between *judgmentalism* and *intuitionism*—just at the moment when the two are supposed to be epigenetically combined, the absence of a principled knowledge of the origins of the form of the understanding—that is, the absence of a genetic account of the categories themselves—undermines the argument at its apex. If we go the intuitionist route, we get the so-called "schematism" of the next section of the Analytic, in which the arguments of the Aesthetic cash out: the categories, as they are combined with the conditions of time especially, give rise to fundamental "schemata" which are used to make sense of the world's lawfulness. The conceptual contribution to experience is secured, but only for a world in which we are finally beings of intuition.

The attractiveness of this argument notwithstanding, it does not respond to the stated intent of the Deduction. It is not trivial that the latter is meant to prove that the categories apply to all experience of any kind—in fact, this argument grounds the separation between the understanding and reason, since they are on its basis a single capacity applied to different material (intuitions, on the one hand, and judgments themselves on the other). This formal similarity would be undermined if the categories were only valid under the conditions of intuition (and this would also arouse the suspicion that the categories are empirically derived, which is clearly unallowable). Thus Kant's opposing argument, the one which comes closest to judgmentalism: that even the synthetic unity of intuitions is ultimately grounded in the transcendental unity of apperception. This argument, running as it does the other way, is a restatement of the early concern that, whatever the content and rules of material judgment (that is, whatever kind of world passes before us and whatever categories we have to apply to it), ultimately, our paradigmatically rational minds control cognition. When we see that our capacity in this respect is essentially a matter of judgment, we see that judgment itself must play the role of truth-warrant if the understanding is to remain autonomous (according to the epigenesis metaphor). Again, however, this leaves us in danger of "spinning in a void," a void that cannot be filled out with some source of factual warrant without, as Kant had observed from the beginning, stripping us of our autonomous truth-perceiving faculty, the understanding itself.

The *Critique* is literally meant to to hang in suspension, as it were, between the organic autonomy of the metaphor of epigenesis and the fixed forms of the categories. The balance is meant to provide a rational system to explore the empirical world, and to orient ourselves in the nonempirical. Its focus on judgment as the central authority in methodology, however, bore different fruit in the next generation. Thus we can say that, running against the textual grain of Kantian history, the Romantics actually analyzed out the impulse toward judgment-based idealism from the *Critique of Pure Reason*, and abandoned the finalism of intuitionism. In Kantian terms this represents a return not to one of the sources of cognition (concepts and intuitions) but to the methodological impulse which treats primarily their unification in judgment. It meant reaching backward to the impulse that resulted in the specific form of the Critical system, to mine this impulse for new theoretical and pragmatic insights. It meant finding, in the methodological version of the metaphorology of the organ, a metaphysics of organological judgment. The Romantics despecified the Critical system and located the methodological motivator which had instigated Kant's life-work. This was because they saw that judgment was a medium, and that this judgment-internal model operated according to the general semantics of the concept of the organ.

THE ORGAN OF JUDGMENT: ON THE SPLIT BETWEEN THE UNDERSTANDING AND REASON

Kant left a clue to the constitution of the mind's organs, seated precisely in the forms of facultative activity—paradigmatically in judgment. Before turning to Herder, I want to spell out this latent possibility, since it lends metaphorological precedent to Romantic organology.[55]

We have now explored the metaphorology of the organ in two respects, both of which restrict the possibility of organology. In Leibniz, we saw a metaphysical use opening to a metaphor, both of which keep the technological term in the boundaries of a preestablished order, a mere instrument. In Kant, we saw a methodological flirtation with the term *organon*, combining in its very rejection in the *Critique of Pure Reason* with the metaphor of epigenesis to describe the intellect itself. In its restriction to intuitive experience and its autonomous yet fixedly formed categorical apparatus, this understanding was awarded only a canon, not, as Lambert had wanted, an *organon*. Being was proscribed from judgment, literally written out of cognition itself.

Or was it? On the one hand, Kant had thrown out the traditional category of being as the "thing in itself" as a mere putative assertion, the essence, in fact, of dogmatism.[56] The world in its materiality was unrationalizeable, while our cognition was, up to the generic limitation of the epigenetic understanding, fully rational. On the other hand, the very categories that made up that rationality had to include at least the assertion of being. Kant brought this element into his system under the heading of the categories of "modality," and they were treated as a special case from the very beginning.

Recall that the categories are not concepts (as in the somewhat misleading phrase, "pure concepts of the understanding"), but rather functions of unification (of concepts and intuitions) in judgment. The categories are literally the instruments of that judgmental apparatus which Kant had found so woefully undertreated in the philosophies he had been weaned on. In the Introduction's famous distinction between analytic and synthetic judgments, judgment itself is first characterized. Analytic judgments merely analyze their concepts—Kant calls them "explicative." Synthetic judgments deserve their name because they add information to their components, furthering our knowledge. (Thus the famous formulation of the metaphysical judgment: *synthetic a priori*, or *both* informative and necessary.) Here, a surprising formulation occurs: describing the synthetic judgment *a posteriori* that "bodies are heavy," Kant points out that "Thus is it experience on which the possibility of the synthesis of the predicate of gravity with the concept of the body is founded, *because both concepts, although the one is not contained in the other, nevertheless belong as parts of a whole—namely, of experience, which is itself a synthetic connection of intuitions—to each other, albeit accidentally.*"[57] If we abstract momentarily from the doctrinal point Kant is making (that the *tertium comparationis* of *a posteriori* predication is experience itself), we can notice that the concept of experience in this passage is (1) a matter of judgment and (2) that that judgment operates as a synthetic whole. A synthetic whole, however, is no mere "aggregate": it does not operate according to the mechanical laws of juxtaposition or mere parataxis. Instead, this synthetic whole itself shares a quality with the concept of the organic whole we above saw treated in Kant's analysis in the *Critique of the Power of Judgment* of the epigenesis debate: true synthetic holism attaches to both concepts. A characteristic of internal equilibrium, of necessary connection obtains in both cases. And while the causal model projected within constitutive judgments about the world is paradigmatically mechanical, the relation of parts to wholes within the judgment cannot be. If it were, no synthesis would take place: our "knowledge" of the world would have to be preimplanted, as it was for Leibniz (so Kant) and for Crusius. That this is not so gives us the hint of a methodology immanent to judgment.

This turn of phrase—it is little more—becomes something like an undercurrent in the argument of the text when Kant approaches the problem of judgment-internal assertion of being, or the categories grouped under the heading "modality." We saw the roots of this approach in *The Only Possible Proof*, but here the argument is put in quasi-metaphorical terms that point up a possible *organ*icity of judgment itself.

The categories of modality are "problematic, assertoric, apodictic."[58] Kant immediately marks them as separate: "The modality of judgments is a quite special function of them, which is distinctive in that it contributes nothing to the content of the judgment (for besides quantity, quality, and relation there is more that constitutes the content of a judgment), but rather concerns only the value of the copula in relation to thinking in general."[59] Judgments as to the possibility, existence, or necessity of those things asserted in judgment are both circumscribed as to their dogmatic contents (things-in-themselves are not determined by such judgments) and also express a relationship to the whole of thought. This is (yet) another way of expressing the basic reduction of metaphysical claims that the *Critiques* are meant to systematize: while "being" cannot be determined in any way by cognition, cognition's rules determine, at some level, the manner of assertion of being itself. Two mereologies, then: that of the synthesis of judgments themselves, and that of the relation of a set of their unifying functions (categories) to the totality of human cognitive activity.

This line of thinking continues when Kant comes to apply the categories to empirical experience in the section on "schematism." The notion here is that each category generates a set of "schemata" in combination with the strictures of the *a priori* form of intuition for the inner sense (time). These schemata are transcendental: without this overlay of pictures, recognition of individual objects (and judgmental subsumption of the same under concepts—"this is a plate") could not occur. With respect to modality, the schemata are: (1) possibility as "the agreement of the synthesis of differing representations with the overall conditions of time"; (2) existence as "being in a determinate time"; and (3) necessity as "being (of an object) at all times."[60] These schemata are meant to determine the *Zeitinbegriff* (roughly, the "essence of time-content")[61] of all possible objects. They express the determining and determinate set of possible relations of judgmental assertion to the whole of cognition (here restricted to intuitive cognition).

Recall that we are, for judgmentalist purposes, abstracting from the content of "time" that here takes the lead in determining the form of the categories. As Kant's argument continues, he deepens this relationship, establishing a "system of principles" on the basis of the transcendental schematism. These principles (of judgment) fall into "mathematical" (or intuitively determinative)—the categories of quantity and quality—and "dynamic"—the categories of relation and modality. The latter are dynamic because they do not determine the form of our experience but are applied according to rules (regulatively) as the flow of our experience demands. They go, as Kant puts it, to the existence (*das Dasein*) of the objects of experience. The categories of modality are, in this context, the basis for the "postulates of empricial thinking in general." That is to say, we assert, more or less in a void, the possibility, reality, and necessity of the objects of experience according to the rules of the *a priori* form of time.[62] In this context again, Kant writes: "The categories of modality have this peculiarity:

as a determination of the object they do not augment the concept to which they are ascribed in the least, but rather express only the relation to the faculty of cognition."[63] Even in the most intuitionist portion of the *Critique of Pure Reason*, then, Kant continues at the level of his text to assert the doubly mereological nature of judgment and its cognizing agent (its "faculty"), asserting thereby the liminal organicity of his synthetic model of judgment itself.

This argument—or discursive strategy—cashes out only in the opening of the "*Transcendental Dialectic*," where Kant explores the transition of cognitive function from constitutive, world-based, intuitively secure judgments to those of a more reflexive nature, judgments about the nonintuitable "absolute wholes" of special metaphysics: the soul, the cosmos, and God.

Where the understanding was defined as the faculty of "rules," reason is defined as the unifying faculty of "principles." And indeed, the unity it imposes is not on nature (the understanding has already done this), but on the understanding itself. Its action is to unify and systematize, and its form of knowledge is always proposed as a "cognition from principles . . . because I here know the particular in the general through concepts."[64] This faculty is the metaphysical faculty, then, but it is fundamentally severed from the content it nevertheless works upon in its object (simple judgments of the understanding). Its always failing function is to propose the known (or conceptual) encounter between the particular and the general—which would have been the function of the *organon*, had it not been restricted by intuition. Kant nevertheless proposes that what replaces this desired synthesis of reason (that is, "dialectical" or illusory knowing) retains the form of metaphysical knowledge, applying itself epiphenomenally to the understanding and thinking *in* the understanding (and not beyond it) to the general in the particular. This is possible, indeed, because it is the action of judgment upon judgment: so far as we do not propose the noumenal truth of these second-level judgments, we remain within the "discipline of pure reason." It is precisely the limited extension of judgments of reason to judgments of the understanding that allows for this curtailed form of "metaphysical" knowledge.[65] Concepts of reason thereby become what Kant dubs "ideas"—necessary unities reason imposes on the rules of the understanding, without any experiential component. Without intuition and the possibility of synthesis, these ideas follow strictly from the form of reason as it is (correctly) disciplined to act upon the understanding, not producing cognition but acting formally metaphysically within that cognition, unifying it and pushing it toward the absolute. In this manner—which opens onto the practical, since it connects intentional, rule-based, and potentially free volition with quasi-cognitive acts—reason serves the understanding, in an expanded sense, as a canon:

> Although we have to say of the transcendental concepts of reason: *they are only ideas*, we will by no means regard them as superfluous and nugatory. For even if no object can be determined through them they can still, in a fundamental and unnoticed way, serve the standing as a canon for its extended self-consistent use, through which it cognizes no more objects than it would cognize through its concepts, yet in this cognition it will be guided better and further. Not to mention the fact that perhaps the ideas make possible a transition from concepts of nature to the practical, and themselves generate support for the moral ideas and connection with speculative cognitions of reason.[66]

The promised "transition" occurs first in the "doctrine of method" of the *Critique of Pure Reason* (and then in the *Critique of Practical Reason*). Here, however, I am interested in the transition proposed from "concept" to "idea," both occurring in judgmental form (as proposition and syllogism). That there is an expanded *canon of orientation within cognition* already shows us what the critique of general metaphysics (canon of the understanding) implies for special metaphysics: relegation to orienting but necessary unities we must reflectively impose upon our empirical experience. This is the transition to the "practical," which will be partially rewritten in the *Critique of the Power of Judgment*.

Kant here states that the form of reason's judgment is the syllogism, which imposes a kind of logic on propositions. These can proceed toward the particular (*en*syllogisms) or toward the conditions of the ensyllogism (*pro*syllogisms). Kant proposes that reason strives to find, in all cases, the "totality of conditions," or the final prosyllogism.[67] Reason thus formally unifies the understanding in the direction of an "absolute whole."[68]

The nonsynthetic action of reason on the understanding—its imposition of unity—is nevertheless not the mere "juxtaposition" of connection. Recall that a preformationist metaphor is used to characterize the "system of pure reason." *At all levels and in spite of the projected constitutivity of the causal nexus effectivus*, judgment follows a mereological model of metaphorically organic heritage. The *nexus effectivus* is the world-picture belonging to an organized reason guided by the *nexus finalis* in its very form and self-relation. The Romantics, sensitive to potentially organic metaphorical models for reason itself, will not have missed this. They did not have to reach forward exclusively to the *Critique of the Power of Judgment* for their notion of an *organ*. They found, in the very rejection of the *organon*, the desired etymological and metaphorological *basis in one thread of Kant's metaphysical methodology itself.* Wholes and parts, and their mutual interaction, formed a judgmentalist research program in the next generation. That this was so, of course, implied a need for (1) an organic and developmental model of metaphysics, and (2) a notion of perception (what Kant dubbed intuition) that was not finally separate from conceptuality. They thus returned to the Rationalist unity of knowledge. But this unity was attended by historicity and the possibility of categorial innovation, and those elements could be supplied only by the last and most dedicated metaphorizer of the organ: Johann Gottfried Herder.

Genesis of and by Organs: Herder's Analogies of Nature and Reason

If Leibniz contributed the metaphysical sense of the word "organ," and Kant supplied a methodological critique of that metaphysics that nevertheless retained and developed its key metaphorical terminology, it was Johann Gottfried Herder who supplied its metaphorical content. He did this by globalizing the term's use throughout his writings, speaking of organs analogically in metaphysical, epistemological, and cultural spheres indifferently.

Leibniz and Kant contributed to the metaphorology of the word, but Herder did far more: he seems to have introduced the systemic use of the word into the German-speaking public sphere.[69] But not, as we might expect, into the French public sphere, from which Herder himself seems to have borrowed the term.

The entry in Diderot's *Encyclopedia* for *organe* is of unknown authorship. It covers, however, the etymological background, but then opens the term up to a recognizably modern use: "ordinary use" is that of a functional part of the body, any part that carries out an operation. The article goes on to divide them between primary organs (built for a single function of similar materials—the veins, arteries, and nerves) and secondary organs composed of the primary—the hand, the fingers, etc.[70] A further entry on "organs of sense" clarifies that these are the parts of the animal "by means of which it is influenced [*affecté*] by external objects"[71] These are sometimes divided into an "internal" organ (the brain) and "external" organs—the ears, eyes, etc.

Thus a recognizably literal sense of the term was given as early as 1765 in the French world. Yet this common usage had not passed into Germany—for that, the young Herder would need Kant's gentle push—also in the mid-1760s—to internationalize his reading habits. And it was indeed, it seems, in some back alleys of his French itinerary that Herder culled the word for his own use.

Herder's *Ideas*, which he wrote in the early 1780s and began to publish in 1784, is a revision of precisely the genre of writing from which he seems to have taken the term *Organ*—*Histoire naturelle*, *Naturgeschichte*, Natural History, the Enlightenment genre of narrating nature's whole course of development. The Comte de Buffon's monumental *Histoire naturelle* (from 1749) was a European best seller and paradigm for the genre for more than a century.[72] Herder's entry into this scene[73] was marked by his fascination with the problems of force and structure in nature's development. If Herder's signal contribution was the historicization of nature and culture,[74] it will have been in the productive tension between force's developmental self-expression and the structure that force took on and imposed that this conception was developed.[75]

It was in the interstices between these two concepts—static structure and dynamic force—that Herder also placed his half-borrowed, half-invented *Organ*. The *Ideas*' narration of formation—*Bildung* as the combination of structure and force—begins with matter itself, and ends with the plurality of particular human cultures. This all-embracing monism, however, needed a way for the apparent contradiction of the static and the dynamic to be put into motion, literally to develop mutually. The system needed to allow for the concrete to emerge from an apparently unsynthesizable duality. The instrument of synthesis was given the name *das Organ*. Thus, at the end of the first volume (1784), Herder inserted a general chapter on natural form and natural-historical method, giving it the title "Kraft und Organ."

Summarizing his doctrine with respect to the debate on preformation and epigenesis, Herder wrote:

> If one looks at these transformations, these living effects in the egg of the bird just as in the womb of the animal that bears living being: then it appears to me that one speaks metaphoricaly [*uneigentlich*] when one talks of seeds that are merely developed, or of an *epigenesis* according to which parts grow onto the individual from the outside. It's formation (*genesis*), an effect of internal forces, for which nature had prepared material to form into itself, in which to make itself visible. This is the experience of nature: this is confirmed by the period of formation in the different species of more or less organic species-variety and fullness of life-forces: only through this do the misformations of creatures through sickness, accident or through the mixture of different species permit of explanation, and this is the only path that the force- and life-rich nature as it were forces on us in all its work through a continuing analogy.[76]

This passage contains the kernel of Herder's ontological teaching. With concrete reference to the empirical details of the epigenesis debate, he makes clear the philosophical stakes of that debate. Force is at issue, and the debate—as the historiography also confirms—goes to the philosophical issue of the constitution of force itself, and therefore of matter. Up to Herder, epigeneticists had tended to defend the division between dead and living matter. As Herder notes here, this conception is etymologically proper to *epi-genesis*, formation "on top of." So Blumenbach's determination, but not Herder's.[77] Herder's doctrine should be called "geneticism" or "generationism," and must be philosophically separated from its natural-scientific counterparts in Wolff and Blumenbach. As Herder would go on to make clear in his *God* (1787), apparently lifeless matter was just a particularly durable structure in a fundamentally organic universe. And "organic" always meant *filled with organs*.

This concept, developed both in the *Ideas* and in *God*, was that of a self-expressive universe, an organic Being in constant unfolding, always in motion and always taking on form, continuous and perfectible. The human stood at the crossroads between its natural and spiritual expressions—for Herder, the problem of *commercium mentis et corporis* was resolved analogically. The fifth book of the *Ideas* is both ontological and methodological, and this cross-disciplinary simultaneity can be seen at the end of the passage above: the "only way" nature shows herself to us is through this penetrating analogy which forces itself upon us in all her works.[78]

If analogy was the answer to the problem with which Kant had confronted the young Herder, it was simultaneously the answer to the difficulty of nature's structure and nature's history. Herder had received this problem from his French reading, perhaps most decisively from Jean-Baptiste Robinet. In his *On Nature*, volume 4 (*On the Animal*) (1766), the third book is entitled "On the Universal Organism." Chapter 1 is poignantly called "On Organization: What an Organ Is." This chapter is a polemic against the notion of lifeless matter—its title's qualifier is to be taken literally. The atomists conceive a brute nature which cannot possibly exist: "Nothing is simple, everything is composed in a material world; a single atom of simple material contradicts like extension without extension."[79] The monism here cuts against the atomists and the spiritualists at the same time: extension without extension articulates contradictions attaching to the soul-body question.

But in a world that is therefore "organic" (in the sense of being *organs* all the way down),[80] what is then the constitutive part which receives the *name* organ? Robinet's contribution—which has earned him much ridicule[81]—is that the organ is the prototype of all being, and that scattered organs constitute the anthropomorphic universe. And Robinet does not shy away from supplying a concrete vision of this essential part: "An organ is an elongated hole, a hollowed cylinder, naturally active: the most complicated organized being can be reduced to this simple idea. The human body, the masterpiece of organization, is nothing but a system of entwined folded tubes endowed with an intrinsic force that results from their structure."[82] With the human body as its finest production, organization is literally simply the universal existence of organs—hollow cylindrical tubes which are fundamentally active. Intertwined and endowed with instrinsic force, they proliferate through the plenum of being and are its basic type.

Herder's own fascination with the problem of type may have been partially inspired by his early reading of Robinet,[83] although the precise provenance of the term is probably not determinable. Yet the organ both held the universe together and progressively broke it down. The term's meaning came to stretch not only analogically across the matter-spirit divide, but also temporally beyond the limits of Leibnizian figure of progress within ultimate stasis.[84]

Herder's first sketch of this problematic takes its cue primarily from the animal. Basic principles are set forth: any effect in nature must be accompanied by force; where there is an excitation (*Reiz*), this must be felt "internally." Should these two principles not obtain, "then . . . the entire analogy of nature stops."[85] Where there is artifice, there must be an artificer—this statement of the *techne-phusis* analogy is rooted in the ontological premise of analogy itself. Where any creature shows intention in its movement, there must be an "internal sense, an *organ*, a medium of this anticipation."[86] There may be more media than we have organs, and thus the world of invisible forces is stranger than our organological sensibility. Nevertheless, "the whole of creation should be thoroughly enjoyed, felt through, worked through: at every new point therefore there must be creatures to enjoy it, organs, to sense it, forces, to enliven it in accordance with each place. No point of the creation is without enjoyment, without organ, without inhabitants: *every creature thus has its own, new world.*"[87] A world of force is analogized—the sense is ontological—into an "internal" resonance of that force through the creature and its organ. This complex of organs determines the *world* of the creature, fundamentally new because immersed in an expansive force always differently concentrated in its particularizing organs.

This sense of the term is broadened in the fifth book. Force is incorruptible, and only its instrument can degenerate. Here organ and tool are equated etymologically: life cannot pass out of existence, although a flower or a tree (life's organs) can. Herder echoes the tradition in which the body is the soul's organ: soul is force, incorruptible.[88] The general point is drawn: "Where we see a force at work, it always does that work in an organ and in harmony with this: without the organ it would at the least never become visible to our senses: the force is simultaneously with it and if we are permitted to believe the continuing analogy of nature, then the force has *formed* the organ *unto itself*. Preformed seeds, lying ready since the creation, have been seen by no eye: what we notice from the first glimpse of the becoming of a creature, are effective *organic forces*."[89] The harmony between force and organ is thus not static—whatever necessity obtains in nature must be united with a dynamic field of organic forces. Herder uses the term in the sense it had in the early eighteenth century: "having organs." Force is essentially organic. And yet, organ is not force—a duality obtains between them which makes their harmony possible. In the basis of nature are light, ether, and heat, a welter of organs in genesis. Herder calls this a "divine current of fire" which pours its forces down to us through a *Vehikulum*. This word—taken from physiological descriptions of the nerves and their electricity—recalls Robinet: concrete form is resisting the chaos of expansive forces. Indeed, their self-restriction is indicated by their name: *organic*. The human body is the result of this organological restriction, this essentially dual force which is yet more essentially *one*. As the organization becomes more complex, its media and organs become finer, eventually resulting in sensation. Finally, the organs complete an internalization or reversal, establishing sensibility.[90]

> Either the effect of my soul has no *analogon* here below: and thus it's neither graspable, how it influences the body, nor how other objects are able to influence it. Or it is this invisible heavenly spirit of light and fire that flows through everything and unites all the forces of nature. In human organization it has achieved a fineness that the earth's structure can guarantee: by means of it the soul works nearly omnipotently in its organs and shines back into itself with a consciousness that excites its inner side.[91]

Again, the analogical nature of the organic universe is seen in its genesis of its highest (physical) form: the human body in its relation to spirit. Consciousness itself is *organ*ic, filled with organs, and this organicity responds to the problem of body-soul interaction. In the crucible of living forces, the soul meets its organs through an *analogon*. Reason is organic, literally without reason (*ana-logon* means "upon" reason, but could easily be taken to mean "without" reason) at the moment of its genesis, and yet determinative of its counterparts (*organa*) as it both is informed by them and returns to itself. This dual gesture stands on the brink of Romantic organology, yet founds, instead of a metaphysics, an anthropology, even an anthropomorphic metaphysics. The organ is not the method: it is ontological, sensible, and spiritual, but its very being and our very knowing are rooted in *analogon*, the unified point where all manner of force meets, underpinned by God.

God, for Herder, may well be *Urkraft*, expressive force, but he also has organs, and the true center of the analogy of nature is the inner human as the final natural organ built (through love) into the divine organ. This is the final sense of "analogy" in Herder's work, the theological end of his naturalistic game. In this vision, God and man share an organ—indeed, the human becomes the analogical organ of the godhead. This unification has been the reality and the goal of history.

Any dynamic force in nature "functioned as an organ of the divine power, as an active idea of the eternal draft of creation."[92] This passing mention of a divine organ recalls Herder's radical assertion in his *On the Cognition and Sensation of the Human Soul* (1778) that the human and his world are God's *sensorium* (the traditional Latin translation of Greek *aistheterion*, the seat of the sense we have come to call "organ"), expressed through a love described well by John the Evangelist yet better by the "even more divine" Spinoza.[93] Herder here takes the terms of the Leibniz-Clarke debate and inserts them into preorganological development of his ontology. When Leibniz had attacked Newton through his proxy Clarke, he had objected that Newton's God was like a watchmaker who needed to intervene periodically to reset the watch. He based this assertion on Newton's claim that (absolute) space and time were God's *sensorium*.[94] That debate—which embittered relations between already hostile camps—stood at the beginning of the confrontation about forces in natural science and philosophy which has accompanied us through this chapter. Herder, by referring to this debate and activating the Newtonian phrase for a Leibnizian force-conception, presents us with the metaphorology of the organ in its highest instance. God's image is his very organ: the human exists in a divine *milieu* which is in constant progress toward the *new*, his experience fundamentally a matter of analogy within an analogical welter of organs across a single substance—God in his expression—which is love. The human is perfectible: his inversion into spirit continues until a full *"inner spiritual human"* is formed, "who is of his own nature and uses the body merely as a tool, indeed who acts according to his own nature even in the case of the most severe derangement of the organs."[95] The organs are corruptible, yet formed in organic necessity. The human forms himself slowly into the essence which is the divine organ, the godly analogy of nature. The human becomes *humanity*.[96]

It is from this point—the self-genesis of the human as godly organ and analogy—that Herder's cultural analysis takes off. Cultures themselves become organic in this literal sense: they are complexes of organs, traditions of individual organismal formation.[97]

The analogy also provides the basis for the final step in Herder's terminological innovation: the rational and linguistic senses of the organ. That there should be an *organ of reason* is already a matter of controversy. That Herder should have thought seriously about one seems implausible. And yet he did; and his thinking on this issue, based in his theory of language, displays the same methodology as his ontological speculation. Reason has an *organ*—language—which provides for its self-genesis through inversion and a resulting consciousness. This genesis reveals the location of reason's encounter with the nonrational—in an *ana-logon*, without or around reason—and makes that encounter productive and innovative for reason itself. And yet the notion of real contradiction, actual duality—so important to the Romantics—is never included in this sliding analogical scale. Herder contributed not only the term and the picture of the universe, not only the ontology and the theology—he also gave the Romantics a picture of reason as itself organic, rooted in a dynamic or organological relation of language and reason. But he did not supply the methodology, for he rejected the Kantian problematic in favor of his analogical overlap between knowing and being.

In his *Meta-Critique* (1799), Herder definitively rejected his former teacher's notion of Critique. The sprawling text takes issue, among other things, with the fixity of Kant's categories. Recall that Kant's conception of judgment was an important anticipation of Romantic organology: the mereology of the judgment, split between constitutive and regulative use by the intuition, is the instrument of the canon of the understanding. This canon is the categories themselves, and Herder's sense that experience provides ontological innovation runs counter to this legalization of our cognition. Thus, when Herder writes that language is the *organon* of reason,[98] he means to organicize experience itself, to give the human a *milieu* that is nevertheless characterized by generality. He is rejecting Kant's division between the *a priori* and the *a posteriori*. And in doing so, he makes the special mark of the human its ability to name especially itself. He writes, "Thus language becomes a *natural organ of the understanding*, such a *sense of the human soul* as that which the power of sight of that sensitive soul of the ancients builds for itself in the eye, like the instinct of the bees builds its cells."[99] Objecting to Süßmilch's notion of a divine origin of languages and to the empiricist derivation of language from noise simultaneously, Herder asserts the self-genesis of reason through human language as a "natural organ." Yet this organ only obtains for humans. Indeed, Herder's concern here for the autonomous validity of meaning is parallel to Kant's concern about truth in the passage on the *epigenesis* of reason. Locke serves as the empiricist foil for both—he imagines a parthenogenetic origin of consciousness and of language. For Kant, representation itself would be undercut in this picture, while for Herder, language would simply not exist—sound and word are separated by the organic development of the human species *as speaking*. Süßmilch's divine gift of language, on the other hand, is like Crusius's (or Leibniz's) preestablishment of truth for Kant. If truth, or language, were given, we would have no way to determine their correctness. In the same passage, Herder notes that *logos* means word and reason, concept and word, and language and cause, all at once. *Alogos* means dumb just as much as nonrational. The human is *homo loquens*: as speaker he is able to *name himself speaker*, to see into his species-essence because he possesses an organ which, inverting the ontological proliferation of organic forces, serves as the instrument of reason.[100]

In the *Ideas*, Herder went on to make this notion work for psychology, describing the psyche's laws as *organ*ic: "That [the connection of concepts] always occurs in accordance with its organ and in harmony with it, that when the tool does not suit, even the artist can do nothing and so on; this suffers no doubt, but also changes nothing in the concept of the thing. The *manner* [*Art*] in which the soul functions, the *essence of its concepts* comes into consideration here."[101] Herder puns on "manner" (*Art* as "kind" but also as "species") to characterize the self-knowing human organ. The essence of the soul's operations—and the essence of its concepts—actually is the trademark of the human (where bleating, so far from mere noise, is the essence of the sheep for the human). The species which can name itself performs this separation—literally passes from animality to spirituality—by way of its organ. That origin of that organ—language—is therefore the moment at which the *a priori* and the *a posteriori* are first, and genetically always—separated. Our hope for knowledge is in this always available separating origin. And that origin is a matter of universal narration, or organic natural history. The human *analogy*, nested in the divine *sensorium* and emerging from first a concretizing proliferation and then an inversion of organs which gives rise to the self-naming organ, is the basis of human history. Herder fulfills his promise to offer a *philosophy* of the history of humanity. This history finds its necessary home in the overlap between ontology and methodology, in the "grand analogy of nature."

It was in fact Herder's reliance on analogy that Kant attacked in his review of the *Ideas* in 1785. As pointed out above, Kant had asked if the *formative drive* was not merely a name for a problem rather than an explanation: was this name not "the explanation of one unknown by another"?[102] He pushed back against Herder, translating his Latin into German and accusing his former student of confusing preformation and epigenesis[103] and ultimately resting his doctrine on unexperienceable forces "as simple limitations, not further explicable, of a self-forming faculty, which latter we can just as little explain or make thinkable."[104] In this key confrontation, Kant slips back into his terminology: Herder's organ becomes a self-generating faculty (*Vermögen*). The organ's delimiting role in being is just as obscure as the force which makes use of it.

To bring his polemic to its methodological point, Kant writes that "the rational use of experience has its own limits . . . and no analogy can fill the immeasurable gap between the accidental and the necessary."[105] The human, who would *be* that analogy, cannot reconcile being's structure with history's development. For us, contingency attaches to the flow of phenomena while the canonical rules of their presentation are absolutely necessary. The necessity of history is in no way graspable—indeed, whether there is necessity is the object of the insoluble Third Antinomy (solved in the "other order of things," practical reason, by postulate). Rejecting both Herder's method (analogy) and his terms (above all "organ"), Kant effectively silenced Herder's philosophical voice, which increasingly took the form of what seemed an outdated polemic against his former teacher. And yet, Herder's contribution to the thought of the next generation was not only oblique, it was also terminological. He donated the decisive chapter of our metaphorology, supplying the content if not the method of Romantic organology. The ontological sense already covered the basic semantic field—in human exploration of Being, the question of force's development in the space between the general or structural and the concrete or developmental was concentrated in the Greek and French borrowing. But Herder also started the trend of engaging physiology contemplatively, and indeed spoke of an "organ of reason," language, by which the human made itself into humanity by giving itself a species name. When Kant attacked Herder—and Herder responded—the full palette of the philosophical and natural-scientific stakes of the imminent organological debate was given. In a metaphysical context first established by Leibniz, the question of methodology was brought to philosophical center by Kant in conversation with Newton. Finally, the richness of the empirical and its potential organic newness—the becoming of the world and of reason—were put in place by Herder analogically. Romanticism inherited the term "organ" from eighteenth-century natural-scientific and metaphysical crises which dovetailed in their production of antinomies with the political upheaval following the Republican revolutions. As they saw the fragmentation of the political world reflected in the increasing isolation of the disciplines, they turned to this terminological context to respond.

THREE

The Organ of the Soul: Vitalist Metaphysics and the Literalization of the Organ

In 1795, Alexander von Humboldt (1769–1859) published a short story called "Life-Force, or the Genius of Rhodes." Appearing in the same issue of the *Horae* as Schiller's "Letters on the Aesthetic Education of Mankind," the story tells of two paintings that appear in the despotic kingdom of Syracuse during the reign of the Dionysiuses in the fourth century BCE. One painting depicts a torchbearer holding his torch high, with throngs of children surrounding him, pressed into order by the attention the central figure commands. The other painting, arriving after years of popular fascination with the first, depicts the same figure, head downcast, torch extinguished. The orders of children fall upon each other, as if they had merely waited for the light to go out. The two paintings are brought to the home of the aging philosopher Epimarchus, who interprets the central figure as the "symbol of *life-force*": "It is as though the earthly elements at his feet strive to follow their own lusts and to commingle with each other. The Genius threatens them commandingly with his raised, blazing torch, and forces them to follow his law, dismissing their old rights."[1] The central figure is that which divides anorganic from organic nature, which causes the organization of orders in nature. Its commanding figure is contrasted with the pacificist and populist Epimarchus, who avoids the despot leaders of Syracuse. Epimarchus, himself passing into eternal rest, conceives of nature as the balance between utter stillness and forced organization, between the tendency to strive to couple and to dissolve:

> Everything strives from the moment of its genesis towards new connections; and only the human's dividing art can present, as if it were uncoupled, what you vainly seek inside the earth and in the dynamic oceans of water and air. There is inert rest in dead anorganic material as long as the bands of relation are not dissolved, as long as no third matter penetrates in order to commingle with those already there.[2]

This definition of life as the constant addition of a "third" that strikes the balance is part of a reconception of life and its cause that spread through German publications in the mid-1790s, raising again metaphysical questions that the Enlightenment and especially Kant had attempted to do away with. Humboldt himself would later renounce this vision of differentiated forces, seeking instead for causal explanations in physical arrangements of the parts of the living.[3] But here, he claimed not only that a life-force separated different spheres or orders of nature, but in fact that only human representation could show this to be the case. Epimarchus's insight is won only as the result of the painting—symbolic representation intercedes where abstract thought and empirical observation could not. The narrative thus suggests that the seat of metaphysical speculation in the 1790s was not some add-on to an otherwise empirically developing set of concerns in physiology, nosology, and other disciplines, but instead, the *instance of consciousness* as a part of science itself. Life-force is intimately connected to that consciousness, and the project that unfolded under various names—transcendental morphology, *Naturphilosophie*, Romantic medicine—was an attempt to include that instance in an account of the world's laws. In one sense, this was a rejection of a kind of Enlightenment empiricism, even of the limitations Kant had and would place on cognition and disciplinary investigations. On the other hand, it shifted the sense of what was metaphysical away from the "totality of being" and toward the very dividing line between consciousness and world. This remained a point of contention for some three decades: should the scientific endeavor include an account of its genesis in human consciousness and culture? The term "organ" was literalized under the aegis of this question, and this allowed for a generalization of its meaning beyond the physical (or cognitive) boundaries of its normal use. Neither merely instrument nor simply a mental faculty, the organ became a part of the body just as it became a universal notion of function.

The physician and philosopher Ernst Platner announced the program of a "new anthropology" in 1772.[4] If the old, Cartesian study of the human was broken into the two substances, that is, into physiology or medicine and psychology, Platner wanted his anthropology to be focused on a unified human being. For this, the discipline would need to focus on the interaction of the substance: "In the end one can observe body and soul in their mutual interactions, limitations, and relations, and that is what I call anthropology."[5] Platner stands at the end of the Enlightenment anthropological tradition and at the beginning of Romanticism. By uniting the human into one form, he follows the tendency of the humanism of the German Enlightenment before him. When Fichte lectured from his aphorisms in Jena in the 1790s, Hölderlin was not the only member of the young generation to take note. Aesthetics and anthropology were co-original in the Enlightenment, and Platner stood at the threshold to the similar discursive intimacy of aesthetics and metaphysics in Romanticism. Anthropology would base itself on the commerce between body and soul, and Platner popularized the term for the instrument of that commerce: *the organ of the soul* [Seelenorgan]. The organ of the soul would in turn provide the fulcrum for Romantic organology.

Disciplinary Organs: Kant after the Inauguration of Criticism

The metaphorical, philosophical, and natural-scientific groundwork for organology had been laid by 1790 at the latest. Herder's *God* had supplied a cosmology of organs, while Kant had supplied the methodological uses of the terms "organ" and *organon*. The concretizing effect of the *organ* on force in Herder was complemented by the concretizing rhetorical choice of *canon* in Kant's notion of the understanding. Meanwhile, Blumenbach's argument for epigenesis in *On the Formative Drive* had won the upper hand. By the mid-1790s, however, the organ's strange absence from or eccentricity to biological debate in Germany gave way to a sudden proliferation: "organ" was now on everyone's lips. Herder was central to this shift. It was in fact Herder who introduced the "generationist" position into the debate on preformation and epigenesis. Herder advocated, in an influential and yet (still) neglected way, giving up the notion of simple mechanical force altogether. A "generationist" was someone who took a position on the implicit debate between Newtonian mechanicists and quasi-Newtonian vitalists. This position never denied the phenomenon of mechanical causality, but rather denied it fundamental status. Herder introduced a differentiated Spinozan Being into the debate, a being that pullulated forces and their specifiying organs. "Organ" came into its own as a scientific term in Germany—in the writings of Carl Friedrich Kielmeyer (1765–1844), Joachim Dietrich Brandis (1762–1846), Johann Christian Reil (1759–1813), Alexander von Humboldt, Johann Wilhelm Ritter (1776–1810), and Samuel Thomas Soemmerring (1755–1830)—on the basis of a renewal of the debate about the nature of force that Newtonians (mechanicist or vitalist) felt had been put to rest. At just the moment Kant's system seemed to foreclose on any future metaphysics (excepting the special metaphysics of the Critical system), the struggle between epigenesis and preformation gave way to a struggle for the soul of the life sciences, a debate between the epigenetic model and a "generationist" or organicist model of being itself. It was in this context that "organ" came, in the mid-1790s, to its full range of literal expression in the German language. The metaphorical uses I examined above were largely borrowed from French, English, and Latin sources. But as the word came to have its now intuitive meaning in German—"functional part," an element in a complex system both existing in mutual determination with the system as a whole—its metaphorical range broadened. And it was in the interplay between an emergent, broad literal sense and a new field of potential metaphor that Romantic organology made its home. As metaphysics entered a new crisis—one determined by the disciplinization of the very knowledge it sought in a profoundly uncertain political Europe—"organ" would come to have a sliding scale of literal values. Hölderlin, Schelling, and Novalis entered the terminological fray, intentionally shifting the open fields of literal and metaphorical meanings of *organ* in order to build a modern metaphysics.

"Organ" came quickly to be associated, in proto-biological discourse, with the developmental aspect of life.[6] Thus Johann Christian Reil could write, in 1795, that "organ and organization is [sic] thus formation and structure of enlivened bodies."[7] Reil returns in this moment to the Aristotelian question of the homogeneous parts and heterogeneous parts—and we can recall that Haller's physiology had also addressed this issue, without using the term "organ." Here even homogeneous parts (like *Faser*—the fibers of the body) are named "organs," insofar as they are complex. Thus Reil establishes an organic system—as a living combination of "matter and form"—in which we have organs all the way down to the divide with "dead" matter. It is as though Leibniz's system has been tempered, included in it a purely homogeneous substance which cannot be organ-filled.

Reil's approach to the problem of matter is Kantian; he tries to integrate the Newtonian attempt not to define forces with Kant's agnosticism about "supersensible substrates." The problem is that we definitely know that we are dealing with two unknowables in dead and living matter—a *contradictio in adiecto* which Reil, on my reading, does not resolve. He writes:

> Representations are possible, according to experience, only in connection with organs. From an immediate effect of the soul independent of organs we have no experience, and thus no real concept. And the capacity of a soul which forms organs for itself before organs are present would have to be of this kind. Matter, one says, is a dead being, insofar as we know it from experience, and we can derive no life from it. But doesn't daily experience teach us that there is one kind of matter (animal) that has life? Why do we not want to attribute the appearances of the same to matter in living nature, as we do in dead nature? Possibly because we cannot know the absolute ground of appearances of living being from their matter? But we can't do that in the case of dead bodies either.[8]

He then flips this argument on its head, emphasizing the complexity of "dead" matter—its formal qualities (crystals, magnets, etc.). The primary representable quality of live matter is, so Reil, that it is "plastic." That plasticity, as the passage above intimates, is a matter of the development and contribution to further development of organs. This somewhat confused passage is an excellent example of the metaphysics which, after Kant and Herder, became unavoidable even among the most sober practitioners of the life sciences. Live matter is purely organic, even where it is homogeneous, since it already there involves structured development. Organs, then, interact with substances outside the live body on a scale from *coarse* to *fine*. With the introduction of the interactive *fine organ*, we can see Herder's influence—but here there is no shift into talk of cognitive organs. Reil also relies on Brandis's 1794 writing of nearly the same name—*Essay on the Life Force*. Soemmerring, as we will see below, will also make use of Brandis, as he will of Platner's 1772 *Anthropology*, which Reil also cites here. Perhaps the most interesting citation, however, is from Gotthelf Fischer's translation of Humboldt's plant-physiological aphorisms. In the notes to this edition, Fischer writes:

> The basis of those bodies that receive life through generation is always extremely small compared to the grandiosity consecrated to its perfection. All this addition and perfecting must effect life from the body itself, from things taken from outside and included in that body, through all the steps accruing to its kind [*Art*]. To effect this inclusion of foreign things into the body, to arrange them suitably for it, there must be workshops and tools that are different and manifold in each case according to the different and manifold determination and needs. These are particularly called organs.[9]

Thus the entire spectrum of processes and parts discussed so widely without a name in the epigenesis/preformation debate became organs. "Organ" has become the point where structure and development meet: organ is formation, but structure is also organization. The term has entered a field of meaning from which it will not again be withdrawn, but which supplies the basis on which the analogy between organic processes and rational forms could become more than metaphorical. In all the talk of formation, drive, and force—which has been so much (and so deservedly) discussed in the literature—a term came into the German public discourse (philosophical, natural-scientific, and literary) to name simultaneously several conceptual pressure points. Gathered around the local problem of structure and development—which itself refers back to the larger problem of generation—were, suddenly, problems of force and expression, cognition and its object, mechanics and organics, *physis* and *techne*.[10]

The crises in school metaphysics from the beginning of the eighteenth century (see chapter 1) seemed, therefore, to haunt its end. The Romantics came of intellectual age in this reiterated crisis-atmosphere, in which the speculative stakes of individual and seemingly isolated areas of knowledge (disciplines) were the explicit topic of debate.

If there is any doubt that this struggle over the general and the particular—this crisis about emergent institutions playing out in metaphysical discourse—was also felt to be political, we need look no further than Kant's 1798 *Conflict of the Faculties*. Originally written in 1794 and withheld due to the Prussian censors (whom Kant had crossed in 1793 with his *Religion within the Boundaries of Mere Reason*), the two halves of this writing tie university reform to the notion of republican government after the French Revolution. Arguing that the regulative, moral teleology of the human race must ultimately recognize its progress in the republicanism, Kant defends—as he had more generally in 1784 in *"What Is Called Enlightenment?"*—the notion of a "philosophical faculty" which represents *public reason* and the interest of the truth alone. This traditionally "lower" faculty—which would replace the theological faculty at the academic helm of the university[11]—necessarily occupies a privileged position in a university appropriate to republican government. Its privilege results from its distance from the state's interests, and that very autonomy comes from its relationship to truth. Here Kant returns to the terms I investigated in chapter 2 of this study: the "laws" which are set forward as a canon of each faculty with respect to government—one thinks of the modern use of the Hippocratic Oath—are distinguished from the rules which bodies of professors give themselves for the practice of their individual disciplines. Although these self-regulatory statutes seem to be essential to the concept of each faculty, they have no governmental authority. Indeed, writes Kant, each can serve merely as an *organon*—an aid—for the present state of each faculty's practice:

> Those books that were authored by the faculty as (supposedly) complete excerpts of the spirit of the law-book for the convenient understanding and sure use of the common institution (of the learned and the unlearned), for example, the symbolic books, are completely different from the law-book as the canon. They can only demand to be seen as the organon, to ease entry to the same, and have no authority; not even if the most noble learned men of a particular faculty have agreed on letting such a book be valid for their faculty as a norm—something they have no authority to do—rather they sometimes introduce them as a method of teaching that remains changeable according to

the circumstances of the time and in any case can only touch on the formal elements of presentation but makes absolutely no difference in the material elements of legislating.[12]

This division is then used again to distinguish within the theological faculty between *pure religious doctrine* (the philosophical canon) and *church doctrine* (the theological organon).[13] Autonomous, public reason is identified with the canon, which should produce disciplinary *organa* according to the heteronomous functions of each faculty. The philosophical faculty's antagonism to the others lies in its ability to make this distinction: it *disciplines the disciplines* by referring them always to the most general human interest. Looking back to the *Critique of Pure Reason*, we can see that "canon" is used univocally across these texts: a body of positive law establishing the legitimate use of judgment for the constitution of our world here finds its social expression in application to the practice of the socially embedded faculties. *Organon*, on the other hand, is demoted to the useful dogmatism of a disciplinary practice outside the philosophical faculty. Above all, Novalis would take direct aim at this distinction in his reading of Kant and his construction of a universal organology (see chapter 6).

The *Conflict* thus takes pride of place in a particular genre of writing Kant engages in after establishing Criticism, one we might call "disciplinary." The review of Herder's *Ideas* is only among the more polemical of these writings.[14] They take their cue from the last sentence of the *Critique of Pure Reason* ("only the Critical way remains open . . . "), and they rigorously apply the notion of *discipline* worked out in the "Doctrine of Method." Those uses of reason which lead away from the canon of the understanding—in fact, those uses which allow *organa* to become statements about truth rather than about disciplinary practice (constitutive rather than regulative)—are to be excluded from philosophy. Indeed, they are to be excluded from the university, and ultimately from the public sphere altogether. After all, the inaugural disciplinary writing had been directed at Herder, and its point of contention had been a metaphysical problem (that of force).

The word *organon*, then, is demoted to its nonmetaphysical use in this context. Kant has now dismissed the problem that had exercised him for several decades, that of an *organon* for metaphysics, or formal rules for general cognition. Simultaneously, however, the very problem which articulated for him the heart of a conflict of the faculties—the problem of the cognition of force—came into natural-scientific prominence and brought the word "organ" to the fore. In this context, the word was literal, but its meaning was not yet fully determined. For that determination, some settling of accounts in the metaphysical debate would be necessary. *The Conflict of the Faculties* socialized Criticism even as it intentionally left out the metaphysical conflict associated with Kant's own earlier use of the term *organon*. But between the essay's composition (1794) and its publication (1798), the issue would be forced into public view by the Göttingen physiologist Samuel Thomas Soemmerring with his 1796 *On the Organ of the Soul*. This literary-scientific event—reactions to his writing, mostly negative, came from all sectors and faculties—effectively brought organ and organon into the same discourse, quite without any intention on the part of Soemmerring or the writer of the afterword to his study: Immanuel Kant.

Toward a New Metaphysics: From the Soul's Organ to Organology

Soemmerring's *On the Organ of the Soul* was published under these disciplinary conditions, which dovetailed with the renaissance of metaphysical problems described above. This conjuncture set the stage for Hölderlin's combination of monism and dialectical thought—the work was pressed upon him by his self-designated literary father, Wilhelm—which found final expression in his theory of tragedy. Among the seemingly *passé* problems that found renewed interest was that of the location of the spirit in the body. The three systems of body-soul commerce—*influxus physicus*, Leibniz's preestablished harmony, and the post-Cartesian occasionalist system—had driven this discourse from the collapse of scholastic Aristotelianism in the late seventeenth century through the middle of the eighteenth. The question had not gone away—nor has it—but the circulating answers in German public discourse had shifted significantly by 1790.

There were primarily two kinds of responses in Germany as the nineteenth century approached. First, there was an anthropological answer, one emphasized by writers like Ernst Platner and Herder. Second, there was a critical answer, emphasized by Kant and his followers, especially Reinhold and then Fichte. In a sense, this latter discourse was *not* an answer, since it took its departure from the logical deadlock of the terms of the question. The increasingly bitter relations between the founders of these discourses[15]—especially as Kant became more polemical in his writings in the 1790s, and Herder began to attack his former teacher—gave way to more reconciliatory gestures with Fichte's 1794 lecture-courses on Platner's aphorisms, which Hölderlin attended.

Scholastic frameworks had fallen away, and the question of locality—especially with respect to the brain—had gained traction as the physiological enterprise became both technologically and analytically more complex.[16] This shift meant that the terms *Werkzeug* and *Organ* were used to describe functional locations in the body, and especially in the brain. If the body was the soul's organ, this meant something immediately material and functional. And the soul's organ could now be investigated for information about that functioning—if only there could be agreement about what that organ *was*.

Kant had addressed the locality question in his *Dreams of a Spirit-Seer* in 1766, arguing that the soul cannot be conceived in the same terms as the body.[17] It was, however, Ernst Platner who spearheaded the reawakening of the question of the "soul's organ," with his 1790 *New Anthropology for Doctors and Philosophers*.[18] This reworked and amplified version of the 1772 *Anthropology for Doctors and Philosophers* introduced the term "organ of the soul" firmly into the public sphere.[19] The organ of the soul is twofold in two separate senses, folding an animal and a spiritual nature into the essence of the human, which is itself both recognitive and active, the body its tool.[20] The body—even its organic parts—cannot be counted as the seat of the soul, and Platner polemicizes against those who misrecognize the nerves themselves as this locus. The body is metaphorically the organ of the soul, nothing more.[21] This presents the reader with a strange dualism: Platner, who insists for holistic reasons on the unity both of the organ of the soul and the human more generally, also points to competing and even seemingly contradictory structures in the body, in the soul, and in their relation.[22] Thus Platner—especially through his continued use of the term *influxus physicus*—introduces not the general "organ" but the *organ of the soul* in a metaphysically discursive manner. And it is precisely the problematic of the dual and the unified that Hölderlin would take up and radicalize.

If "organ" had been literalized, it had also been, in a parallel discourse, metaphysicized again. Where the conflict between physics and biology had gained obliquely metaphysical proportions, the traditional problem of *commercium mentis et corporis* had been remetaphysicized just as Kant sought polemically to force a different (canonical, disciplinary) solution.[23]

Platner set the tone for the entire interaction Hölderlin would have later in the decade with Soemmerring's writing:

> Because the human soul, like in all probability all finite spirits, can neither produce the material of the representations appropriate to its manner of existing and so determined by God, nor can it take them immediately from the world in front of it, and that much less can it produce an immediate influence in that world: so it needed a tool as means [*Mittelwerkzeug*] by which it could partly represent the world according to its relation to it, and partly according to these representations influence that world. *Organ of the soul*.[24]

Not only the structure of Hölderlin's dialectical metaphysics (see chapter 4 of the present study), and in fact not only its terms were given here, but in fact the definitional task of the Romantic organ was described by Platner, perhaps for the first time. For we can see here, in an argumentative style that attracted neither Fichte nor the Romantics, both the metaphysical and the political or ethical dimensions of the interventionary tool of Romantic organology. The *Mittelwerkzeug*, which both cognizes and acts and is the principle of any possible unity between these two human activities, is also defined as the possibility of human cognition of the world itself. The division in cognition which prevented this unified instrument in Kant's criticism is here itself criticized. The organ of the soul was, as we shall see, the predecessor of the Romantic attempt to systematize the absolute knowledge that Kant had called "intellectual intuition."

Soemmerring was not exactly alone, then, when he proposed that the "fluid" in the nervous system was the *sensorium* Descartes had once located in the pineal gland. Soemmerring did, as Michael Hagner has pointed out, recall the literal terms of the Cartesian debate, by citing liberally from authors from Descartes himself to the novelist (and his friend) Wilhelm Heinse (1746–1803). If it seemed that the question of the seat of the senses could be separated from that of the location of the soul, Soemmerring incautiously equated the two.[25] He thus broke the discursive rules of the resurgent metaphysics, making its terms explicit and violating Kant's warnings about the separation of the faculties. As we have seen, he was not alone in continuing to ask about their connection. Yet his literary archive, a kind of panorama of "old" metaphysics, provoked vehement rejections of his proposal from perhaps the two most dominant cultural figures at the time: Goethe and Kant. It took Hölderlin's eye to recognize, through the prism of Soemmerring's citations of Heinse—as we shall see—the connection between post-Kantian philosophical concerns and post-Spinozan theological anxiety (especially in the wake of the Lessing controversy).

Dividing the nerves into twelve pairs, Soemmerring focuses on the interface of the "fluid" in the "cavities" of the nerves (*Nervenhöhlen*) with the ends of those nerves in the brain.[26] Detailed drawings and descriptions of especially the sensory nerve-pairs lead to the larger question: How can sensation arise from hard matter? Soemmerring's answer is that it is precisely a soft kind of matter that allows this: water. Citing widely from Descartes to Kant to Wilhelm Heinse, Soemmerring asks after a *medium uniens* that connects soul to matter in perception. His answer is the dynamic fluid in the nerves. Where Platner had rejected a "visible" locale of this interaction, Soemmerring incautiously proposed a point where physiology and metaphysics could meet.[27] In doing so, he prepared not only the term "organ" for a more complex use but also gathered a citational web which Hölderlin and Novalis went on to exploit. Important among his sources are not only Descartes and Kant, but also Platner, Brandis, and Herder. The argument culminates in a rhetorical question ("But then why should an apparently homogeneous fluid not contain our spirit, not be able to serve it as an organ?"),[28] which is then followed by a citation of Herder ("no force of nature is without organ").[29] Herder's use of the term, as the passage suggests, was the most present for the shift that occurred in the 1790s. And yet, as Soemmerring makes clear, he is not addressing the question about force directly: he is interested to show where it is appropriate to talk of the soul's instrumentalization of the body. That interface—the location of the *sensorium commune*, the *proton aistheterion*, and the *sedes animae*—is in the literal waters of the nerves.

Soemmerring's predominant interest in the physiological question about the soul—its organs, not its status as or as interacting with force—did not prevent his treatise from bearing on the debate about the unity of nature and the plurality of forces. Indeed, his writing is singular among those I have identified as shifting the discursive use of "organ" in the 1790s. This is because he thematizes the mediation of the complex physical organ (rather than the question of the organ's relation to force). He emphasizes the *organ* itself, the medium of communication, rather than the Newtonian problem of force.

The responses to Soemmerring were largely negative.[30] Goethe wrote to his erstwhile friend[31] on 28 August 1796, complaining that he had mixed up the duties of physiologist and philosopher. He would have done better to "leave the philosophers completely out of the game" and to have ended with §26, the last "empirical" section of the writing. Ideas about nature were mere instruments for research, organs for appropriating from nature an understanding which might never demonstrably correspond to the reality it attempts to encompass. This organ is quite clearly the limited *organon* which Kant had envisioned for practice in and out of the university disciplines. Organs of the soul stood in stark and public contradiction to the use-value of concepts as metaphorical instruments in empirical research.

The treatise received an afterword from "our Kant," to whom Soemmerring dedicated the treatise.[32] Kant's assessment of the work is divided into two points: a rejection of the metaphysical pretensions of the physiological enterprise, and a welcoming of the possibility of discovering where sensation occurs.[33] Relying on his earlier position in the *Dreams* and its development in the *Paralogisms* of the *Critique of Pure Reason*,[34] Kant rejects the notion of a locale for the soul, but finds the notion of an organ comparatively interesting. The notion of water as the element which makes sensation possible is intriguing, according to Kant, because the fluid dynamics of the medium indeed suggest the alternatingly binding and separating factor that is required for such sensation to arise.[35] Here "organ" gains its literal meaning, intentionally reduced to its material element (from its Aristotelian ambivalence—see the introduction above) and regarded as the dynamic source of empirical perception. The rejection of the locality of the soul is, however, not as straightforward as it seems. It relies on two notions, which together present a clarifying conjuncture of the disciplinary and theoretical programs of the late Kant. First, the notion of the locality of the soul is said to be an "error of subreption"[36] in which a confusion of faculties (*Vermögen*) occurs. What the *Dreams* had discussed as *subrepted concepts* had been revealed in the *Critique of*

Pure Reason as moments of the failure of critique, moments where a concept (such as "soul") is combined in a judgment with an intuition (such as "locale") in such a way that there is a genuine conflict between them. What is here called "subreption" is simply a hurried combination of judgments which, when their origins in their relative faculties are properly revealed, can be shown to conflict so basically that the judgment must be relinquished. This theoretical objection is reflected at the level of university politics: Soemmerring's proposal pushes the medical and philosophical faculties into conflict through the attempt at a coalition where none is possible. The second half of the afterword becomes more interesting precisely through this lens, since Kant speaks there not merely as a layman, but as the voice of the philosophical faculty and its public reason. The philosophical and medical faculties are divided by their relative positions of autonomy with respect to the state—and by their internal rules for practice—but also have the right to organize the results of the other faculties in keeping with truth itself. The task of philosophy, which is founded on an essential separation from the other disciplines, reflects its metaphysical pretensions on the social level. This is one way to describe the disciplinary genre itself. After dividing the faculties and singling out the understanding as the world-constituting canon in the *Critique of Pure Reason*, Kant moves from the notion of "regulative ideas"—areas where we can have orientation but no determination—to the regulating of just those ideas. But in regulation comes a different sort of determination—social prescription—which in turn requires such regulative beliefs as the progress of society, the amenability of nature to human aesthetic and scientific purposes, etc. Thus Kant's later works present us with a kind of social mirror of the former metaphysical enterprise. If there was any Kantian conclusion which the Romantics wanted to move beyond, it was this one. But the disciplinary and metaphysical stakes had been given, precisely at the moment when the term became literal. The Romantics began their intellectual work in this semantic confusion, and inherited its conceptual problems. They recrafted their inheritance into a modern metaphysics.

PART II

Romantic Organology

Toward a Technological Metaphysics of Judgment

The layer of Romantic discourse centering on the word "organ" reveals a different Romanticism—one focused on Critical methodology, on a new metaphysical approach to concrete scientific and political topics. The doctrine that emerges—reconstructed here from its incomplete textual expression—can be called "organology." It is the project of the coming chapters to present this neglected doctrine. Originally metaphysical in its pretensions in Leibniz, the word "organ" had been introduced fully in German only in the 1780s, as we have seen in part 1. Under the Critical system of Kant and the analogical system of Herder, it had gained a complex universalizing overtone (as the "*organon* for metaphysical knowledge") and a historical depth (as the developmental connector of orders). From a strange absence in biological theory, it had, on the other hand, been literalized into what is for us an intuitive sense ("functional part of a living body"), albeit in newly metaphysicizing projects of natural philosophers in the 1790s (the term "biology" was first introduced in 1800). There were, then, three basic meanings in fairly chaotic play when Hölderlin, Novalis, Schelling, and others adopted the term. Each of these figures found the biological term useful as a concept that combined both location (often "material location"—see the introduction above) and function. The eye, for example, is the name for the physical location of an "organ" that is really the combination of that location and its set of rules: the ability to see. The "organ"—as this combination—is thus the concretion of a field of possibility (in this case, the field of the visible). Dovetailing with this problematic was that of the "*organon* of metaphysics"—although Kant had rejected such a tool (and, as we shall see, the Romantics took his methodological considerations very seriously). Because the metaphysical was "general" (metaphysics treats "being insofar as it is being," *ens inquantum ens*), a tool—which is formed for a specific function—would delimit that generality and obscure the task. But an *organon*, the Romantics would reason, could be general enough if we follow the second-order insight that the complex organ—as a location and a rule, or a specified function—is itself fundamental. That all things could be, or could be *made*, organs, was Novalis's central insight. But Hölderlin's, Schelling's, and Novalis's efforts did not consist in projecting the structure of an organ onto being. As I will be arguing, their engagements with the emergent doctrines of German Idealism led them away from a "metaphorics" of either mechanism or organicism, and to seek a higher methodological core for their metaphysics in the seemingly abstract concept "organ." Organology is not the same as "organicism" (which is a fair label for Herder's cosmology and metaphysics). The notion of the organ includes a negativity fundamental to the kind of systemic and cognitive syntheses the Romantics produced. The organ was always an enabling and a restricting factor, and its elevation to a (perhaps *the*) methodological concept involved not merely projection, but systemic synthesis in a nonsystemic world. It allowed for theoretical efficacy in the contingent flow of phenomena and history. The concept thus allowed them to develop what one could call a "technological metaphysics," one that is infinitely open-ended to radical possibility both in the material or empirical world and in the history of human cognition. To tell the intertwining histories of nature and human cognition as a single story was the common project of Novalis, Schlegel, and the scientists who wrote what they called *Naturphilosophie*, the philosophy of nature. These "histories" sought to confer necessity upon contingent structures, not as a way of demonstrating their internal rationality, but as a way of gaining a theoretical toehold for intervening in these histories, for reproducing them as provisionally rational, in order to produce something radically new—a different order of things. "Organs," then, were the locus of this philosophical hope, which combined methodology with application to the "real." Rather than producing a representation of a world, a Nature, or a history, the Romantics' focus on organs was meant to recast any possible sense of "world," "nature," or "being." By naming the parts of any of these wholes "organs," the Romantics did not wish to replace one picture of the world with another—they meant to undermine the merely pictorial or representational sense of the world. Organs delimit and open the space between the possible and the actual—as Aristotle had shown— and thus change the metaphysical question about the primacy of thought or being from one having to do with representation (thought "pictures" being; an ultimately represented world produces the thought that does the representing) to one centered on function and reproduction. No simple knowledge of "being" was the object of this metaphysics, but instead a discrete kind of knowledge, a discontinuous and developmental model of cognition meant to respond to being as it is grasped and can be grasped by that cognition. I call this "technological" because it is literally based in human *techne (in fact, it construes thinking as techne)*, and consists in aesthetic, kairological, and constructive attempts to deal with historical being, both at the natural and social levels. For this task, something more than a new representation was needed. Rather than a metaphor of synthesis (the state as a body) or of automation (the body as a machine), what was needed was a system that focused always simultaneously on the mysteriously constituted real area of application (say, the physical for physics, or governance in statecraft) *and* on the methodological interface between knowledge and action inside this area. Knowledge of and action in a field were, according to the definitions in the discursive air at the time, organs. These connecting pieces between parts and wholes, between the possible and the actual, needed to be made into tools, or *organa*, in order to be useful. And reciprocally, the *organon* of metaphysics was only conceivable under the condition that it become a variable organ, a general term for the connection between parts in an order, and ultimately for connections between orders. This philosophy, as paradigmatic as I think it for post-Revolutionary European thought, is also of potential use in any world marked by the uncertainty of its technological and political circumstance. To address and redress such a circumstance, the Romantics thought, one needed a flexible metaphysics. That is what I call "Romantic organology."

Being was put in brackets by Kant.[1] Whether one takes his fundamental position to be that we "color" our experience of a real but unreachable world, or rather that there is a world fundamentally different from our particular colorations—in either case, the real is excluded from the procedure of philosophy, makes up only the horizon of that procedure. Being can very well be as it may: it has little to do with philosophical method. Indeed, it is for just this reason that Kant does not create what the next mini-generation—especially Reinhold and Fichte—wanted from him, namely: a highest cognitive principle for the deduction of other forms of knowledge. This mini-generation largely ignored the question of being altogether (with the exceptions of Maimon and Jacobi, who understood that Kant could not leave this question out, although he also could not solve it), focusing on the possibility of a new faculty (*Vorstellungsvermögen*) or a different theory of self-consciousness as the source of legitimation of claims (*Ich*). This is why the cluster of the earliest Romantic philosophical writings (Hölderlin's *Judgment and Being* and Novalis's *Fichte-Studies*) so forcefully reraise the question of being.[2] Indeed, both of these seminal texts not only ask after the traditional topic of metaphysics (being as being), but they also propose cognitive categories that correspond to being in general ("intellectual intuition"). And, as Manfred Frank has shown, they both reject the possibility that those types of cognition are available to us.[3] Being had become alien, beyond the reach of cognition, but not, strictly speaking, "outside" it—it was, as Frank and Dieter Henrich have both emphasized, the ground of consciousness, the enabler of but also the the unknowable and contradictory root of self-consciousness.[4] This is why "being" appears in Romanticism alternately as exalted and dark, as the highest source of reveling or revelation and the destructive "night," the blindness of finite human existence. This, too, is why Romanticism has been taken as both mysticism and proto-existentialism—"being" is alien to what is human in both philosophies. The Romantics had learned from Kant that being was not *ipso facto* cognizable, and then, having learned from Fichte that the tools of cognition contained synthetic forms on the basis of this alien "being" (perhaps) inside of but contradicting consciousness, they saw themselves as free to experiment in and out of the metaphysical tradition. This experimentation,[5] I will argue, took the form of a methodologization of metaphysics. Where Kant had made metaphysics a matter of epistemology,[6] the Romantics made theoretically good on Newton's (and Wolff's) reduction of being to method: for them it was not merely that we could "only know" what was available to our methods of knowing (the tautology that papered over the empiricism of Newton and the idealism of Berkeley), but that, lacking knowledge of a being *other than* the one our categories would allow us to know, we would have to experiment with the "being" projected from those categories, and then allow *both* the being and the categories to change synthetically—to interact historically. But the guiding light of this approach had to change very quickly: for this form of playful synthesis to be meaningful, being had to be fundamentally *composite* rather than fundamentally "alien."[7] And indeed, the foreignness of Being, for Hölderlin as for Novalis, did not stem from the predicament of Kant's proscription, but instead from a quality attaching to the conception of being "in general," whether taken as "outside" or "inside": its wholeness. It was not that individual beings were brute "things" that could not be approached by thought, as though these were two different orders in the Cartesian tradition. It was that the putatively other order (being) and the proper order of our consciousness both posed the problem of totality. Being seemed to be explicable only through its totality, and consciousness seemed deducible only through its relation to that totality. The vertiginous aspect of these early texts is driven by this problem (what Frank calls the "ordo inversus") of the seemingly impossible but absolutely necessary question of the interface between being and thought, where this interface itself requires a totalizing form of cognition not available to Kantian or Fichtean categorical synthesis. The first move, then, was to recoup the sense of what was needed (albeit under the mark of its impossibility): a method for knowing the totality of being, or: *an organon of metaphysics*. That tool, as I shall argue below, is a kind of transcendental tool, a tool that makes more tools out of the material on which it works. It efficacy is based on the conviction that "being" is general and real and provides the material for the work of the *organon*. This conviction, which always allows being its alien properties, also always insists that being be workable, that it be not cognitively "available" but cognitively "possible" (in the sense that it is possible to work on it, that it is, as Novalis will claim, possible in the second degree). The question of "being," then, is written in multiple ways: for the Romantics, "beings" (*onta*) are not brute, "outside" factors for cognition to passively receive and then reconstruct in representation; "being" in general, while it appears to possess the alien quality of a different order than thought, is also (*perhaps*) the "internal" ground of that thought (as self-consciousness). Taking both of these propositions provisionally and improvising new and real relations between their objects, the project of Romantic metaphysics become "technological," that is, it becomes a matter of constructing working systems that can produce and reproduce (rather than merely represent) in what appear to be multiple orders of things (as we shall see paradigmatically with Hölderlin: thought, being; mechanics, organics; *physis, techne*). Reproduction of orders is "transcendental technology" because it presents the condition of possibility for an order of things and aims to make that order available not only to cognition but also to action. Romantic metaphysics continually reproduces the point of interface between thought and being, and thus makes the totality (*Seyn*) of orders flowing from that point available for alteration. This point is where the problem of writing and metaphysics encounter one another. Winfried Menninghaus has argued that the Romantics hold a positive notion of "the letter and representation as *medium* and *production*."[8] What is always already inscribed, however, is function in the process of becoming—being is written, to be sure, but this means that its inverse order is always available for real interventions between orders, at the level of natural and historical laws. Writing and manipulation are already possible—what remains is to discover the limit and the method of metaphysical innovation.

Hölderlin discovered the organ in the work of Samuel Thomas Soemmerring (1755–1830), who claimed that the organ of the soul lay in the waters of the nerves. Schelling united physiological debates on specific forces and their locales in the animal body to construct an electric organ at the base of human cognition, an organ containing a true contradiction, as would cognition, action, and the artifacts of the human world. Novalis's

Romanticism is organology—for him, anything can be made into an organ, one derived from polemics with Kant, borrowings from Franz Hemsterhuis, and ultimately imagining the cosmos as populated by little more than infinite potential organs. These Romantics borrowed from and focused their analytical energies on a kind of imaginary science, a supradisciplinary conspectus of method that constitutes a metaphysics for an always contentious modernity in which neither cosmos nor society can form the suggestion of a stable imprint of the other, or of some putatively higher order.

FOUR

The Tragic Task: Dialectical Organs and the Metaphysics of Judgment (Hölderlin)

> A human being should act with calm;
> We should reflect, we should unfold
> Enhancing, cheering all that lives about us
> for full of high significance
> Magnificent nature, bearer of the silent force,
> Surrounds the one who intimates
> That he must shape a world
> That he may call
> Her spirit forth, this human being suffers
> Care within his breast and hope;
> An overwhelming yearning
> Its roots deep within him strives upward;
> For he can do much and lordly is
> His word; the world is then transformed
> And under his hands
>
> —HÖLDERLIN, *The Death of Empedocles*

In June 1796, Friedrich Hölderlin (1770–1843) fled with the Gontard family, for whom he was working as a tutor, into the hills outside Frankfurt to Kassel. The Coalition War had again spilled over the border of the Rhein. Hölderlin was calm—he was close to where the legendary Hermann had conquered Varus,[1] and thus felt positioned between two crucial battles as historical spectator, as he put it to his brother: "For you, my Karl, the proximity of such a prodigious drama [*ungeheures Schauspiel*] as that which the giant steps of the republicans now guarantee can strengthen the soul deeply."[2] The post-Revolutionary scene was a spectacle, indeed a play—he would rewrite that play in the years that followed, as *The Death of Empedo-*

cles. And he would build the theoretical necessity of spectatorship into the drama itself.[3] Hölderlin might waver in his optimism as the Coalition Wars dragged on, but he understood his task quite early: to produce the theoretical counterpart to the Revolution. And theory, he agreed with Schiller even as he tried to surpass him (on which more presently), had to be rooted in concrete production. A generically constrained *techne* was needed to bring a new era into being.

In Kassel, Hölderlin met a man he would call "Father Aether" or sometimes "Zeus." Wilhelm Heinse, the author of the novel *Ardinghello, or the Happy Isles* (1787), was a friend of the Gontard family, and his unequivocal pantheism had inspired Soemmerring to look for the instrument of exchange between body and soul, the organ of the soul rooted in the fluid in the nerves. As we have already seen, Heinse had associated the term "soul" with the functioning part within the one and all, the *hen kai pan*. Soemmerring had then, about a decade later, added an *organ* to the soul. It is likely that the older writer Heinse put Soemmerring's recently published *On the Organ of the Soul* in Hölderlin's hands.[4] If so, it was a fateful moment.[5] Hölderlin was forced into the role of philosophical spectator of the war, conversing about poetic production with "Zeus," even as he read a treatise about the material conditions of cognition along with its disciplinary afterword by Kant. The contradiction of the Revolution was complemented and ramified by the tension between metaphysics and the organ of the soul. The twin antinomies were a spectacle playing out before Hölderlin, and they posed from the start a question about representation, embodied in the encounter with Heinse, who had made metaphysics central to the novel.[6] This was the conjuncture that produced Romantic organology.

Hölderlin is the inaugural thinker of Romantic organology. The story of his intellectual development has been told in rich detail, with twin focuses on his reaction to Kant and Fichte, on the one hand, and the emergence of his poetic theology—starting with faith, proceeding through a pantheism, and emerging with both Christ and Dionysos, devotion to nature's whole and mythology—on the other. The history I construct here unites these diverse elements. Indeed, in a neglected chapter of Hölderlin's development—his reception of the term "organ" from Soemmerring—I find the origins of that historiographical dualism. Hölderlin is and has often been shown to be a thinker of contradictions. But his innovative use of the term "organ" shows that he is this in a far deeper sense than is often recognized. Organs become dialectical in his thought,[7] in precisely the technical sense later conferred upon that term. They are at once structure and development, at once sensual and rational, at once form and content. Precisely because they are this unity of opposites—as Hölderlin puts it, *the one differentiated in itself*[8]—they are also the instruments of structural dissolution and reunification.

Hölderlin's most central philosophical gesture—the dialectical figure of real, developmental contradiction captured in a greater unity—emerged not from his reading of Fichte, but from his realization that Fichte (or Kant) and Spinoza (or Lessing) could be united only by a metaphysics of contradiction: organology.

Organs develop, to be sure, but they do more than that. Where time traditionally resolves the logical issue of the coincidence of opposites,[9] organs confer real contradictions on the world, and also resolve those contradictions. They are antinomial and univocal both simultaneously and in succession. For this reason, they offer a particular kind of answer to the Kantian deadlock about the understanding. Where Kant had rhetorically asked the tradition to justify its application of logical judgments to real states of affairs, Hölderlin responded—unexpectedly—with the notion of an organ of the spirit,[10] explicitly in contradiction to its unified pairing "being," which Hölderlin came to call the *aorgic* (*das Aorgische*) in order to underscore its logical opposition to the organ. By expanding the range of contradictions in our cognition and then focusing those contradictions in the flexible lexeme "organ," he pointed the way toward a metaphysics of judgment, a post-Kantian approach to the problems Kant had left behind when he shifted from a possible *organon* to his canon of the understanding. Hölderlin saw in the form-content unity of judgment a unified frame, a locus of investigation. The structure of the world opened in its very development in that frame— what Kant had called "intellectual intuition" found an organ, again unexpectedly, in the tragedy. The tragedy became the genre of presentation of the totalizing opposition between "organ" and the aorgic in historical form. Tragic writing was an invitation to new forms of absolute cognition. Hölderlin broke the deadlock of post-Kantian theorizing in Jena in two gestures: first, with the notion of a dialectical organ; second, with the turn from pure philosophy to genre theory.[11] This latter move made tragedy a privileged organ of absolute knowledge in concrete and historical form. Literature was instrumentalized, and metaphysics renewed: the first form of Romantic organology took shape between the speculative organ and that instrument of literary expression, genre. Tragedy was both philosophically classical and epistemologically actual—it was meant to provoke a new (organological) form of consciousness into existence. The figure of an organ of intellectual intuition addressed and attempted to resolve three real contradictions: the critical opposition of judgment and being; the biological deadlock between the mechanical and the organic; and an emergent struggle between nature and art, *physis* and *techne*.

The cosmic picture that emerged as the necessary correlate of this complex thought is what is usually referred to as Hölderlin's Classicism, for it was only the Greek tragedy that reflected the one pole of his contradictory picture of developmental ontology. And it was only the possibility of a contemporary tragedy that could fulfill the organological promise, the promise of mutual interpenetration of the absolute and the particular which had eluded him in *Hyperion*.[12] That tragedy would have to build spectatorship into its very thematic, making the observation of Empedocles's return to the "aorgic"—i.e., his suicide by volcano—into the call for a radically new era.

Hölderlin's self-conceived task, then, was to observe the emergence of that era and to provide the theoretical tool for its completion. That tool would have to have the characteristics of the nature and history it would serve to alter, namely contradiction and function. Just before drafting the *Empedocles*, Hölderlin wrote in a letter:

> And with respect to the general, I have a comfort, namely that ferment and dissolution must *necessarily* lead either to destruction or *to new organization* But there is no destruction, so the youth of the world must always return from our rot. One can in fact say with certainty that the world has never looked so colorful as now. It is a *prodigious variety of contradictions and contrasts*. . . . But that's what it should be! This character of the better known part of the human race is certainly an augur of extraordinary things. I believe in *a future revolution of opinions and kinds of representations* [*Revolution der Gesinnungen und Vorstellungsarten*], that will put everything past to shame.[13]

The revolution would be metaphysical. Hölderlin is not only referencing Kant's famous comparison of Critique to Copernicus, but seems to be directly quoting Kant's first popularizer, Karl Leonhard Reinhold, who had written that Kant's system was "[a] shaking of all heretofore known systems, theories, and *kinds of representations* [*Vorstellungsarten*], the extent and depth of which the history of the human spirit has produced no example."[14] Reinhold's earthquake is epistemological and recent; Hölderlin's revolution is metaphysical and still to come. And as we shall see, the very nature of contradiction would have to change for that revolution. Contradictions would have to occur in history, and be resolved there. And for that resolution, the philosopher would need a tool that both resolved and created contradictions, that cut across the apparently resolute dichotomies between being and thought, mechanics and organics, and freedom and necessity. Hölderlin was the first to name that tool an "organ."

Intellectual Intuition: Judgment, Being, and the Beginnings of Hölderlin's Metaphysics

Platner had set the stage for an organ of the soul that would unite real cognition (for him, the passive capacity of the organ) with action (the ability to influence the organ-mediated world). Hölderlin set his sights on just such an epistemological figure, which he found in the Kantian/Fichtean notion of "intellectual intuition." He wrote to the Jena philosopher Friedrich Immanuel Niethammer (1766–1848) on 24 February 1796, only a few months before the encounter with Heinse:

> In [Schiller's *On the Aesthetic Education of the Human*] I want to find the principle that can explain to me the separations in which we think and exist—but it is also capable of making the conflict disappear, the conflict between subject and object, between self and world, even between reason and revelation,—theoretically, in intellectual intuition [*in intellectualer Anschauung*], without any need for our practical reason to come to our aid. For this we need aesthetic sense, and I will call my philosophical letters "New Letters on the Aesthetic Education of the Human." In them I will also go from philosophy to poetry and religion.[15]

The "New Letters" never came to be, but Hölderlin pursued the project laid out here throughout the years leading up to 1800. Hölderlin makes clear that Schiller has laid out the "principle" for resolution of contradictions between subject and object, self and world, theory (philosophy) and practice (poetry and religion)—but Schiller has not elaborated the doctrine which this principle makes possible.[16]

Writing to Schiller five months before his letter to Niethammer, Hölderlin seems to agree:

> The discontent with myself and that which surrounds me has driven me into abstraction; I am trying to develop the idea of an infinite progress of philosophy for myself, to show that the unrelenting demand which we must make of every system, the unification of the subject and the object in an absolute—I, or whatever one wants to call it—must be aesthetic, in intellectual intuition, theoretically however is only possible through an infinite approximation [*unendliche Annäherung*], like the approach [*Annäherung*] of the square to the circle, and that in order to actualize a system of thought, an immortality is just as necessary as it is for a system of action.[17]

Note, however, that Hölderlin shifts the tone out of the anthropological register and cleaves closely to Fichte's vocabulary. In the letter to Niethammer, he has changed his mind: Fichte's "practical" reason should have pride of place over a theoretical intellectual intuition, and Schiller's representational and corruptible reconciliations are not enough. What is needed is a metaphysics to ground artistic practice, and an artistic practice with synthetic (literally, antinomy-resolving) materials to offer that philosophy. Perhaps the most striking element of this exchange is the lack of "organ" in Schiller's mature vocabulary (see the introduction to this study).

Hölderlin, however, wanted to push beyond both Fichte (who rejected absolute knowledge in intellectual intuition) and Schiller, toward the resolution of real contradictions in both thought and in existence. This would require actual intellectual intuition—an instrument with which to resolve real contradictions in both spheres—and "aesthetic sense." The interaction of these two figures—which Hölderlin would come to call "organs"—anchors the remainder of Hölderlin's theoretical efforts.

The concretization of this problematic occurred slowly. Hölderlin had seen the connection of the "soul" in Heinse to the organ of the soul in Soemmerring, but he needed a philosophically sophisticated way to bridge the *hen kai pan* with the functionality of its parts. This led him to the area

Hegel would not incidentally later call the "absolute spirit": Aesthetics, Religion, Philosophy. He wrote on 4 June 1799, when he was preparing the third and final draft of the *Empedocles*, to his brother, who had announced his own intention to found an "aesthetic church." Hölderlin wrote floridly: the human essence that lay at the base of aesthetics was the drive to make things better, to construct anew that which nature had given. This drive was the "artificial or formative drive [*Kunst- oder Bildungstrieb*]."[18] The terms are both biological: formative drive is the force that forms the living being, as Blumenbach held; artificial drive is the animal's ability to construct apparently rational things, explemplarized by the beehive. Yet Hölderlin clearly means "artificial" also as the "aesthetic drive." The human animal is formed to form. The activity of forming cannot exceed the bounds of being, as Hölderlin goes on to clarify—all our activities flow back into the "ocean of nature." And yet the consciousness that philosophy raises, the image that art offers of the "infinite object" of that philosophical drive, and the religion that requires him to seek and believe that that object is to be found in nature itself—these are the disciplines of the spirit.[19] The "infinite object" that comprises their common center is the contradiction that Hölderlin will later crystallize in the phrase "the organ of spirit." The object cannot be infinite—it is determinate, and determination requires negation, as the Pantheism Controversy had taught all new readers of Spinoza. Humans do not make the substrate of that determination; force is not "the work of human hands." And yet the triad of absolute spirit presents concrete tools for precisely the "direction" of that determination: "Philosophy and beautiful art and religion, these priestesses of nature, thus influence most immediately the human, are there for the human, and only give his real activity, which works immediately on nature, the noble direction and force and joy, thus these also work on nature in a mediated real way."[20] The disciplines of the absolute spirit determine the direction of human praxis. What is given to the human is not made by him; but the making is his own, and this indirect praxis—or revolutionary metaphysics—corresponds to the only possible appearance of nature. Nature is beyond judgment, as we shall see in a moment. But the nature that appears is open to determination both theoretical and practical in the judgment itself. To give direction to nature or to determine an object is finite; the possibility and the substrate from which the determination is made is infinite. In this potential is the metaphysical praxis Hölderlin will come to define as *Poesie*, and for which he will need an equally contradictory term: the organ of spirit, which precipitates in his thinking from his first realization of the apparent finality of the contradiction between judgment and being—the excluded middle being "intellectual intuition"—in 1794/5.

The *locus classicus* for consideration of Hölderlin's notion of "intellectual intuition" is a short fragment entitled "Judgment and Being."[21] Here Hölderlin—who had been attending Fichte's lectures in Jena—draws a false etymology of the word *Urteil* [judgment] as *Ur-Teilung*, original separation.[22] Judgment relies on prejudgmental unity, but is itself (first) separative. It is only possible to judge on the basis of an original separation: elements of judgment must be given, and for that, they must not be unified.[23] Their unity preexists the judgment, and is based on the ur-separation "of the subject and the object that are most intimately united in intellectual intuition."[24] Judgment's hidden capacity is to create the separated environment of subjectivity itself, on the basis of which original separation other factors for judgment can be offered to analysis and synthesis.

The counterpoint to the ur-separation is Being. As Hölderlin stresses, "identity" (even that of A = A, the first judgment derived from *self*-identity in Fichte's *Science of Knowledge*) is *not* the intrinsic unity which exists only in intellectual intuition. The latter is a figure of absolute identity, one which forms a true contradiction with the original separation at the basis of human consciousness, judgment. The opposition, we can note, is a really existing contradiction—it occurs at the basis of whatever we might name reality, and it is not a circumstantial opposition, but rather a necessary one (or a contradiction). The structure of self-consciousness—the awareness of the self, where the self is both subject and object of that awareness—directly contradicts the notion of pure unity. Judgment—which Kant had made the universal medium of our cognition—encountered real opposition.[25] And where Kant had called that contradiction "generic preformation" or "epigenesis of pure reason," constituting but unavailable to the understanding itself, Hölderlin shifted the burden of metaphysical inquiry to that contradiction, the interface between judgment and being. "Intellectual intuition" was to become the figure of this mysterious interface.

The term "intellectual intuition" was coined by Kant, and is a terminological key to his revision of Rationalism. Rationalist epistemologies have corresponding figures of absolute (or absolutely certain) knowledge. So, for example, Spinoza's *scientia intuitiva*, in which the rational order of being is reflected in an intuitive leap or immediate presentation of the logical chain of events. The logical conclusion is not arrived at through ratiocination—instead, the rational is immediately presented as truthful.[26] As we saw in chapter 1, Leibniz used a version of this logic to cap his epistemological tree, calling it "intuitive adequate knowledge," or knowledge of the ultimate ingredients and full composition of complex objects. The stakes of this epistemological figure were high. What *God* could know was always part of the question, making the figure of absolute knowledge a theoretical parallel to the *physis-techne* analogy (which is based on the difference between divine and profane making). God's knowledge—the figure of total synthesis and analysis simultaneously—differed, for the Rationalists, in extent but not in kind from ours.

Kant's term—intellectual intuition—is based precisely on a rejection of that parallelism. Our knowledge is not merely different in kind from a putative divine or absolute knowledge—our analysis of our own faculties leads to the conclusion that the figure of such knowledge can only serve as a negative example, as something we can positively see that we do not possess. The canon of the understanding excludes precisely the figure that an *organon* of reason would have given us: metaphysically certain knowledge.[27] This separation of our cognitive capacity from traditional figures of absolute knowledge was based on a specific reaction to Rationalism after Leibniz. Kant rejected the notion that clarity and distinctness existed on the same progressive plane—for him, there were two separate roots of cognition, intuitions and concepts. The separation of these two justified the mysterious characterization of pure reason as "epigenetic"—a passage between two fundamentally different orders had to occur (mechanical and organic matter, literally; the forms of intuition and the pure concepts, metaphorically). The limiting possibility of an "intellectual intuition"—the presentation of an object in the absence of an externally grounded perception—implied a violation of the theoretical *physis-techne* differentiation. If we could intellectual intuit the world, it is hard to see how we would not be then also implicated in its production.[28]

If the stakes were as large as the possibility of human ontological production and the question of the cosmos's proper rules (divine or not), there were two other problems which confronted Hölderlin in his adopted term. The first was that of the quasi-antinomy between mechanical and organic nature. The second he had managed to identify and include in his earliest mention of the term: the relation of subject to object *as a function of* the relation of judgment to being. The first was given in Kant's reverse-coinage of "intuitive understanding" in the *Critique of the Power of Judgment*; the second by Fichte's early flirtations with the term to designate self-consciousness.

In the *Critique of the Power of Judgment*, Kant laid the framework for a philosophical approach to judgment itself. As essential as the doctrine of judgment had been to the *Critique of Pure Reason*, his focus there had been on the twin roots of judgment's cognitive work, intuition and concept. The *Critique of the Power of Judgment*, then, takes its start from a refinement of the model of judgment. The judgment as defined in the *Critique of Pure Reason* had been a matter of "subsumption," literally the subordination of an intuitive complex to a conceptual determination with the aid of overarching forms of that unity (categories). In the *Critique of the Power of Judgment*, Kant calls this subsumptive activity "determinative judgment" (*bestimmende Urteile*), and names a new kind of judgmental activity, "reflecting judgments" (*reflektierende Urteile*). This type of judgment operates in the temporary absence of the correct determinative unifier. While a unity is produced in the judgment and given to the mind (for reflection), its proper status as determined (with respect to our faculties) is not yet included in the judgment. This allows space for reflective activity, for the pursuit of higher unities among complexes of judgments. Indeed, Kant singles out three areas in which reflective judgment is necessary: the progressive determination of scientific laws in general, the judgment of taste (including natural and artistic beauty), and the special case of the organic. It is the last of these that causes the *Critique of the Power of Judgment* to readdress the issue of simultaneously intuitive and conceptual cognition. Hölderlin, while using the term "intellectual intuition," would make use of both conceptual formulations in his notion of an *organ of intellectual intuition*.

Subsumptive judgment, Kant reasons, is good enough for most natural determinations. Even where reflecting judgments are needed, they are often temporary—we need to reflect so long as the higher unity for subsumption is not discovered, but only just that long. The discovery allows the constitutive determination of the natural world to progress. This model encounters an obstacle, however, when it comes to judgments about organized beings. The difficulty stems from the model of judgment in the *Critique of Pure Reason*, which holds to efficient causality as paradigmatic. We understand and constitute the world where cause and effect are unilinear—indeed, that efficient causality is one of the essential conceptual ingredients we supply to nature. When we observe organized beings, however, we notice that their parts—as we have seen above—seem to have a causal feedback loop into their wholes. The form of our understanding does not allow for that type of judgment to constitute our world, however: as Kant claimed, there can be no Newton for even a tiny blade of grass.[29] Reserving subsumption, we can easily produce a reflecting judgment about organic beings, but this is not enough. There must be some analysis of what unity in judgment is appropriate to these mutually causal beings.

Intuition for Kant is the representation of particulars without anything "general." It is therefore always of the senses, because the general can only be contributed to knowledge by the concept. On this basis, Kant polemicizes against any possible "intellectual intuition,"[30] that is, representation of noumena. But Kant makes a famous regulative exception in the name of biology in *Critique of the Power of Judgment* §§76 und 77. Here, Kant describes a mode of knowledge which does not (*a*) conclude to a whole from its particular parts, but (*b*) for which the whole is first *given*, and which can therefore derive the particulars from this whole. Kant calls (*a*) "analytic-general," and it is our type of knowing: we understand the whole from its parts, and we can therefore analyze the conceptual/general infinitely, without being able to clarify the connection in the other direction. We can take the whole apart, but we cannot deduce the parts from the whole—the whole that would determine its parts is not an object for us. And yet, such a determination is precisely what we observe in organic beings: the whole appears to influence and even determine the parts according to some greater concept. Kant calls this parallel mode of knowledge "synthetic-general." Were we able (*b*) to know the whole as determinative of the parts, then we could construe (even "construct," in the specific sense that Kant excludes) the conceptual-general in its synthetic connecting of the parts. Our knowledge would be "synthetic" in precisely this metaphysical sense, reproducing the actual order of connection we observe in the organic world. We have to assume this mode of knowledge for the sake of understanding living beings—but the connection of the "goal" (of life itself, its "concept") of the whole to its parts (in, e.g., the phenomenon of growth) is not in our constitutive ken. Our type of (synthetic) knowledge subsumes particulars (intuitions) under a general concept. This constitutes (*a*) a whole that can be analyzed, but no insight that would (*b*) allow for the synthetic construction of the individual (the connection of the parts according to a rule, or better, a concept). Such a synthetic "construction" would make possible a kind of intuition within the conceptual, an intuition of the parts of the concept-representation as parts of the concept. Kant thus calls this mode of knowledge "intuitive understanding." The problem with intellectual intuition from a critical standpoint—the possibility of our (co)production of the world—is thus different from the problem with intuitive understanding. The latter would give us something else we do not have: a world-constituting form of mutually causal cognition. Recall that the form of reason is organic in Kant, but through its separation from the understanding fails to constitute our experience of the world. That constitution, which dovetails but is not identical with the problem of intellectually intuiting and thereby producing the world's content, would give us something very like what Leibniz had thought as "intuitive adequate knowledge": truthful cognition of the mutuality of the organ-world. We might, in this case, still "make" the world, raising concerns about the *physis-techne* analogy. We would definitely, however, have insight into our own manner of synthesizing knowledge—Kant's primary Critical exclusion. The point at which judgmental forms find rational grounding thus runs into two distinct contradictions: the difference between the order of being and the order of representing, and the line between mechanical and organic matter.

But hidden in these two contradictions is a third, more basic, opposition: that between judgment as already internally contradictory structure and being, as Hölderlin had described in *Judgment and Being*. That the latter fragment is a critique of Fichte is as well historiographically witnessed as it is textually clear: Fichte does not use the figure "intellectual intuition" in the *Science of Knowledge*.[31] He had, however, considered using it both before and after the canonical 1794 *Science of Knowledge*.[32] Fichte's texts invited Hölderlin into this conceptual problematic, and Hölderlin seems to have tarried there at least until about 1800.

The popularization and development of Kantianism was fast.[33] Fichte, publishing in 1794, was already reviewing efforts to refute the second generation of Kantians. Reviewing a skeptical work directed against Kant's popularizer Karl Leonhard Reinhold, Fichte defended the possibility of "closing" Kant's system by supplying a common root for intuition and concept. Reinhold had proposed representation (*Vorstellung*) as a generic category binding the two,[34] but Fichte rejected this solution as a mere description, with no binding argument. Simultaneously rejecting the outright skepticism of the work he was reviewing (Schulze's *Aenesidemus*), he wrote of a different solution: "We [actualize . . . this transcendental idea] through intellectual intuition, through which *I am*, and indeed: *I am simply because I am*."[35] This hermetic combination of the self's putative necessity and intellectual intuition was informed by an insight that had a great impact on the next generation, even if it was not this writing that carried it there. Fichte's move was simple: true criticism was actually "negative dogmatism," the radical bracketing of all "things-in-themselves." In that methodologically thingless space, the possibility of intellectual intuition could not be excluded. Instead, forms of judgment needed to be investigated without the assumption of the source of those forms. The *I* was to be the focus of Fichte's formal investigation, and this investigation reopened the question of an *organon* for metaphysics—because Fichte allowed no elements of the *Not-I* to be determinative methodological factors, the possibility that there was a general tool for knowledge in the *Science of Knowledge* was all too real. In the *Aenesidemus* review, Fichte went on to state a principle that would survive the discontinued use of the term "intellectual intuition" in the *Science of Knowledge*: "The I is, *what* it is, and *because* it is, *for* the I."[36] The formal investigation of this structure is contained in the first three paragraphs of the *Science of Knowledge*. Here we need only see that the entire structure is *for the I*. By connecting the I's consciousness of itself *as I* to the notion of an intellectual intuition, Fichte thought to solve Kant's problem. All knowledge would have to go by way of this recursive and self-supporting structure, and the methodological exclusion of "things" meant that finite knowledge was underpinned by the self's giving of itself—and its actual knowledge—to itself. To be clear, even here Fichte did not go so far as to suggest that we could know intuitively in this sense. That activity was unconscious, reflected in finite knowledge. By creating a sphere of self-justifying knowledge—consciousness *for itself*, as Hegel would later dub this structure—and excluding "things" from its purview, Fichte offered Hölderlin an opening to wade into the cascade of contradictions surrounding the term "intellectual intuition."

In a first step, then, Hölderlin returned the term to its objective provenance: absolute being.[37] But he did so by way of logical contradiction, making the judgment's recursive structure antinomic to that of being itself. He thus placed himself between Kant and Fichte. Returning to the question of being meant regressing behind Fichte's "negative dogmatism," while reintroducing the question of being and judgment meant asking whether there were not other Critical answers to Kant's system-question (how can we judge rationally about things?). Hölderlin thus approached metaphysics with a framework both real and contradictory, addressed not only to the Critical question, but also to the opposition between *physis* and *techne*, and that between mechanics and organics. Such was the anticipatory step taken in *Judgment and Being*—its elaboration would have to wait for Hölderlin's genre-theoretical writings from the late 1790s.

The danger Kant wanted to avoid—and which Fichte had seemed to embrace—was two-sided. Excluding "intellectual intuition" served simultaneously to ward off charges of nihilism (the absolute as knowledge would lead to the lack of a concrete world) and to prevent the possibility of ideal determination of that world.[38] For Kant, the latter concern was determinative: we set the conditions for experience of the object-world, but we do not determine the flow of phenomena out of which we condition those objects.

Hölderlin, by placing himself between Kant and Fichte, managed to inaugurate organology by maintaining both the rationality of the "for itself" in judgment and the open flow of phenomena in its real contradiction, being. By including a cascade of antinomical structures in this single framework, Hölderlin reopened the question of metaphysics in a context where "organ" could suggest itself as the bearer of such structures. Dialectical organs were to come from the confrontation of these post-Kantian concerns with problems about the unity of nature, stemming ultimately from Spinoza and the Greeks but mediated by Heinse and Soemmerring.

Tragic Organs: The Genre-Theoretical Metaphysics of Judgment

THE PROCEDURES OF SPIRIT: JUDGMENT AND BEING

Between philosophy and the life sciences, the eighteenth century had constructed a grand analogy between animal generation and the human mind. Kant had taken this analogy further, speaking of an "epigenesis of pure reason." Reason itself generated according to laws it could not know, could not discover, since there was no Newtonian method for the kind of force that formed Reason's categories. In the analogy, the categories were the functional parts of the body Reason. They were what would come to be known, in biology, as "organs." But it remained for the Romantics to draw the analogy out along these lines, literalizing and expanding the definition of "organ" as they did so. Hölderlin did this work by extending metaphysics into *genre theory*. And indeed, the analogy remains precise: where Kant's discourse requires an *Art*—a species—Hölderlin exploits the alternate sense of this word: genre. In works like *"On the Different Kinds of Poetry"* (*Über den Unterschied der Dichtungsarten*), Hölderlin was taking the first steps toward organology by leaping from speciation to genre theory, and filling in the missing "organ" from an analogy now become literal. The genre as organ made metaphysics literary, because it put the ambitions of the "organ of philosophy" into concrete and restrained written form. It approached internal contradiction with the tools of literary writing, especially tragedy.

The highest contradiction remains that between judgment and being. But if this contradiction is to be more than a permanent antinomy,[39] then figures of its resolution must exist. In Hölderlin's aesthetic and genre-theoretical writings, these figures are "intellectual intuition," the "pure," and "organ."

The essay *"On the Procedure of the Poetic Spirit"* starts from the premise that poetic activity is deeply ontological. Hölderlin marks out a series of basics insights that the poetic spirit must possess before it succeeds in its production. The spirit itself idealizes and generalizes the material it works on, and yet the particularity of that material is itself ideal. The reversal of predicates seems to follow on the derivative nature of the aesthetic task: material is ideal because it is already *in* the spirit of the poet when it first is worked on. Likewise, the activity of the poet is concrete, although his cognitive capacity is discursive, original separation, *ur-teilen*. What becomes clear over the course of the essay, however, is that this first impression is incorrect. The activity of the poet is rather placed in the order of being itself. This placement does not resolve the central contradiction already outlined in *Judgment and Being*, however. Indeed, the problem outline here is precisely the problem associated with intellectual intuition in general. In Hölderlin's terms, the difficulty is that, while the spirit both splits being and unites it—and while being itself is both one and differentiated—the means of true interaction between these analogous forms (*the harmoniously opposed one*) is not only uncertain, but quite apparently impossible. The problem is that the oneness of the conscious apparatus cannot be presented to itself, since presentation is discursive, is based on *ur-teilen*. Representations of all kinds of structures are possible, but access to the original unity—intellectual intuition—seems impossible. This impasse had been noted by Kant, of course, and subsequently by Fichte. Hölderlin puts it this way:

> [Infinite unity] is thus never merely opposition of the unified, and also never merely relation unification of the opposed and shifting, opposed and unified is inseparable in that unity. If this is the case, then it can be passive in its purity and subjective totality, as original sense, indeed in the acts of opposition and unification, with which it is effective in the harmoniously opposed life, but in its last act, where the harmoniously opposed as opposed harmoniously, the unified as interaction is captured as one in it, in this act it simply cannot and may not be captured by itself, become an object to itself, if it should not be a dead and lethal unity a being become infinitely positive instead of an infinitely unified and living unity.[40]

The activity of the poet is infinite unification, a process underlying both judgment and being. And yet this unification is not merely the synthesis of judgment, the joining of juxtaposed elements, however deeply. Its infinite quality makes it positive in a strong sense, one we should call dialectical. For its unity includes both a contradictory duality and a further unity. Infinite unity swallows the negative in a further positivity. As it occurs in the procedure of the poet, this unity is an impossibility. This is because the object of consciousness—any consciousness—is precisely that: an object. The fixity of the object's unity contradicts, however, the harmony of opposites in the greater, infinite unity. Lest this talk of infinite unity sound too mystical, we should say that the problem is not that some asserted "higher unity" is at stake. Instead, the problem is that the structure of the harmoni-

The Tragic Task 147

ously opposed *one* is not merely taken as the structure of both consciousness and for being. The separative-yet-unifying unity, which is indeed valid for both judgment and being, is, at the moment of the poet's reproduction of the order of being, *reproduced* (not merely represented). Thus we have something like a contradictory unity uniting two separate contradictory unities. *Mimesis* is the technical reproduction of unity-in-duality such that this structure reflects itself literally infinitely. The task of the poet is a reproduction of the real contradiction between judgment and being which is at the root of all cognition. To complete that task, he must objectify in representation the simple yet infinite interface between judgment and being. The objectification of that process is contradictory—it is an impossible task, or an infinite approximation[41]—because the limited unity which can only be its result cannot contain the infinite unity from which it will be torn. The object of the aesthetic unification of opposites is not a simulacrum of human freedom, *pace* Schiller, but the affirmation of absolute cognition. It is only that affirmation which can confirm the desired freedom.

In that actual freedom, Hölderlin finds the metaphysical determination he had confirmed as early as his poems to Soemmerring. Hölderlin's imperative for the poet reads: "Set yourself in free decision in harmonious opposition with an external sphere, just as you are in yourself in harmonious opposition, but unrecognizably so as long as you remain in yourself."[42] The actuality of the infinite reflection between judgment and being—the dialectical determination that unity-in-duality is reproduced as an infinite unity, or a higher contradiction—can only be realized in the cognitive act of the reproduction of that infinite unity. In other words, *mimesis* is the moment where reproduction and representation are identical. And as long as they are not in that unity—as long as representation is merely a reflection of one or the other of these unities—an infinitely discursive oscillation is the basically human property. Thus the product of objectification does not count as the poet's essential task, since this product is merely an indeterminate admixture of these two perspectives. Instead, the procedure of that objectification reacts on the objectifying consciousness. The choice to produce an aesthetic object thus passes through the contradictory structure of being itself—its ideal particularity and its real generality—in order not merely to produce but to capture its own process as reproduction of the infinite task of aesthetic production. Freedom emerges in this process, and points simultaneously to its theoretical solution—in an organ of intellectual intuition—and to its genre-theoretical home in the tragedy.

"Organ" appears suddenly in the *Procedure*. Hölderlin states that the poetic effort to grasp life is characterized by an immediate conflict (*Widerstreit*) between the individual (or material), the general (or the formal), and the "pure." He continues:

> The pure captured in every particular mood conflicts with the organ in which it is grasped, it conflicts with the pure element of the other organ, it conflicts with alteration. The universal conflicts as particular organ (form), as the characteristic mood, with the pure, which it grasps in this mood, it conflicts as a striving in the whole against the pure element which is grasped in it, it conflicts as characteristic mood that which lies closest. The individual conflicts with the pure element that grasps it, it conflicts with the form lying closest, it conflicts as individual with the universal that is alteration.[43]

This triple conflict is "life," which Hölderlin equates here with both the object of poetic representation and being insofar as it is organized. The problem arises not from the conflict within that order of being, but in the poet's attempt to work on his subject, on life. The first opposition is that between the "pure" and the organ, which, in the next clause, gains its own "pure" element. Then the general is also named "organ," such that the antinomic conflict between the general and the particular—the expected opposition, metaphysically—is shifted into a conflict between grasping or conceptualizing organs (*begreifende Organe*) and change or alteration (*Wechsel*). The various ontological determinations—pure, general, individual—are shifted into a struggle between *organs* as forms and the development from one form to the other, living change. Reil's determination that organs were developmental and organization structural is here reversed—the organ is the formal structure, and is caught in a contradiction with change. Since the result of change is new organs, the organ's struggle as it grasps the various categorical levels of being and represents them is a struggle with more organs. The task is infinite, and is constituted by the synthesis of the finite and the infinite: the organ is the name for the impossibility of producing the absolutely new in the absence of absolute power. It is thus a name for *techne*, and designates the post-Revolutionary metaphysical task.[44]

This is, however, far from the end of the poet's task. Indeed, his task will be the presentation—in fact, reproduction—of intellectual intuition. Hölderlin now sets a condition on the organ. The organ is "directly opposed to spirit," but is also the container of that spirit, and that which makes all opposition (*Entgegensetzung*) possible.[45] This organ must now be grasped as having several opposed functions. First, it is definitionally that which allows formal opposition to be introduced into harmonies. This is an organological gloss on the doctrine of the *ur-theil*: the organ is assigned the judgment's function of analyzing or separating within larger connected wholes. The obverse of this function also belongs to judgment: it binds this second, representational whole together formally. Having analyzed, it synthesizes; from the *ur-theil* comes the *Urteil*.

Hölderlin now adds new conditions. The organ must also be grasped as materially opposing disharmonious moods (*Stimmungen*) while formally binding them together. This function seems to mirror the notion of synthetic judgments *a posteriori* in Kant—taking arbitrarily occurring phenomena and binding them through a disjunctive judgment (corresponding to the negative of the category of interaction). The organ sets up a frame in which various elements are opposed materially, dissociating them in terms of content but joining them in the judgment—in the organ—itself. The obverse of this capability is in turn added to the organ's definition: the organ also materially binds these moods while formally opposing them. This is, then, the true metaphysical organ in a post-Kantian mood: the *organ* synthesizes the very material of its object while introducing formal opposition within it. The totality of the synthesis is the object of analysis, and at the basis of its activity is a generalized but concrete—*organological*—synthetic judgment *a priori*, the production of cognition itself. The organ becomes the basis of what we can call a metaphysics of judgment, or organology. Hölderlin places the frame of philosophical inquiry in the forms of judgment, rejecting the premise that the "material" elements of that judgment should be treated as external to those forms. This is, to be sure, an idealism—indeed, it is the beginning of *German* Idealism—but it does not commit the sin Kant wanted to avoid. The organ of judgment does not determine the world it perceives, but exists in a dialectical codetermination out of which autonomous and truthful cognition can emerge. The task of the poet is fundamentally connected to this larger metaphysical task: the concrete production of consciousness through the organ's conceptualizing (*begreifende*) objectification, which is freedom.[46]

As a binding element in cognition, the organ thus produces "formal life," while as a separative element, it cognizes that production. This leads Hölderlin to the most general requirement for the organ:

> If the *organ of spirit* could be regarded as that which must be *receptive* in order to make the harmoniously opposed possible, then this must be so for the one as for the other harmoniously opposed, that it thus, insofar as it is a formal opposition for the poetic life, it must also be formal connection, that it, insofar as it is materially opposing for the determinate poetic life, must also be materially binding, that the bordering and determining is not merely negative, that it is also positive, that it must to be sure be regarded in the case of harmoniously connected elements as separated, the one opposed to the other, but both thought together is the unification of both, then the act of spirit which with respect to meaning has a thorough conflict as its consequence, is *just as* unifying as it was opposing.[47]

The organ, that which had been "directly opposed to the spirit," is that very spirit's property. The organ of spirit—the inauguration of Romantic organology, rejecting centuries of opposition to this conceptual possibility—must be receptive, and this even as it is formal. This dual quality is the basic dialectical property of the organ. As a cognitive function, it binds the opposed notions given as "organ" and "organization" in the life sciences, problematizing development within itself. The term serves as a contradictory conceptual unity, but this literal *contradictio in adiecto* is cast as philosophically salutary. But it is not merely biological development that is problematized. Instead, it is the representation of being—poetic and cognitive activity—which reflects the problem of change. "Poetic life" is determined, both generally and concretely, as this chiasm of separation and synthesis both formally and materially. The organ of spirit captures the problem of the location of the soul in terms of judgment's forms and activity. Like Kant, Hölderlin can point here to contradictions arising from the forms of judgment. Intellectual intuition, as we saw above, is for example a seeming impossibility because of the way in which objects are presented in judgments, with reference to an infinite unity not included in their determination in representation. Yet Hölderlin transforms the *disciplinary organ* into a *dialectical organ*. This transformation needed only the recognition that the conflict between the philosophical and medical faculties—the determination of the nature of body and soul—is a real contradiction which is represented and reproduced in the organ, in judgment. This is why Hölderlin goes on to characterize the limiting activity of organ of spirit as "positive." The passage from organ to organ goes by way of the positive production of limitation—of represented objects or events.[48] This act of the spirit is thus the point of coincidence of synthetic and analytic cognitive activities. That point is called *organ*.

Point is indeed Hölderlin's gloss on this formal yet receptive possibility of absolute knowledge. Proceeding from the conditions of the organ to its concrete elaboration, he writes—and I quote—at length:

> But how is it grasped in this quality? as possible and as necessary? Not merely through life in general, for in this way it is this indeed, insofar it is regarded merely as materially opposing and formally connecting, determining life directly. And not just merely through unification in general, for this is what it is so far as it is regarded as formally opposing, but in the concept of the unity of the unified, so that of the harmoniously connected the one as the other is present in the point of opposition and unification, and that *in this point the spirit can be felt in its infinity*, so that it appears through the opposition as finite, that the pure element, that conflicted with the organ, is present to itself in just this organ and is thus first a living being, that, where it is present in different moods, which immediately following on the basic mood is merely the extended point, which leads there, namely to the middle point, where the harmoniously opposed moods encounter one another . . . and the [contradiction's] meeting in the point replaces the simultaneous inwardness and differentiation of the sensation at the basis of the harmoniously opposed living and at the same time becomes more clearly depicted by the free consciousness and more formed, more universal, as a world in the world, and thus as the voice of the eternal to the eternal.[49]

The goal is clear: the contradiction between the "pure" and the organ should be unified in a feeling of spirit's own infinity. This can only be achieved organologically—the realization of the human task is only possibly through tools which are not incidental to that human. The organ is the essence of the antinomy between judgment and being. Every rhetorical turn, each reversal of terms in this passage is merely one more fold in the organ's capacity to know and to produce knowledge. And it is through the acts of the spirit, occurring in its organs, that "mere life" is exceeded, the mirror of the contradiction of organic developing forming a hyperbola, occurring on opposite sides of the y- and x-axes simultaneously, and each containing two infinite approximations or asymptotes. This figure includes an "extended middlepoint," where finally spirit and life coincide. But this coincidence is itself strictly identical with the moment where the unifying and separating activities of the organ are also identical, and only where these identical activities are most material. Finally, this "simultaneous inwardness and differentiation" is felt as the development of the spirit itself in its chosen material—its freedom. The poetic spirit's activity is, then, not only mimetic reproduction of the rules of an order, of a world. This activity literally redounds onto the spirit itself, making it a *world within the world*. In other words, as material (both ideal and material) is given poetic form (both concrete and abstract), the organs that make up consciousness become active in both of their contradictory activities—synthesis and analysis—simultaneously. Organology's first notion of *mimesis* comes into view. This is a notion of the production of an order of things, the invention of rules. Let us bracket the issue of the origin of those rules, and focus only on the point that is made here for organology as a metaphysics of judgment.

The metaphysics of judgment does not exclude other capacities, just as Kant's critical focus on the understanding did not exclude other faculties. In Kant, we see the priority of the faculty of the understanding providing a set of rules for the use of the other faculties. If I am right to point to organology as a metaphysics of judgment in Hölderlin, then it is not a matter of excluding, for example, sensation. In the passage above, it is a certain senstation or feeling (*Empfindung*) at which the poet aims, and it is this production which results in the religious feeling described, "the voice of the eternal to the eternal." The point is not to take away from that goal—ultimately the new mythology—but to see that it is rooted in an overlap between speculative and aesthetic tasks which alters both discourses. Organological procedures have always been involved in the preparation of what is felt, and organs provide the conceptual basis on which to make use of generically different discourses (metaphysics, aesthetics, poetic form, genre theory) and the "facultative" (really organological) bridges between them.

On the metaphysical side of this question, Hölderlin certainly aims at a knowledge which Kant had rejected (but not before he had defined and suggested it). The overlap between aesthetic and metaphysical activity is the moment where representation becomes (re-)production. The whole procedure of the poetic spirit is the re-production of the infinite interactive capacity of the spirit in its organs. Rather than a representation, one gets a reproduction in the sense that a world is constructed, a world in the world.[50] We might call this metaphysics of judgment *technological*, since it defends the notion of work on being—on the "most material"—both metaphysically and aesthetically.[51] Organology makes the cognizer and the poet equally participants in the dialectic of history, which is dialectical precisely because of that participation, because of those human organs.

Yet the metaphysical problem remains. The greatest concern for Kant—as for Kantians—must be that this type of participation lacks a legitimation, an argumentation for its connection of judgment and being—in short, a deduction. The focus on judgment as form and content allows this argument; its operational term is "organ." The worry remains, however, that in such a system judgment determines its phenomena, that the world is simply now dependent on the spirit, with all the attendant problems attaching to idealism. Let us take just the problem of determination, leaving the problems of organic judgment and world-production for the following sections of this chapter.

Kant defends a robust conditioning contribution of conceptuality to experience, as we have seen, but is careful to disallow the determination of the dynamic of experience by that conceptuality. Indeed, this is what the phrase "transcendental idealism and empirical realism" is meant to capture. Any metaphysics lacking the external conditioning provided, in Kant's case, by the material of sensation combined with the forms of and then actual intuitions, runs the risk of just such a determination. In the case of organology, the fear must be that, by placing the ontological and epistemological burdens simultaneously in a single term, we speculatively free the human to determine a world in which he then lives wrongly. We produce the illusion of control.

The organ, however, should do precisely the opposite of this. Its cognitive and representational abilities are traced back, in the *Procedure*, to a productive ability which can only exist because of its simultaneous implication in two systems of contradiction-in-development. Ultimately, however, the highest contradiction is that of judgment and being. And yet judgments have content—they aim at or allude to being, however partially they fulfill their task. What Hölderlin claims to discover is that the feared "determination" is mutual. But rather than simply conflating two orders of being (or of judgment), Hölderlin is making a deeper claim. He is claiming that the relation established in the overlap between cognition and being is not representational, but instead on the order of ideal production. The production of a world within a world is the signal of this deep-seated capacity, and it cannot occur without the organological structure, without judgment's basic contradiction. In the moment of representation, the organ separates and binds simultaneously, producing a cognitive object. In that very production is contained the infinity of the organ itself, its own self-knowledge as knowing or forming itself in the material it chooses to produce. It is as though Hölderlin has taken the formula for the differentiated monism of the 1780s—Jacobi's "being in all existence"—and transferred it into epistemology. "The knowing in all cognition" would be the formulation of a Critical monism. Judgment does not but can determine its phenomena, but only insofar as it exposes its categorical apparatus to that external sphere—the sphere of its own contradiction, and thus exposes itself to the possibility of development. And that development is a matter of passage—from organ to organ, or from genre to genre.

ORGANIZATION AND ITS DISCONTENTS: GENRE THEORY AND INTUITIVE UNDERSTANDING

An intellectual intuition must lie at the root of every tragedy, according to Hölderlin.[52] This means that the highest contradiction—that between judgment and being—must be presented in the tragic form. And yet, in the late 1790s, the concept *being* came to have a further, specifically tragic, term: *the aorgic* (*das Aorgische*). With this etymological invention, Hölderlin both introduced yet another real contradiction and specified its sphere of application in the tragic form. The aorgic is logically opposed to any organization—the problems of intellectual intuition and intuitive understanding, as they had been given by Kant, thus appear as different versions of the same problem. On the one hand, the contradiction between judgment and being needed the judgmental organ as its frame of developmental and dialectical resolution. On the other hand, the specific contradiction between the mechanical and the organic was already a matter of the forms of judgment for Kant. For Hölderlin with his metaphysics of judgment, it was a matter of sharpening this contradiction—in a first gesture, by making it a *logical* contradiction—and then finding its resolution in the theory of tragedy. And this shift into the genre-theoretical had it stakes in eighteenth-century theory of tragedy.

Classical tragedy theory emerged from midcentury metaphysical debates. Lessing's famous break with Gottsched and the French school was prepared by his collaborations with Moses Mendelssohn in the 1750s. In 1755, Lessing and Mendelssohn had collaborated on a number of writings, prominently the essay "*Pope a Metaphysician!*" There the general question of poetic form and speculative reason had been broached, with the argument that poems cannot be treatises (and thus that Pope was not a "Leibnizian"). Meanwhile, Mendelssohn had published his own *Philosophical Writings* in the same year, which included a consideration of "mixed sentiments" to address the problem of the enjoyment of tragedy. The collaboration reflected a deep intellectual affinity and friendship. When Friedrich Nicolai raised the parallel problem of the morality or pedagogy of poetry—claiming that the intent to teach should be removed from tragic writings in particular—this provoked a disagreement between Mendelssohn and Lessing. Nicolai collected their letters debating the topic and published them as *Correspondence on Tragedy*. The disagreement came down to metaphysics.

The important point for the present context is a difference of task assigned to the tragic form. Mendelssohn held a strictly rationalist line.[53] Tragedy—as all art—is finite imitation of the *perfectiones* (*Vollkommenheiten*) of God's universe. We are presented in the poetic form with human attempts at adequate intuition, pointing toward originary images (*Urbilder*) which underlie the phenomenal world. These images of perfection contain the morality which Nicolai had sought to remove, and thus must play a role in producing what he calls admiration (*Bewunderung*), the uniting of lower and higher sensations in the tragic presentation (*Darstellung*). Thus tragedy occupies a relatively high rung on the "progressively semiotic" ladder, offering a taste in the finite, human order of the adequate intuition that is proper ultimately only to God. United in that intuition would be virtues and the reciprocal perfections of the metaphysical universe. The picture is thoroughly Leibnizian, down to the metaphorology investigated in chapter

1 above. Mendelssohn writes of intuition grasping the perfections of the world in its own reciprocal gesture, mimicking the organicism of the metaphysical universe. This metaphor establishes the organico-rational stakes of the metaphysical debate about tragedy.

Lessing objects by shifting the ground of the debate to tragedy, first removing the metaphysical assumptions. Rather than virtue, tragedy's task is to awaken sympathy (*Mitleid*). Both men had agreed to the proposition that "the best human is the most sympathetic human," and Lessing accordingly makes his theory depend on the technics of producing sympathy. He thus reintroduces pedagogy into tragedy without the Leibnizian background. Indeed, what he advocates could be called *moral technologies*, attempts to intervene in the moral sensibility through poetic representation. Removing the moral-metaphysical image of origin or perfection, he argues that the end—the tragedy should produce the most sympathetic human—must not be conflated with the means. Mendelssohn, it follows from Lessing's presentation, has done just this: the goal is moral perfection, and the example of moral perfection must therefore be contained in the tragedy. Lessing objects: where we have the end, we have no necessary connection to the means. In knowing the goal, we do not know how to get there. If we grasp the means, on the other hand, the possibility that the end will emerge is a real one.[54] This points to the technology of producing morality, to the possible future synthesis of moral capacities. Lessing writes: "If it is then true, that the whole art of the tragic poet is oriented to the sure exciting and continuation of sympathy alone, then I say, the vocation of tragedy is this: it should *extend our ability to feel sympathy*. It should not merely teach us to feel sympathy for this or that sufferer, but instead it should make us broadly vulnerable [*weit fühlbar*] so that the sufferer at any time and in all his forms can move us and win us to his side."[55] For Lessing, then, the human has a future, and this future is in a moral synthesis that must still be produced by art (tragedy) and reason, without a representational model toward which we could work. If we follow the metaphorical stakes in Mendelssohn, we find that Lessing's text suggests a developmental organic model for the pedagogy of tragedy. Rejecting representationalism between means and ends, Lessing makes the task of tragedy open-ended progress: it is literally for moral purposes, and it is obliquely grounded in the metaphorical organicism of Mendelssohn's arguments.

We could say that "to extend a capacity" anticipates the Kantian notion of a synthetic judgment—it extends, gains, wins more unto itself as it progresses. Of course, sympathy will not have been *a priori*, neither for Kant nor for Lessing. But Hölderlin—and I am not claiming that he knew this passage—would find a way to unite this moral technology with absolute knowledge. For that, he would need to characterize the tragedy as a very particular organ: an organ of intellectual intuition. Combining Kant's two prohibited forms of metaphysical knowledge—intellectual intuition and intuitive understanding—in an organ of judgment, Hölderlin could descend from the contradiction of judgment and being to the contradiction between the organic and the mechanical, and from there begin to suggest the way toward a political theory of tragedy.

Hölderlin divides poetry into its classical triad—lyric, epic, and dramatic (which he reduces to tragic). Each has an appearance in productive tension with its "meaning."[56] Tragedy's meaning is rooted in a "metaphor of intellectual intuition," while its appearance is heroic. Thus the narrative aspect of tragic form combines the apparent ideal mood of the lyric poem with the naïve appearance of the epic tale. Underlying this combination, however, is our familiar figure of absolute knowledge. In the elaboration which follows, we can see that Hölderlin is in fact dealing with the alternative figure of the intuitive understanding, the regulative "synthetic-general" judgment of beings with organs (and thus reciprocal causality) described in the *Critique of the Power of Judgment*: "The tragic, in its external appearance heroic poem is, according to its basis tone, ideal, and a single intellectual intuition must lie at the basis of all works of this kind [*Art*], which can be no other than that unification with everything that lives, which of course can't be felt by the more limited mind [*Gemüth*], which only has a presentiment of it in its highest striving, but it can be known by the spirit."[57] We can recognize the shift from intellectual intuition to intuitive understanding both in the mereological characterization and in the immediate reference to "the living." Further, the genre—the specific manner of uniting form and content—of tragedy demands that this knowledge be a unity only cognizable by spirit. As we have seen above—the language follows that of the *Procedure* quite exactly—spirit must perform its dialectical task with the aid of its organs to arrive at the cognitive production that unites judgment and being.

Here, the poet is given the further, specifically tragic task of uniting his presentation with the totality of life.

If it is right to think of Hölderlin as moving to the figure of intuitive understanding in this passage (while retaining the term "intellectual intuition"), then we must note a shift from Kantian doctrine. As Hölderlin continues, his description of the task of tragedy retains the characteristic explored above more generally, that of contradiction. He will etymologize this distinction in the *Ground for Empedokles*, opposing *the organic* to its privative opposite, *the aorgic*. Here we can see this move at the level of tragic form. If the content of the tragedy has to do with elemental organization and disorganization, its form must reflect this in the organs of the poetic spirit. No pause is possible, Hölderlin writes, in the tragedy, because it must be always engaged in material synthesis and separation in order to operate effectively. If the first task—one common to the forms of poetry—is the unification of parts into a greater whole, this whole must, in the tragedy, gain the same concreteness that the individual parts possess. The whole thus gains content while the parts gain inwardness. This process is the literal opposite of the tragic plot. The hero's apparent fall is, in fact, a devolvement into chaos—the parts of his life must be pushed into the greatest possible disarray and tension. The less organized the parts in this sense, the grander the synthesis at the level of the whole. The mood of the lyrical individual—apparently ideal, but based on sensuous suffering—is then felt in the whole. The disorganization is thus guided by an eventually felt whole which produces the tragic effect. Hölderlin points to his conclusion from the *"Procedure"*: intellectual intuition can only exist as the extreme of absolute knowledge *in an organ*, or "insofar as it goes out of itself," because its unity, which must be infinite, cannot brook the contradiction of a limitation, even where it is in fact limited. That process—the general dialectical process of cognition—has a privileged place in the tragedy, which thus takes metaphysical cognition as its hidden ground:

> And here, in the surfeit of spirit in unification, and its striving towards materiality, in the striving of the divisible infinite aorgic, in which everything more organic must be contained, because everything more determinately and more necessarily present makes an indeterminate, unnecessary thing necessary, in this striving of the divisible more infinite towards separation, which communicates itself in the state of the highest unity of all organic being to all

the parts contained in this unity, in this necessary *caprice of Zeus* lies in fact the ideal beginning of all actual separation.[58]

The tragedy as genre—as poetic form, as *meaning*—presents us with the attempt of spirit to concretize itself, just as we saw above. Now, however, it is confronted with the specifics of its nature as organized and organizing, even as the absolute knowledge contained in its efforts to objectify the world push it toward the *an-organic*, to the indeterminate. As Hölderlin continues, he makes clear that the tragedy's form must present continuing separations reaching toward the apex where intellectual intuition—being, the aorgic—originally becomes organized even as it becomes as poem a higher unity. It thus presents dialectical unity in its privileged human or organological form.

This presentation makes clear reference to Kant's passages on intuitive understanding. The problem that makes intuitive understanding a regulative necessity is that of organized beings, literally beings which seem to possess reciprocal causality between their parts and their whole. For Hölderlin, the contradiction between organization and nonorganization plays out in the tragedy. And yet it does so slightly differently than it could for Kant. In the *Critique of the Power of Judgment*, the pair mechanical/organic makes a candidate for an antinomy, a contradiction which occurs because its object is made of parts from different faculties which cannot interact according to the canon of the understanding. Kant's examples—freedom, the infinity of the world, etc.—are generated by reason's drive to the absolute as applied incorrectly to objects of intuition. So, for example, the "world" can have an intuitive sense, but its determination as the totality of phenomena stems from the faculty of reason. When reason tries to answer the question of the size of that world, it should be disciplined by the canon to perceive that this question has two mutually facultative roots. Quantity cannot be applied both infinitely and intuitively at the same time.

The same does not hold, however, for the distinction between the mechanical and the organic. The judgment, which can be applied subsumptively (constitutively, in keeping with the canon) or reflectingly (anticipatorily), cannot produce an antinomy because it does not by itself produce real ontological determinations. The opposition between mechanics and life is thus given at the judgmental level, but made—by the designation of intuitive understanding as regulative—merely oppositional, not truly contradictory. Indeed, the *Critique of the Power of Judgment* is built on this delicate balance between two forms of judgment and their respective sphere of application. The supersensible substrate, and the regulative conclusion to God's existence—Kant's teleology—are founded on that nonlethal divide.

In a founding gesture of Romantic organology, Hölderlin first shifts the critical focus to judgment, bracketing the question of the ontological status of its contents in favor of an exploration of its forms (organs). He thus makes the judgmental opposition mechanical/organic into a contradiction, but one based in the forms of judgment itself. This means that the question of organization is opened up to a new kind of investigation, one which uses the formal-receptive organs of spirit to ask after the point of differentiation of the two orders. The "antinomy"[59] thus becomes resolvable, but can also be generated. Contradiction gains a history; in other words, dialectics come into being.[60] And it is immediately concretized in genre theory, addressing the Kantian problem of judgments of reciprocal causality—of organic beings—in the tragic form. Tragedy becomes the organ of an investigation into the origins of organization out of contradiction, of dialectical determination. Genre theory is, then, given pride of place in this second part of Hölderlin's metaphysics. The genre becomes the organ of metaphysics, not merely its instrument but its living function, the form it demands to achieve its effect, the plastic means of the metaphysical epoch. Genres—or species of representation—are the organs of the mind as animal body. Hölderlin has merely literalized what eighteenth-century biology and Kant offered to him metaphorically.

Epochal Twists: The Death of Empedocles and the Romantic Ethics of Tragedy

Hölderlin chose the philosopher Empedocles as the subject of his own attempt at tragedy. Empedocles was, then as now, a figure of philosophical legend, perhaps most famous for his putative suicide in the active volcano Mt. Aetna. His doctrines, such as Hölderlin could reconstruct them,[61] ran closely parallel to the ancient monism Hölderlin had encountered in Heinse's *Ardinghello*. Empedocles wrote a cosmology[62] based on the four elements (indeed, he is usually considered the founder of the latter doctrine). Separating these elements is one of two basic forces—conflict—while love binds them. The welter of the world is thus what Hölderlin would call *harmonious conflict*, a productive conflict in the elements.

The project is an attempt at a modern tragedy.[63] Hölderlin was translating Sophocles (the *Antigone* and the *Oedipus* cycle) as well, and Empedocles is thus chosen as a paradoxically modernizing figure.[64] Where *Antigone*'s struggle with law and custom reflects a Greek necessity, and where Oedipus's tragedy circles around problems of knowing,[65] Empedocles is singled out for actual tragic production as the representative of Western (non-Greek) modernity.

Or is he? As much as the logic of classicism—to imitate the flourishing of the ancients—seems to drive the choice, within the organological framework we can detect a complex mechanism of that classicism which makes the tragedy formally effective for modern and metaphysical purposes. The *Empedocles* became, as I will now show, the organ of a metaphysical metapolitics based on a break with the *physis-techne* analogy. With modern metaphysical tragedy, the radically new came into view.

The fragments of *The Death of Empedocles*—the tragedy remained incomplete—present the biography rather than the philosophy of Empedocles. The three versions are often taken to present Hölderlin's shifting attitudes toward the Revolutionary project and the Coalition Wars.[66] The political themes are patent: Empedocles's self-identification as nature's god earns him the descriptor "unlimited sense," an echo of the genre-theoretical descriptions of the poet. The plot occurs—in all three versions—between his self-apotheosis and his suicide, in the period of greatest political tension in Agrigentum. And yet the political element of the tragedy is not merely representational—Empedocles is not merely a cipher for the crisis of sovereignty of Hölderlin's own time. It is rather in the *mode* of presentation (and in the reception of the tragedy) that the politics reside.

Poetry and philosophy come too late and reconstruct their object.[67] This reconstruction, however, constitutes an *acceptance* of that history,[68] and its reproduction differs from the initial experience in that it takes on the characteristic of necessity.[69] The necessity, that is, of real historical contradiction and the task of reproducing and altering such contradiction. This is how, as Kurz puts it, the reconstruction becomes "a legal power of reality that sets free actions"[70] in productive contradiction with the "hardened positivity" of the actual. This philosophical schema and political ambition were founded on the most general contradiction, the organ of intellectual intuition. And it was not the representation of this organ that formed the revolutionary kernel of the tragedy, but the making of the tragedy into this very organ itself. For this purpose, Hölderlin thematized the spectatorship of that very contradiction in Empedocles's intradiegetic observers and in the tragic chorus in the second draft. These technical shifts in tragic presentation allowed the *Empedocles* to project the organological task of giving a new "direction" to the forces of history, of bringing forth a "new world."

The material presented in the tragedy—the "heroic" narrative—is focused on the highest point of possible contradictory tension in human life, presented in Empedocles's decision to politicize his total identification with nature as elemental power. The decision is antediegetic, such that the plot is driven by both the theoretical tension of the identification itself and the political consequence. Thus Hölderlin politicizes the description given in the *Ground to Empedocles*:

> where, then, the organic become aorgic appears to find itself again and to return to itself, so far as it holds itself to the individuality of the aorgic, and the object, the aorgic appears to find itself, in that it, in the very moment where it takes on individuality, also simultaneously find the organic at the highest extreme of the aorgic, such that in that moment, in this birth of the highest enmity the highest reconciliation appears to be actual.[71]

We can see here what the "New Letters on Aesthetic Education" would have contained. By allowing the contradiction to emerge in the tragedy and to find its resolution in that organ, Hölderlin demonstrates how theoretical and practical philosophy can be combined in intellectual intuition. The unification of theory and praxis occurs in the cascade of produced real contradictions—starting with that between judgment and being, proceeding to that between the organic and the mechanical (here concretized in Empedocles's self-identification with *physis* itself as the *an-organized*), and ultimately finding its meta-political point in the production of a contradiction between *physis* and *techne*.

Hölderlin writes in the *Ground* that "nature and art are merely mutually and harmoniously opposed to one another in pure life."[72] This opposition is the particular basis of the depiction of Empedocles, and it is this opposition in particular that makes him into the figure of his nation:

> Thus Empedocles is a son of the heavens and of his time, his fatherland, a son of the violent oppositions of nature and art in which the world appeared before his eyes . . . such a human can only grow out of the highest opposition of nature and art . . . so Empedocles is, as I said, the result of his time, and his character refers back to this, just as he emerged from it. His destiny depicts itself in him, as if in a momentary unification that must dissolve itself again in order to become more.[73]

As has often been recognized, Empedocles has a national and epochal task in his theoretical and political contradictions.[74] His destiny, however, is defined by a given contradiction between nature and art, as Hölderlin repeats here. As I have elaborated in the introduction, nature and art were often analogized, the latter contained within the former's rules. Why, then, do they emerge here—in the epochal tragedy of modern destiny—in opposition?

If Empedocles's "unlimited sense" can be read as intellectual intuition, then we can see the emergence of a third contradiction essential to that figure. As I showed above, intellectual intuition both carries with it the danger of an ideal determination of the phenomena and also makes a theoretical parallel to the *physis-techne* analogy. If we possessed intellectual intuition, this would suggest something like an ability to make the world. As we have also seen, however, the first contradiction—between judgment and being—addressed in Hölderlin's philosophical writings shifted from a representational model to a reproductive model of cognition and aesthetic production. In that shift, the possibility of a contradiction between *physis* and *techne* becomes possible. When the poet establishes a *world within a world*, there is no guarantee that the laws of the two worlds will coincide. Indeed, one could argue that they must not, if development—from organ to organ—is to occur.

The epochal destiny of Empedocles is, then, defined by a third contradiction produced by the figure of intellectual intuition, that between nature and art. This firmly confers metaphysical relevance to the poet's project: if nature and art are opposed, and thus exposed to the cycle of separation and unification of dialectical organs, then the possibility of metaphysical innovation is present in the tragic form.

Empedocles—at the level of heroic narrative—must be destroyed by these contradictions, as the *Ground* makes clear. Recall, however, that the heroic narrative of tragedy is its appearance. Its "meaning"—tragedy as organ—must be sought in intellectual intuition. And yet, in *The Death of Empedocles*, the heroic narrative (the content) coincides with the designated meaning of any tragedy (the underlying intellectual intuition). This is captured in the formula of the contradiction between nature and art. What is a given contradiction for Empedocles must be reproduced for Hölderlin's present. Art must be raised into opposition with nature, precisely through the narration of their coincidence—Empedocles's intellectual intuition, replete with political consequences. The tragedy's task is to produce this contradiction, and it can do this only by becoming the means of cognition in the present of intellectual intuition proper, epochally, to ancient Greece. The tragedy, as the finite bearer of the dialectical process of ancient cognition, becomes the desired organ of intellectual intuition. Tragedy literally objectifies the infinite, and in doing so presents the present (around 1800) with its destiny. More than merely the immanence of absolute knowledge in every objectification is thus presented. The *Empedocles* becomes the chosen sphere of self-objectification of intellectual intuition itself. It raises the possibility of the application of metaphysical knowledge to concrete situations. And it does this formally yet with openness to any possible arising phenomena: it does this organologically. The *Empedocles* presents us with the possibility of the intentional, systemic yet aesthetic intervention into the order of beings, the production of new categories. These are conjured by the figures in the play who seek to understand Empedocles. It is their limited sensibilities, and that of the chorus, that present the audience with the clarion call for dialectical thinking—as we shall see. Furthermore, the *Empedocles* crosses, as it were, from a metaphysics into a metapolitics, maintaining a theoretical space for necessary contradictory encounters between spirit's highest knowledge and history's most concrete contingencies. It marks out the tragedy as the beginning of Romantic organology.

Before leaving Hölderlin and turning to Schelling, I want to point to a technique Hölderlin uses to realize the epochal significance of his *Empedocles*. This technique involves the concrete presentation of the contradiction between organs and intellectual intuition—the epochal struggle of the tragedy—in the figures surrounding Empedocles. I read this technique as a final reaction to Schiller, one which takes the latter's own theory of tragedy into account, anticipating the terms of Schiller's 1803 writings on the tragic chorus.

If Empedocles has an "unlimited sense," those around him do not. Even those sympathetic to him—his disciple Pausanias and the devoted youths Delia and Panthea—are presented as severely limited in their understanding of his internal process. Especially Delia and Panthea, who never actually speak with Empedocles, emphasize this limited relation, and thus the status of the contemporary observer of the *Empedocles* as distantiated from the narrative hero. These middle figures, these sympathetic spectators, are the aesthetic means by which the underlying organological message of the play comes to view. This message, I contend, runs directly counter to that narrative witnessed by the spectator, and thus makes a concrete contradiction with epochal significance the very fabric of the play.

Schiller's theory of tragedy is, like his more general aesthetic theory, in deep dialogue with Kantian categories. In tragedy, Schiller finds a path to the socialization of Kant's theories of freedom,[75] both at the level of narrative and at the level of reception. The narrative contradiction—the suffering of the good—in the tragedy is reflected in our mixed pleasure in that which is not "purposive" (*zweckmäßig*), in "that which is contrary to (its) purpose" (*das Zweckwidrige*). Following Kant, Schiller claims that the presentation of natural teleology confirms our sense of the true moral teleology.[76] Tragedy does this by allowing pleasure to emerge from the presentation of the non-natural.[77] Schiller addressed one technique of this presentation of contradiction (with the goal of the presentation, as always, of our freedom as humans in reason) in the preface to *The Bride of Messina*, "*On the Use of the Chorus in Tragedy*" (1803).

The chorus of ancient tragedy is the means by which the poet finds the proper balance between sense and reason, between *telos* and chaos. Schiller writes:

> The ancient tragedy, which originally was satisfied merely with gods, heroes, and kings, needed the chorus as a necessary accompaniment, it found this in nature and needed it because it found it. The actions and destinies of the heroes and kings are in themselves already public and were so in the simple times of old [*Urzeit*] even more. The chorus was consequently in the ancient tragedy more of a natural organ, it followed directly from the poetic form of

actual life. In the new tragedy it becomes an artificial organ [*Kunstorgan*]; it helps *produce* the poetry.⁷⁸

The chorus as natural organ existed in the ancient tragedy to balance the violent effects of the plot. Schiller suggests that this organ was given by the social form of the *polis*, itself a representative of that aesthetico-political balance. In the modern tragedy, the poet (one imagines the *sentimental* poet, in his terms) must create this organ, which, by providing the means to the desired balance, helps the poet in turn create his poem. The distance created by the chorus is one of framing. As a quasi-spectator—but one in possession of full knowledge of the tragedy—the chorus rescues the spectator from any possible excess of violence, guiding him toward his realization of his own freedom. From a production angle, this device ensures the poet a first distantiated balance.

It is possible for the poet to misplace this balancing device, however: "Getting rid of the choir and the combination of a sensible powerful organ into the characterless boringly recurring figure of a poor confidant was thus not such a great improvement of the tragedy as the French and their worshippers have imagined to themselves."⁷⁹ The figure of the confidant—also a kind of ersatz spectator—is a poor replacement, for Schiller, for the chorus. Perhaps this is because of the chorus's expected bird's-eye view of the plot and the fate of the hero. However that may be, Hölderlin was unsure about which organ to use.

Only the third version of the tragedy contains hints about how Hölderlin might have used an actual chorus. In a plan for the continuation of the third version, we find *Chorus* written throughout, always accompanied by an enigmatic question mark, and once with the laconic "future."⁸⁰

In the third version itself, a sketch of the chorus's role follows the discussion between Empedocles and Manes. Indeed, the third version has only three scenes. First, Empedocles soliloquizes. Next, he confides in Pausanias, rejecting his confidant's desire to follow him into death.⁸¹ Finally, the conversation with Manes determines Empedocles's epochal destiny, justifying his suicide.⁸² Only at this point does the chorus enter—and Hölderlin wrote only a sketch of their contribution. Their words run:

> New world
> and it looms, a brazen vault

the sky above us, curse lames
the limbs of humankind, and the nourishing, gladdening
gifts of earth are like chaff, she
mocks us with her presents, our mother
and all is semblance—
Oh, when, when will it open up
the flood across the barren plain.
But where is he?
That he conjure the living spirit[83]

The chorus fills the role later ascribed by Schiller. It sings the generality of Empedocles's destiny, helping the spectator to the tragic insight. So far from the aesthetic realization of Kantian freedom, however, the chorus as organ here offers the epochal destiny of Empedocles as the unification of the general and the particular. The mockery of the world lies in light of Empdocles like chaff before us, and the world disappears into illusion. An aching for the fullness of time is present in the metaphor of the flood washing away aridity, and the chorus finishes with the subjunctive command that Empedocles (?) conjure the figure of dialectical unity, neither mere life nor mere consciousness, but the living spirit.

The passage is prefaced by by the words "new world." It is unclear whether we should read this as an impossible stage-direction or as a note for earlier lines from the chorus. What is clear, however, is that the highest unity or reconciliation is here given in the organ of the tragedy, itself an organ of intellectual intuition. Here the higher contradictions of that problematic are encapsulated: if the intuitive understanding achieves its conjuring of living spirit in the resolution of the contradiction between the *an-organic* and the organized, and if the condition of that resolution is the metaphysical production of the *Procedure*, then the *world within the world* here produced is fundamentally new. The chorus points up the tragedy's task in metaphysical innovation. It suggests—barely—that cooperation in the production of a new order of things is the intellectual intuition at the basis of the attempted modern tragedy. It suggests that the produced contradiction between *physis* and *techne* can be resolved only by the re-production and alteration of both orders. The world within the world alters the former, framing world.

This alteration and intervention is not Empedocles's destiny, but ours, for Hölderlin. The *meaning (Bedeutung)* of the modern tragedy is its contribution to the ability to change the world systemically. Tragedy is the organ of a radical metapolitics. This becomes even clearer in Hölderlin's proleptic violation of Schiller's rule about the chorus, in the formal presentation of the tragedy's meaning in the sympathetic character Panthea.

Panthea is the daughter of Agrigentum's leading politician, Kritias, whose hostility toward Empedocles is barely tempered by the latter's antediegetic medical treatment of his daughter.[84]

In the opening scene, Panthea describes Empedocles, whom she worships, to Delia, the daughter of the visiting priest Hermokrates. The two find themselves in Empedocles's garden, but in his absence. Panthea opines about the unlimited sensibility of the prophet, describing his spirit in very much the terms of the genre-theoretical works. He binds the organic world together and to himself, and his potion has awakened a similar feeling in Panthea herself:

> The sounds that surged from his breast! in each syllable
> of his, every melody sang out to me! and
> the spirit in his words!—at his feet
> I'd sit, for hours at a time, as his pupil
> his child, gazing out into the ether that is all his own
> and, clambering joyously to his own heaven's
> height, my senses fairly wandered.[85]

In the course of the scene, Delia expresses reservation at Panthea's apparent desire to imitate Empedocles[86]—the impending suicide strikes the reader or spectator immediately. Panthea distances herself[87] from this danger with heavily philosophical, indeed organological, consequences. Panthea's distance from Empedocles is the distance from *mimesis* to representation.

Empedocles, as should be clear by now, is involved in the productive devolvement into the *an-organized*. His destiny is the heroic dissolution of himself through identification with the elements, ostracization from his society, and physical disintegration in lava. Read with the genre-theoretical writings, this highest, heroic reconciliation is a regression into the common root of consciousness, into the intellectual intuition that allows cognitive activity to be more than representational. Empedocles reaches into the depths of consciousness to find the productive root of mimetic activity itself.

Not so Panthea, who is caught in the organs of finite consciousness. She expresses the deadlock as follows:

> O eternal secret, what we are
> And seek, that we cannot find; what
> We find, that we are not.[88]

Her finite attempt to grasp the unlimited sensibility which has partially lit her consciousness meets an impasse, the impasse of self-cognition in intellectual intuition. We find only that which we are not, while we strive for that which we cannot find. Indeed, this creates a contradiction in the finite organs of consciousness:

> I think after him [*ich sinn ihm nach*]—how much still
> Must I think on him? oh and if I have grasped
> Him; what would that mean? To be him, that is
> Life and we are merely its dream.[89]

Mimesis (*Nachahmung*) here becomes imaginative reflection (*Nachsinnen*)—mimetic reproduction and its organological possibilities are reduced by the spectator-consciousness of the adoring Panthea to reflection. She thinks him in every way possible, but even if she grasps him, she says, it does not matter. The figure she represents gives only the feeling of not existing, of being an illusion derivative of Empedocles. There can be no representation of intellectual intuition.

Panthea's predicament is the organological key to the *Empedocles*. For it is in her speech that the contradiction between organs and absolute knowledge is generated for the spectator. The sympathetic figure—precisely the organ which does not, like the chorus, guide the audience through the narrative—reflectively pursuing but failing to grasp the infinity of Empedocles's destiny, becomes the epochal signifier in the modern tragedy. The contradiction produced, that of an organ of intellectual intuition, is the desideratum of the organological tragedy. The play opens with the generation of this contradiction (in the first version). If it also closes (in the third version) with the choral reassurance of the newness of the world, then we can see the tragedy's *meaning* (in Hölderlin's terms) as the creation of an organ for the third task of intellectual intuition: metaphysical innovation.

The mission of Platner's *Seelenorgan* had been simultaneous receptivity and external influence. This dual task is conferred, after detours through Heinse and Soemmerring's interventions in late-eighteenth-century pantheism and (post-)Kantian epistemology, onto the Romantic organ. That concept here comes into its transcendental, literal own: it makes possible an absolute knowledge as a tool of historical change. It introduces the notion of transcendental technology into the European theater. The metaphysics of judgment, or organology, thus created a new genre of post-Kantian metaphysics. This metaphysics was based in the notion of a *dialectical organ*, defined as an ability to generate and resolve real contradiction. In philosophy, this meant tracing cognition to its reproductive roots; in genre theory, finding a home in tragedy for the unification of mechanical and organic orders of being; and in the tragedy itself, generating actual new contradictions to construct a new world. Organology was inaugurated as the metaphysics of judgment in passage to a radical metapolitics, a theoretical or spectatorial Jacobinism.

FIVE

Electric and Ideal Organs: Schelling and the Program of Organology

> What then is that secret band that ties our spirit to nature, or that hidden organ through which nature speaks to our spirit or our spirit to nature?
>
> —FRIEDRICH WILHELM JOSEPH SCHELLING,
> *Ideas for a Philosophy of Nature*

Friedrich Schelling (1775–1854) was the biggest risk-taker among the German Idealists, doggedly exposing Idealist methodology to that which it might not be able to internalize—nature, things in themselves, the beautiful, the divine.[1] This is why, although he remains a leading figure in the German Idealist pantheon, he was also able to set the direction of a major movement in the history of science—*Naturphilosophie*. This movement, which is often cast as the fringe of the royal road, was the most likely site for organology to find a home in the mainstream of scientific development, a notion against which I argue in what follows. But Schelling's elaboration of the key insights of organology cannot be reduced to what came after him, whether that be his sometime collaborator Hegel or his enthusiasts in the sciences. Schelling combined his risky Idealist methodology with a philosophy of nature to produce a Romantic metaphysics rooted in the term "organ." Combining the problem of specific forces—organic, chemical, physical—with the physiological exploration of electricity, Schelling developed a model of the "organ" that mediated between different spheres of nature, and between the body and cognition. This model resulted in an aesthetic doctrine that anticipates the problematics of technological development later articulated by Karl Marx.

Schelling did not merely adopt the term "organ" from physiological discourse: he played a central role in the idiosyncratic mix of idealist philosophy and natural-scientific speculation that came to be known as *Naturphilosophie*.[2] Schelling's attempts to unite his brand of Romantic Idealism with his wide-ranging knowledge and exploration of the natural sciences mark his philosophical production in the years leading up to 1800. The organ was a central term in both projects, and it is their intersection—the point at which, as Schlegel would demand, the ideal meets the real—that makes up Schelling's contribution to organology. By constructing a schematic organ in nature's orders (as we shall shortly see), and by allowing this organ to emerge into perception and cognition in the physiological processes of animal electricity, Schelling tried to make good on his promise to "idealize" nature. Looking back on this project, barely a year old, in 1800, he would develop the notion of the philosophy of art as the "organ of philosophy." The result was that the scientist's and artist's world were analogously filled with organs, and this term thus became the crux of a large-scale conceptual shift. Where the tradition had held, even up through Kant and Fichte, that nature and self, object and subject, were the loci of the eternal antinomy between the necessary and the free, Schelling proposed a metaphysics that started in the middle, where the two were already joined in individuals. He proposed a world populated by "organs," the point of indifference, as he put it, between the law and the individual. The touch of human construction thus lay on all things. Where Marx would see the factory as the accumulation of human constructions reversing the very progressive purpose for which they were built, Schelling reserves this notion of acting on the already-created as the core of a philosophical optimism about a modernity where necessity and freedom are always intermixed. In this way, he is responding to the same anxiety as Hölderlin: metaphysics must be a matter of historical epoch, and that epoch must be concretized in some object of analysis and action. But where the tragedy formed that hope for Hölderlin, for Schelling it was a matter of joining cultural and scientific endeavor—a point on which Goethe and Schelling agreed. To represent the world as an infinite series of organs was to allow for a new metaphysics that could hope to engage that world, even where its artifacts seemed to take on unexpected characteristics. It was to allow for being to emerge in contours around metaphysical method, and to preserve the possibility of constructing and reconstructing whatever should emerge. Schelling's ideal physiology thus formed the programmatic chapter of Romantic organology.

If Romantic organology was inaugurated by Hölderlin, it was brought into its classical form by his roommate from the Tübinger Stift, Schelling. The full confrontation between metaphysics, science, and disciplinarity would have to wait, however, for Novalis's version. But where Hölderlin had drawn his term from physiological and metaphysical debates (and applied it to genre theory and to the meaning of tragedy), Schelling found himself an essential part in shaping those debates. The debate recognized and recuperated the return of metaphysics in the re-iterated confrontation between biology and physics. Schelling's dialectics of nature, derived from his early readings of Kant and Fichte, came to be the basis of a European *Naturphilosophie*—a dedicated mixture of speculation and empiricism—which died out in Europe only in the 1830s and 1840s. Taking the logic of the instrument to its transcendental limit, Schelling's organ offers a different take on his series of early systems than has otherwise been possible. By returning to Kant's rejection of the *organon* for metaphysics, Schelling's texts reveal an etymological "indifference point" (*Indifferenzpunkt*) between knowledge and being (Nature), one which ultimately is the fabric of our world, not divided into subject/object or necessity/freedom, but populated by syntheses of these antinomies in development.

With Schelling, the instrumentalization that organology made possible applies itself across three separate fields—science (and electricity specifically), metaphysics (the explicit reintroduction of an *organon* now in interaction with ideal organs), and theology (reaching into the metaphorical register of the body as mind's *organ* and the world and its humans as *organs* of the divine). "Organ" becomes an analytical tool capable of synthetic intervention in its field of potential. The concept of the organ—by this time fixed in a relatively stable sense in biology—is thus returned to its not-so-distant etymological past. The metaphorology of the organ becomes organology by way of transcendentalization and simultaneous application.

Schelling's organology makes use of science for the reestablishment of metaphysics. The organ as Schelling makes use of it allows us to see that his *Naturphilosophie*, his metaphysics, and his theology result in, but do not presume, the metaphysical picture he reawakens. In the first step of Schelling's organology, a dialectical picture of nature itself (Nature) is developed.[3] This picture is meant to serve as a "proof of idealism"[4]—Schelling's conviction is that a truly scientific natural science should arrive at reason as the result of its activity. Idealism, or the real inclusion of reason in Nature, is the result of doing science. That is because true natural science is rooted in a general *organ* (that which makes the organism organic, mereologically), but plays out in the constructed natural world in a specific organ (actual organs as bearers of electrical qualities). Schelling thus reconstructs Leibniz's physics across the Kantian divide and in a post-Herderian context (especially in dialogue with Kielmeyer). By doubling organs, making them internally oppositional, and by reducing the logic of organic judgment and redoubling its representational activity, Schelling redevelops the Leibnizian fundamental, metaphysical organ on the other side of Criticism. This is only possible in the post-Herderian context because histories of organs come into view, not in a general sense but as the resolutions of rolling sequences of oppositions and resolutions.

The second step—*ideal organs*—submerges the mind in nature by analogizing nature's most general and most specific reciprocal cause ("organ," or, as in the epigraph above, "band") with the tools of philosophy. This analogical reasoning, which is meant to recuperate and include all of Fichte's achievements in a new metaphysics, results in a system (the *System of transcendental Idealism*) that seeks to objectify consciousness, producing itself as aesthetic product. That process—intellectual intuition—can certainly never come to an end, and thus a fundamental agreement between Schelling and Hölderlin underscores their differences from the 1790s.[5] Where Hölderlin had turned to the figures of absolute knowledge in their tense but productive relation to finite literary form, Schelling turned to the finite products of scientific knowledge, and ultimately to aesthetics as well. The theoretical punch of his speculation moves in the same direction that Hölderlin's had: toward the possibility of intervention in the mutual and analogical rules that support the increasingly dialectical double-system of *Naturphilosophie* and transcendental idealism. Aesthetic intuition becomes ontologically relevant because it points toward metaphysical innovation. Indeed, the admixture of organ and *organon* in a productively aesthetic intuition was the discursive moment that allowed organology to emerge as a doctrine. It thus undermines the traditional pairings nature/necessity and soul/freedom, replacing them with a synthetic world in developmental flow.

The Organs of Organization: Intuitive Understanding

Between 1797 and 1799, Schelling drafted no fewer than three systems of the philosophy of nature, the *Ideas for a Philosophy of Nature* (1797), *Of the World Soul* (1798), and *First Outline of a System of the Philosophy of Nature* (1799). These efforts constituted a singular systemic move in the rapid development of German Idealism. Inspired by Kant, Fichte, and the immediate natural-scientific context (especially the work of Kielmeyer and Alexander von Humboldt), Schelling was attempting to fill in perceived gaps on both sides: Idealism needed a stronger doctrine of nature, and science needed idealist underpinnings if it was to be called science at all. In elaborating these interventions, Schelling made organs general—the most literal terms in his "organic" system, returned to their Leibnizian metaphysical provenance—and particular, coagulations of electrical qualities in the developing world. In both uses, he emphasized dialectical structures of duality and triplicity, preparing the analogical and isomorphic way for his eventual combination of organs and *organon*.[6]

In the *First Outline*, Schelling throws his lot in with the epigenesis movement: *"all formation occurs through epigenesis."*[7] The developmentalist perspective[8] should allow, for Schelling, a comparative physiology. Comparative *anatomy* merely attends to the structures of various organs, not to their role in the formative process (*Bildung*).[9] The new science would be based on the specifics of the organ, defined once as the general relation of the universe to itself (nature), and again as the functional part caught in the larger forcefield of that Nature. The conceptual basis of this program is captured as follows:

> They would relate simultaneously to the whole organism as cause and effect of its activity. That which so relates itself to the organism (as a whole) is called an *organ*. Where opposed functions are united in one organism, these functions

must be split up into various organs. Therefore, the more the multiplicity of the functions increases in the organic domain of nature, the more complexly must the system of the organs develop (in part called "system of vessels," which is completely wrong, for within the organism nothing is merely a *vessel*). Insofar as each organ exercises its special function, it would receive a life of its own (*vita propria*)—to the extent, however, that the exercise of this function is still possible only within the bounds of the whole organism, it would only receive a borrowed life and it must be so in accordance with the concept of organization. If the manifold possible proportions of organic functions can be deduced a priori, the whole multiplicity of possible organisms would also be deduced, because the organic structure depends on this proportion.[10]

This passage not only lays out the fundamentals of Schelling's system of nature, but also clarifies his relationship to Kant in that context, and that on three separate levels. Moving backward through the quotation, we can see that the first point is that Schelling opens the possibility here that life science could be constitutive, that is, that the prospective comparative physiology could proceed *a priori*, making up an essential cognitive condition for humans. That is, of course, precisely what Kant had rejected in premise and tried to recuperate in the "Critique of Teleological Judgment," as we have seen.[11] He does this for reasons profoundly in keeping with Hölderlin's establishment of a metaphysics of judgment. Comparative physiology would be based on the notion that the world in organic development (development of and through organs) could be known *a priori*, where this *a priori* takes on an adverbial sense to the productive judgment of the scientist/philosopher. This would only be possible with the eventual correct sense of the *organ* itself, which Schelling sketches here. The second point, then, is that the *organ* is dual—its has its own life insofar as it has its own functional unity, but also a "borrowed" life, insofar as that unity *is* the direction of a force's contribution to the great organismal unity.[12] If we return to Kant's definition of the organ, we can see the full impact of this point:

> In such a product of nature each part is conceived as if it exists only through all the others, thus as if existing for the sake of the others and on account of the whole, i.e., as an tool (organ), which is, however, not sufficient (for it could also be an tool of art, and thus represented as possible at all only as an end); rather it must be thought of as an organ that produces the other part (consequently each produces the others reciprocally), which cannot be the case in any

instrument of art, but only of nature, which provides all the matter for instruments (even those of art): only then and on that account can such a product, as an organized and self-organizing being, be called a natural end.[13]

Kant's interest, here in the *Critique of the Power of Judgment*, is in objects in which an idea (*Begriff*) appears to precede or determine the order of the parts (thus the juxtaposition of art and organisms). Schelling's first step was to challenge the notion that these objects are somehow second-order. In a framework of judgments as basic elements of investigation, what could lead us to assume that status? Bracketing that question,[14] Schelling develops the picture of organicity from Kant. Where Kant's focus falls on the problem of mereological precedence, Schelling starts *medias in res*. Given a reciprocally acting whole, the name for the agent of that reciprocity is organ. And it follows from that slight definitional shift that the organ must be dual, that is, it must be both a unity (functioning and in itself) and a difference (from the greater whole for which it serves that purpose). The third point thus emerges from the second and the first: if there is the possibility of making judgments *a priori* through experimentation and experience, and if further there is no reason to assume the cognitive status of the organism as secondary to begin with, then the organ becomes the object of investigation. Its process, which we encounter in a divided form, as a duality in development, follows from its definition. And beyond this mere conceptual conclusion, the organ also serves as a first analogue to judgment itself. After all, judgment is precisely that function in cognition which produces ambivalent unities, by Kant's own lights. That these unities then come to be analyzed into regulative and constitutive is the result of Kant's version of Criticism. Judgment's function with respect to the fourth modality makes up a rather large part of the organ-metaphorical strain of the *Critique of Pure Reason*. Schelling is merely drawing a consequence from a different set of assumptions within a generically identical methodology. The preliminary result—and I emphasize *result*, where this is often taken to be the premise of Schelling's *Naturphilosophie*—is that the reciprocal causality of the organism mirrors the form of judgment insofar as it is abstracted from the other faculties. This conclusion set Schelling up to break with the Critical system in a Critical manner, a topic I shall consider below.

The beginnings of this break are clear in a section of *Of the World Soul* that treats the cognitive status of the organism. The stated intention of the section is to investigate the origin of the concept of organization, and Schelling proceeds according to Kant's initial findings in the *Critique of the Power of Judgment*. Kant had written that the notion of an "end" (*Zweck*) is "the object of a concept insofar as the latter is regarded the cause of the former (the real ground of its possibility).... Where not merely the cognition of an object, but the object itself (the form or existence of the same) as an effect is conceived as possible only through a concept of the latter—here one is thinking an end."[15] The notion is that, in the encounter with phenomena that seem to display objective autonomy or formal independence, the end suggested by the encounter is merely a concept that appears to exist and determine the object. This concept—*thought* to be existing—differs from the concepts at work in the canon of the understanding, in which the faculty provides and never asserts the reality of the concept (except as a condition of possible experience). The encounter with such phenomena receives the name *Zweckmäßigkeit* (purposiveness), a term that implies the extra work that judgment does in this case. The spontaneous ascription of reality to the concept—which Kant, as we have seen, goes on to regulate and discipline—is a case study in the separation of subsumptive judgment from reflecting judgment, and grounds the separation of the life sciences from physics. Yet it also is a necessary part of the judgmental apparatus, without which we would have no recourse to analyze aesthetic and organismal objects.

Schelling writes that, in the mechanisms of nature, we never perceive anything that "forms its own world."[16] Where we encounter what must eventually be called an organic being, we note the reciprocal causality in our judgment and its tendency to assert the reality of causality. Schelling literalizes the mereological element of this thought: for a whole which appears to subsist independently, the concept which appears to guide it is like a sphere in which the parts interact. The sphere is perennial (at least locally): it forms a kind of substrate of the phenomenon.[17] In keeping with Kant's analysis, Schelling then asserts that this sphere cannot itself appear. This is because the sphere is not the intuited part of the phenomenon—it is the concept, "the *monument* of disappearing appearances."[18] What Schelling does with this gloss on Kant's conception of cognition of organisms anticipates his future work with the term "intellectual intuition." He reduces the framework for investigation of this phenomenon to its judgmental basis, that is, he attends to both the intuition and the concept at work without making an assumption about their respective ontological statuses: "Since the concept of this product expresses nothing *actual*, insofar as it is the concept of mutually operating *appearances*, and since on the other hand these appearances are nothing *that remains* (nothing fixed), insofar as they operate within this *concept*, therefore *appearance* and *concept* must *be inseparably united* in each product."[19] The concept, the substrate, the monument of appearances, is not actual, but makes up a real cognitive condition for the play of those appearances. Only within that concept can these appearances be such as they are, unified as a field of intuitions. The conclusion—that in the organic product, appearance and concept are inseparable—slides subtly out of the Kantian conclusion it appears to mirror. Kant agrees—for regulative reasons—that intuitions and concepts are united at a higher level in the cognition of the organism. But for Schelling, it is not intuition (*Anschauung*) but appearance (*Erscheinung*) which unites with the concept. The concept's status as nonactual is partially contradicted by its necessity in not merely unifying appearance (that is what intuition does), but in providing a field of possible play for those appearances, a substrate. This first conclusion leads Schelling to differentiate types of matter. Matter—the proposed substrate—must be of different kinds if it is to support different sorts of appearance-fields (different *Potenzen*, potentialities). But the concept of matter requires it to be permanent and divisible. Schelling's investigation thus leads to a conception of different matters divisible infinitely in themselves but qualitatively whole and distinct from one another: "It must be divisible like every other matter, ad infinitum, indivisible, as this particular matter, similarly ad infinitum, i.e. such that through infinite division no part that does not yet represent the whole, refers back to the whole, can be met with in it. The distinguishing character of this product (that which removes it from the sphere of mere appearances) is thus its absolute individuality."[20] The point for Schelling, at one level, is that matter itself must be organized in order to be valuable for natural-scientific investigation. The concept of matter must be differentiated according to its quality, which he here calls absolute individuality (see below for more on quality). But the passage also has indirect implications for the picture of cognition of organisms. The judgment which applies must be capable of switching between conceptions of matter, that is, between concepts which determine the field of play for appearances. This means that the nonactual status of the concept is not attached to a subordination to another type of judgment, but instead to the framework of unity necessitated by experience. Schelling will now immediately refer to the *Critique of Pure Reason* and ensure that "at least" a regulative use of this conception of matter is necessary. And yet the concept itself, changing as it must in the reflective process of investigation, undergoes a process of redoubling which allows for alteration—just the condition that obtains for all organisms. Schelling concludes:

> Thus the double aspect of every organization that as an *ideal* whole is the cause of all *parts* (i.e. of itself as *real* whole), and as *real whole* (so far as it has parts) is the *cause of itself as ideal* whole, wherein one can then sees without difficulty the absolute unificiation, proposed above, of *concept* and *appearance* (of the ideal and the real) in every product of nature, and depends on the finite determination *that every truly individual being is simultaneously both effect and cause of itself.*[21]

Intuitive understanding—since that is clearly what Schelling is analyzing—is not only a higher unity of concept and intuition, but also a duality within that higher unity that allows for the reflective process of concept-determination to occur in any experiential context. The reciprocal causality of the organism is literally taken up into the relation between concept and intuition in the intuitive judgment. The unity of the real and the ideal in any cognition of nature is both absolute and iterable. It is both determinative of the appearances and alterable in the face of experience. It unites, in a way that Schelling did not see in 1798, empirical realism and transcendental idealism in a single judgmental style, one that paradoxically determines the phenomena and remains open to their contingent flow. At this point in Schelling's career, this implicit revision of Kant's doctrine of the intuitive understanding is little more than a promise, on which he first makes good in 1800, in the *System of Transcendental Idealism*.

General and Particular Organs: Comparative Physiology and the System of Forces

In the *First Outline*, Schelling gives a "general schematic of the sequence of stages" in his system of Nature:

Electric and Ideal Organs 181

Organic	General[22]	Inorganic Nature
Formative Drive	Light	Chemical Process
Irritability	Electricity	Electrical Process
Sensibility	Cause of Magnetism?	Magnetism?[23]

The task of the *Outline* is to deduce this system, which, as noted above, includes separate but interconnected orders, here specified as *anorganic*, *organic*, and *universal* (Nature). In each order, quantitative and qualitative balances of forces interact, even as the whole interacts with the other orders. As we can now see, at each level, the points of interaction are called "organs," which are the *referenda* both in general and in particular. Schelling invented neither the system of forces nor the use of "organ" within that system, but his contribution to both helped him to develop *Naturphilosophie* as a branch of Romantic organology. The system of forces had to be bound by a general concept of the organ, and complemented by the particularity of the electrical organ. Force had to be made dialectical, and the organ had to be transcendentalized. The result was that organology could instrumentalize natural science, as a preface to its intention of intervening in nature itself.

The notion of the system of forces was always bound, for Schelling, to the possibility of a comparative physiology. And he explicitly recognized his forebears in both projects.[24] The system of forces had been given first by Herder, and then been expanded and altered by Blumenbach and Kielmeyer.

Herder's *Ideas* had been the *locus classicus* for the systems of forces that emerged in the vitalist natural-scientific scene in the 1790s. The third book of that work had proposed, in keeping with the general cosmology analyzed in chapter 2 of this volume, a comparative physiology[25] on the basis of the interaction of basic forces in the animal. For Herder, the formative drive included, as it had for Blumenbach, both nutritive drive and reproductive drive (the three elements ultimately united in Wolff's *vis essentialis*). In addition to these, Herder included Haller's irritability and sensibility. As in his earlier work, *On the Cognition and Sensation of the Human Soul*, irritability and sensibility were made into progressively concrete versions of a single force. Indeed, in the *Ideas*, all of these forces were made part of a single organic force-complex: "A single principle of life seems to rule in nature:

this is the *aetheric* or *electric current*, which is worked every finer in the pipes of plants, in the veins and muscles of the animals, finally even in the nerve-structure and finally kindles all the wondrous drive and forces of the soul at which we marvel in animals and humans."[26] The relation of forces was, for the monist Herder, a proliferation more than a combination. A single, ultimately divine force, held the cosmos together and displayed itself in the animal individual and species. As we shall see, the suggestion that aether or electricity was the basic characteristic of this internally differentiated force reached both back to the aether-tradition and forward to Schelling's instrumentalization of electricity. As Herder continued his speculation, he tied the system of forces and the promise of a comparative physiology to his cosmology. The effects of the force visible in the external world were bound to their ability to find organs of mediation. Without those organs, the various media of the qualitatively different force-fields could find no external expression.[27] Individual animals, including the human, might not have the right organs for the perception of a given medium, but in principle, the cosmos is self-perceiving (it is, indeed, God's self-perception), and thus suffers no discrete breaks in its force-organ relationship. The cosmos is thus made up of force-organ relations, built (in this respect) as complex units of animal experience or *worlds*.[28] The physiology which prospectively follows on this system of forces differentiates animals by the complex of their organs,[29] with the endgame of the human standing erect, open to sensing, cognizing, and speaking of his world.[30] The basis of this system was in the differentiation of force and organ, in the moment of encounter of the general and the particular, which, Herder noted, first created the difference between *inner* and *outer*.[31] This designation, too, would bear fruit in the 1790s, and in particular in Schelling's *Naturphilosophie*.

Schelling marked out another significant predecessor to both projects in the Stuttgart biologist Carl Friedrich Kielmeyer's *On the Relations of the Organic Forces amongst Themselves in the Series of the Different Organizations*.[32] Kielmeyer had given this speech on the occasion of the duke's birthday at the Hohe Karlsschule in Stuttgart in 1793.[33] Kielmeyer marked out five different forces that were distributed for interaction among animal *organs*: sensibility, irritability, and forces of reproduction, secretion, and propulsion. He treats these and their relations, trying to derive general laws of their interactions. Indeed, as Schelling noted, these were laws of compensation:³⁴ where one force preponderated, another was repressed partially. The basis of that effect was the differing qualities of the forces, but the resulting relations were at least vaguely quantitative. Ultimately, an inverse proportion between the force of reproduction and that of sensibility was the result of Kielmeyer's investigation: where sensibility is afforded the organism, fewer offspring are possible, and where dozens of offspring are possible, less force is devoted to the organs of reproduction.³⁵

Blumenbach follows Kant in making organisms reciprocally causal, and dubs this causality a "system of *organs*."³⁶ The organs are the locus of the relations between forces, which are divided up and conferred upon (*verteilt an*) these organs. The physiological struggle between irritability and sensibility noted by Schelling³⁷ plays a role in the very formation of those organs: "and what before was irritability develops itself in the end to a capacity for representation, or at least to its invisible most immediate material organ."³⁸ Organs were not merely passive receptors of their forces, but actually formed by them. Where Herder had seen a single force differentiating itself, Kielmeyer saw unified organs differentiated according to a series of forces. And this enabled him to anticipate another key element of organology: the possible voluntarism of any organ. Writing of humans, Kielmeyer states that "[with] the Reason that arose in its organization [the human] gained the faculty that he has in common with the other animals to change the relation of the other forces arbitrarily, within certain boundaries."³⁹ This capacity or faculty gave humans the ability to instrumentalize other species to make room for themselves: "It is more than probable that [the human] will force many [species] to disappear completely, in order to make space for himself as an organ performing its own replace in the grand machine."⁴⁰ Thus the human became the organ of a violent but rational evolution, one which would sweep away entire species to make room for the faculties and organs of reason. The ability identified here would become an essential part of organology, as we shall now see.

The division of forces to different organs had been a problem in physiology at least since Haller, but it exercised Schelling especially strongly. Indeed, it was in this problematic that he saw the positive need to go beyond eclectic observations of nature and proceed to a systematic philosophy, a metaphysics, of that nature. The mission statement of Schelling's still-prospective comparative physiology (as quoted above) had read, in part: "Insofar as these organs performed each its own peculiar function, they had a *proper life* (*vita propria*)—but as far as the performance of this function was only possible inside this whole organism, it was only a *borrowed life*, and that is what it must be according to the concept of organization."[41] He would go on to call this division a duality of all organs, and it was the spur to his dialectical organology. Its source, however, was in the third of his named predecessors in the system of forces and comparative physiology: Blumenbach. Blumenbach's *Physiological Institutions* [*Institutiones physiologicae*] (1786, translated as *Anfangsgründe der Physiologie*, 1789) had separated between five forces at work in formation and preservation of organisms: *formative drive*, "contractility," irritability (*vis muscularis*), a proper function of individual *organ*-formations (*vita propria*), and sensibility (*vis nervea*).[42] The standout here is the *vita propria*, a crucial term in Enlightenment physiology that goes back to Albrecht von Haller. The sense of this term is the action of the functional part in and of itself, the way the organ moves according to its own arrangement and capacity. Lacking electrical apparatuses for experimentation, physiologists needed this concept to differentiate between different motive actions. Blumenbach assigns to the *vita propria* functions which cannot be attributed to simple contraction, excitation, or sensing (like the womb's efforts during labor, the excitability of the nipples, and the dropping of the testicles). The problematic he raises is precisely that which fascinated Schelling, and indeed, Schelling appears to be very close to Blumenbach's formulation: "But other than these still a fourth life-force comes into consideration, namely the particular life under which I grasp those forces which one perceives in particular organs [*quae singularibus quibusdam corporis partibus, peculiaribus functionibus destinatis*] specified for particular performances, and cannot obviously be brought under the previous classes."[43]

But if Schelling took the term *vita propria* from Blumenbach, he followed the 1789 translation by Eyerel in using the term "organ." *Einzelnen Organen* here translates *singularibus partibus corporis*, and the extended modifier in the German (*zu einzelnen Verrichtungen bestimmten*) translates the Latin needed to define "organ" (*peculiaribus functionibus destinatis*). This collates the problematic of the organ as it presented itself to Schelling. Caught up in a system of forces, the organ was the site of concretion of those forces, a flexible receptor of a various fields of potentiality referring to those forces. Further, organ was the *referendum* and *relatum* of the organized being as such. Its general determination was in that relation between forces in their concrete effects—matters are nothing but expressions of "living copulas," organs, bands[44]—and its particularity would have to be discovered by finding a consistent quality through which those effects were related to each other and to the larger system. The organ had a dual task, and a dual structure—but it would quickly come to have a triple task and structure in Schelling's work. For that, its electricity had to be taken into account.

Quality,[45] for Schelling in the 1790s, was increasingly electric.[46] His turn to electricity followed developments in experiments especially with "animal electricity."[47] In 1791, Luigi Galvani published the results of his experimentation with severed frogs' legs, believing he had discovered electrical energy—cast in general at the time as "fluid"—in the tissue of the animal. His discovery had an effect on the lettered spheres in Germany that can only be described in the metaphorical terms given to his name because of that very effect: the young generation was galvanized. The galvanic result, however—that is, the very existence of what is now called bioelectricity—remained a topic of scientific debate until the 1840s.[48] Alessandro Volta attacked the results of Galvani's experiments, claiming that the various metals used to produce the reaction were the source of the electrical impulse—"animal electricity" was a nonstarter, produced only as a result of the production of "heterogeneity" in metals. As Karl Rothschuh has shown, the reason for the confusion was partly an inability to produce a strong enough electric field to demonstrate the internal electrical current in the muscle-nerve complex. Particularly in the experiments undertaken by Alexander von Humboldt, however, both purely metal electricities and internally "animal" electricity were present.[49] Volta invented the battery on the basis of the former; Romanticism dug its heels into its scientific interests, trying to unite metaphysics with physics, biology, and now chemistry, on the basis of the latter.

Talk of "electric organs" extended back into Enlightenment science; the British scientists Henry Cavendish (1730–1810) and John Walsh (1726–1795) debated whether the torpedo fish was possessed of such "organs" in the 1770s. Galvani's discovery, however, intensified the physiological part of the debate, and when the controversy was at its most heated, in the 1790s, it dovetailed with the debates on life-force to produce a shift in the very concept of the living. Central to this shift was Alexander von Humboldt.

Volta had already directly disputed the possibility of an electrical organ[50] when Humboldt weighed in with a lengthy work on the "excited fibers of the nerves and muscles" in 1797. The experimental work he had done, he claimed, was meant to explore the difference between physical and organic matter—the very terms of the debate on life-force were at issue in the locality of the electric organ. In the two years since his publication of *The Genius of Rhodes*, Humboldt had turned his narrative into a scientific program. Having worked on this problem for some years, Humboldt now claimed to see an at least partial answer in the excited organ, which behaved much as a musical instrument's string did for the artist.[51] Humboldt showed that both "galvanic" and "electric" activities were possible—the interaction between muscle and nerve was itself a relation of local excitability. The question of where the excitability was located—whether it was flowing through or originally "in" the muscle or the nerve—became the central question. Haller's "irritability" was thus pinpointed for research. Humboldt dubbed the Galvanic potential "excitability" (*Erregbarkeit*), using the fashionable terminology of John Brown's medical system (on which more in chapter 6). A separate fluid—Galvanic fluid—carried impulses through the nerves and to the muscles. Crucially, the Galvanic response could occur only where a certain *nexus*, or polarity, of muscle and nerve—a pair of organs—was present.[52] The notion was that experiment could locate different forces in the organization of bodies. Schelling would adopt this notion but reverse it: the organ was necessary for the expression of any force, its specificity the basis of different kinds of matter, with animal life essentially electric.

Johann Wilhelm Ritter supplied the final step for Schelling, in a popular 1798 lecture entitled *Proof, That a Constant Galvanism Accompanies the Life-Process in the Kindgom of Animals*.[53] Heterogeneity of the conductors was necessary for both electric and Galvanic reactions, as Volta had realized early on in the debates. But Ritter was disatisifed with the polarity of muscle and nerve, and claimed that the nerve itself was polarized.[54] Partly through experimentation on his own senses—positive and negative electricity would produce polar tastes, acidic and alkaline, for example—Ritter showed the heterogeneity of the nerves. By experimenting on nerves that did not combine with muscles to produce motion, but led directly to sensation, Ritter narrowed the sense in which organs were electric. He concluded by construing that narrow sense as universal: both the anorganic realm and the animal world could, as long as they contained sufficient potential heterogeneity, produce electrical phenomena. The electrical organ, Ritter had implied, was itself in a state of polarity or indifference, waiting for an impulse. Such was the entire universe, the "great animal."[55]

Schelling adopted this model but extended it to the problem of cognition, casting life, sense, and ultimately reason, as we shall see, in terms of radical heterogeneity and the potential it generated. Certainly organs bound together the world of forces; and certainly organs were the operative locales of sense. But organs were also at the basis of rational cognition, of the mind in general, and of the possibility for construing and remaking the world. Because humans knew the world through syntheses of heterogeneous states, they made the world that way too.

Schelling had used the system of forces and the problem of their expression to reconfigure the notion of an organ. As we have seen, this organ was, in a first step, entirely general, following Herder's and Blumenbach's uses. It was the universal *referendum*, applicable to any situation of mereological closure. Those situations were, of course, legion for Schelling's concept of Nature. In fact, "Nature" itself was, as I have been arguing, only legible through the organological cipher. Organs—as they had for Leibniz—went all the way down. They were the ontological crux of a Critical *Naturphilosophie*, one which focuses on judgment in the absence of the assumption of other fundamental faculties. Schelling's move here required them to be dual (and eventually triple), dialectical doctrines he developed out of the textual complexes I have laid out so far in this chapter. As he particularized these organs, he electrified them. In doing so, he refined the notion of the natural organ and allowed for a mutual instrumentalization between philosophy and science.

In *Of the World Soul*, Schelling re-poses the problem of force, now as the problem of organs of all kinds, especially electrified organs.[56] The system is dynamic, but where should the force reside? Is it inside the organized body, or outside? Those who believe in the pure immanence of force in bodies are physiological materialists (the main object of polemic seems to me to be Reil), and for Schelling, they cannot answer the key question of the emergence of quality, or of the differentiation of forces from each other. Those who hold that force is entirely outside of bodies are chemical immaterialists, and their permanently dualist position cannot explain the interactions of forces and bodies.[57] For Schelling, neither position is acceptable, and only one which insists on both conditions—a simultaneous inner and outer relation of forces to bodies—can form the correct beginning for investigation. He moves immediately to the organ, which is the locus of that simultaneity. And his source combined with contemporary discourse puts it in turn immediately in the context of electricity, since Haller's notion of irritability was now being investigated in the terms of the debate on galvanism.[58]

In *Of the World Soul*, the particular and electrical organ is dual and bound to the question of irritability: "If *Haller* had conceived of a *construction* of the concept of *irritability*, then undoubtedly he would have seen that it is unthinkable without a *dualism of opposed principles*, and thus a *dualism of the organs of life*."[59] The organ[60] is the point of differentiation between inner and outer,[61] the point of manifestation of force. Without it, no Nature—and no experience. The *a priori* construction Schelling is seeking finds a terminological basis in the duality of the organ. In *On the World Soul*, "organ" is associated primarily with irritability (as we shall see in a moment, the term is transferred to *sensibility* in the *Outline*). The two forces are opposed to each other in a quasi-compensation system, as in Herder and Kielmeyer. The two are active and passive relations respectively to the dual (positive and negative) action of the World Soul itself: "The organs on which the *positive first cause of life continuously* and *immediately* works, these organs must be represented as *active*; those which, however, its only influences in *mediated* way (through the first), as *passive* organs (nerves and muscles)."[62] The internally divided first principle expresses itself in a particular order in the excitable and sensible organs, which, dual in themselves in order to receive the dual impulses which delimit their functions, together form a unity-in-duality. That dialectical unity obtains only for animals, and most particularly for humans. In a fundamental sense, the organs here are those that relate the most particular (and most complex) products of Nature to Nature's original organization. Their local functional capacities simply are the expression of force, the differentiation of which is a matter of organization, a matter of the internal arrangement of organs. Electric and cosmological at once, these are metaphysical organs in the most literal sense.

Electric and Ideal Organs   189

But they are also transcendental organs. Schelling makes this clear by explaining what he thinks Haller missed: "*Excitability is a synthetic concept*, it expresses a manifold of negative principles."[63] The organs which bear the weight of relating Nature to itself at the general and particular levels are implicated by a synthetic judgment which is a condition of that system. Schelling is concerned here to derive that system *a priori*, or rather, in keeping with my conclusion from above, to *make that judgment a priori*. The passage in which Schelling confronts and alters the Kantian notion of intellectual intuition (see above in this chapter) in fact follows this confrontation with Haller. The internally dual *organs* of irritability lie at the basis of the conception or organicism, and not *vice versa*. By starting with those judgmental and electrical organs, Schelling can conclude to a dialectical organicism:

> And such a dualism lies in the first principles of the philosophy of nature [*Naturphilosophie*]; for that only beings that belong to *a single physical species* are fertile with one another—which is the highest principle of all natural history . . . follow simply from the general fundamental proposition of dualism (which is confirmed in organic as in anorganic nature), that only between principles of *a single species is there real opposition*. Where there is no *unity of species*, there is also no *real opposition*, and where this is *no real opposition*, there is no *generative force*.[64]

As in the generation of animals, so in the most general and most particular organs: a dualism based on real repugnance, on the actuality of opposing forces expressed only in those organs, is the principle of Nature. The unity that runs through this passage, and the conception of Nature in the *Naturphilosophie* more generally, must itself ultimately be differentiated into specific unities that make a *third* for each such duality. That third is the capstone on the dialectic of Nature, and emerges in the *Outline*.[65]

In the most general sense, that third is already the unity proposed as *aether* in the *Ideas*, and metaphorically dubbed the *World Soul* in the work by that name. In order for the truly dialectical picture to emerge, however, it must be specified at different levels of organization. So, in the most general sense:

> Thus, the anorganic external world again presupposes another external world in relation to which it would be an inner. Now, since the activity of the original

organic being is aroused only by the antithetic activity of its external world, this is again itself sustained through an external activity (in relation to it). Together with the external world that it immediately opposes to itself the original organic being would then be again jointly opposed to a third, i.e., again mutually an inner in relation to a third outer.[66]

The liminal space between inner and outer is always organ, and this is no exception. Organs are simultaneously the separative and binding elements of natural systems. They form the (general and particular) third, the legitimation of *Naturphilosophie*: "The composition of organic matter proceeds to infinity because every organ organizes again into infinity, is again mixed and formed in a peculiar way, each distinguishing itself from the other by means of particular qualities.—But what is quality itself?"[67] Quality, as we have seen, is electricity. We can now add that it is electricity in organological expression. General electrical quality exists only through the mediation of the specified third between inner and outer, between force and product: the organ. And the locus of that organ's operation is now shifted from the excitable formation to the realm of sensibility.

Sensibility and the reproductive force are opposed to each other, as they were germinally in Herder, and doctrinally in Kielmeyer. Sensibility marks the freeing of the organs of the animal for the purposes of perception of the environment, and must be compensated by redirection of the productive force. It also opens up the animal to the other realms or orders—and it is from that openness that reason should emerge, on the basis of organs of sensibility. Because sensibility and the reproductive force form an opposition mediated by the already-dual irritability, it is easy to see why triplicity becomes Schelling's focus.[68] The triple structure emerges from dualities in single points of contact. Reproduction and sensibility both rely on triplicity, while irritability, which is dual, ties the two together into a larger triad. If we focus on sensibility (which is where organs occur in the text), we can see why Schelling's particular organs are paradigmatically those of (animal) electricity or galvanism.

For Schelling, there are three levels of interaction: the organ, the field of differentiated forces, and the single organic force itself.[69] The single force and the differences are simply the material fields of interlevel interaction according to the schema of the universe given above. The contemporary status of natural-scientific investigation allows only some certainty about these interactions. The organ, however, is the relay between the levels, and between any whole and any part: it makes the system work. And indeed, the electrical or more properly galvanic organ clarifies the relation of general to particular in this dialectical Nature.[70] This is because the establishment of qualities requires the interaction of general organs with particular organs. Nature as a whole—which ultimately includes human reason and allows that reason to operate successfully—must display, for the *Naturphilosoph*, the very connections that make *Naturphilosophie* possible. Schelling thus shifts the term "organ" from its physical meaning in previous writings on electricity to one in keeping with his view of Nature. The organ becomes the central third that makes natural development, but also cognition of Nature, possible.

The local organ of animal electricity is, then, triple, or it is the third in a triadic structure. The debate between Galvani and Volta had been on just this problem. Volta had proposed a dual structure or "principle of heterogeneity" for electrical phenomena, which he limited to metals. However, that principle could only operate with three bodies (two differently charged metals in a conducting medium).[71] The operation of this triadic excitation-system is clear for irritability, which reacts according to its dual nature. In the *Outline*, irritability is no longer the bearer of the problematic of life, of organs. This is because Schelling has now located the problematic of life in sensibility and its multiply triadic structure.

Sensibility crowns the triad of the animal order,[72] as we have seen, and it is expressed in one way through the triadic structure of the galvanic reaction. Yet it is also the third between the general and the specific organs: it is that which relates those two organs to each other. It is the organ of organs in the system of Nature.

An organ may have a specific function, but it can be put to either negative or positive use. Formation is clearly positive: it involves the combinative creation of a new being. Excitability (*Erregbarkeit*) for Schelling, is negative, because it is passive—it records shocks absorbed into the animal's body. *Sensibility*—and this is the breakthrough in the *Outline*—*combines these two*, synthesizes the two electrical poles of the organ's capacity. What should be impossible is really the foundation of the order (the organic order, here), and it is also what allows that order to interact with another one. This means, for this case, that the isolated and dumb forces of reproduction and muscle contraction, when they are combined by their simultaneity in the organ (in this case, the brain), make possible a reference to another order: the inorganic, and even the "general." In other words, it is the indifference of the organ to its positive and negative charges, or the combination of passivity and activity, that constitutes the beginnings of cognition as it emerges from animal life. This is in keeping with Haller: *sensibilitas* entails *representatio*; *Vorstellung* is nothing other than neutral states of electricity in the nerves. But the organ's duality is what allows it to refer to something else, that is, to an order of being that is not contained within the body: to the world. But in the very moment of reference, the knowing organ is identical with the known force from another order. Cognition is materially based, and we can cognize material (of any order) only through the fact of this structure, which is common to all the orders. Subject and object are identical both physically and ideally, and neither is reducible to the other.

Sensibility, for Schelling, disappears behind its appearance. It can only truly be known internally. The sensitive part of the reaction to Galvanic excitation can in no way be observed but only inferred in another being. The irritable reaction makes both reproduction (eventually) and sensibility possible, by providing initial movement and "protection" respectively.[73] Sensibility is thus an internally oppositional field of potential within certain irritable beings,[74] the organ of consciousness. Sensibility—this final stage of the organ before its ideality emerges from the "proof of idealism"—provides the final natural relay, the final organ of nature. It produces Nature—and as Schelling is thus done constructing that Nature, it produces the need for an organ of reflection on that construction, one which can ensure the initial identity of the production and the construction. Construction would have to rely on a mirror-organ (or more accurately, a system of *organa* and organs for philosophy), which Schelling came to articulate only in 1800.

Ideal Organs: Intellectual Intuition and the Organ-World

Schelling's organology came to full expression only in 1800, in his *System of Transcendental Idealism*. In this work, Schelling shifted the very terms of the antinomy freedom/necessity away from the subject/object divide and onto forms of synthesis that combined both freedom and necessity: art, and all human-made things that contained contradiction. His argument anticipated the cultural critique of technology that would arise in the middle of the nineteenth century, but cast the contradictions of *techne* in optimistic terms derived from the organological project.

Where Schelling had literalized the philosophico-biological organ of the eighteenth-century philosophical metaphorology, he transcendentalized the metaphor as it applied to human reason. His talk of a "spiritual *organ*" began as early as the *Ideas* (1797), and he developed it into a complex condition of human cognition in general. This transcendental use of the term allowed Schelling to revise his early commitment to Kant's canon of criticism, and to expand Idealism into a metaphysics based on an *organon* of reason (the philosophy of art) with the organs of philosophy (the I) and of that very philosophy of art (the genius). By 1800, Schelling articulated this program as a Critical, historical metaphysics. The Idealism that the *Naturphilosophie* was supposed to prove was complemented by the self-objectifying intellectual intuition of the transcendental philosopher, in whose possession Schelling found a tool that needed realization: the figure of the third, the *Band*, copula, or organ of simultaneous knowledge of the self and of the world. The question was only in what cognitive activity this ground could be brought to consciousness. Schelling answered this question with the notion that "aesthetic intuition" was the conscious version of the spiritual organ, and made the philosophy of art into the core of organological metaphysics. As it had for Hölderlin, art as a mode of cognition offered Schelling a way of conceiving a world within a world—and it was from the flexibility of the organ that he first conceived of a nonfit between an encapsulated world (and art object) and its encapsulating world (the universe). If the laws of the one did not correspond to those of the other, this could suggest the possibility of theoretical alteration of the given.

For Kant, the question of synthetic judgments *a priori* had been about a "third" which could perform the synthesis. In the case of empirical judgments for Kant, the third is simply empirical intuition: predication to a subject occurs in a verifiable sphere that can be checked (*is* that jacket blue?) at will. Where judgments both extend our knowledge and do not derive from empirically intuited contents, we need a reliable source of synthesis.

As we saw in chapter 2, for Kant the two sources of such synthesis are the *a priori* forms of intuition (space and time) and the categories. By the time Schelling's publishing career had hit its stride in the late 1790s, the possibility that a *single third*, a generalized basis of real cognition, could be found, had become the theme of a movement we now call German Idealism.

Schelling calls that third thing in cognition "organ": "What then is that secret band that ties our spirit to nature, or that hidden organ through which nature speaks to our spirit or our spirit to nature?"[75] This was indeed precisely the possibility that Kant had come to reject. In the *Amphibolies of Reflexive Concepts*, an appendix to the *Analytic of the Understanding*, Kant had attacked Leibnizian metaphysical concepts on the basis that they conflate the two sources of cognition. The dogmatism of the Leibniz school is here formulated as lack of attention to the boundaries between the sources of judgments. The attribution of intensive magnitude to monads (with, for example, the consequence that they are absolutely isolated) is a confusion of extension (based on the *a priori* form of space, thus an intuition) with pure intension (which is conferred on objects by the categories, and is thus ultimately a conceptual quality). Giving his architectonic one more Aristotelian name, Kant showed how Criticism could operate to elucidate the errors of former systems by securing conditionals for different types of assertions. Even the self, as Kant would go on to argue, is not both source and object of a single kind of knowledge. Nature (our nature) does not offer us even the traditional security of the Cartesian *ergo sum*. The self and nature both belong to the single plane of judgment, composed of categorically formed concept-intuition unities. The possibility of knowing either root of cognition nonjudgmentally is denied us. Kant writes:

> Observation and analysis of the appearances penetrate into what is inner in nature, and one cannot know how far this will go in time. Those transcendental questions, however, that go beyond nature, we will never be able to answer, even if all of nature is revealed to us, since it is never given to us to observe our own mind [*Gemüt*] with any other intuition than that of our inner sense. For in that lies mystery of origin of our sensibility. Its relation to an object, and what might be the transcendental ground of this unity, undoubtedly lie too deeply hidden for us, who know even ourselves only through inner sense, thus as appearance, to be able to use such an *unsuitable tool of investigation* [*ein so unschickliches Werkzeug*

unserer Nachforschung] to find out anything except always more appearances, even though we gladly investigate their non-sensible cause.[76]

Schelling's *hidden organ* is a direct answer to Kant's anthropological charge that the ultimate ground of sensibility's reference to objects is *hidden too deep* for our *unsuitable tool*. "Organ" is thus the (at least) definitional locus of a principled insight into the common root of cognition. A rather stringent definition of Criticism is at the basis of this hope. For Schelling, any notion of Nature external to our mind is simply nonexplanatory. We can posit natures and substances arbitrarily, but the emergence of the idea of such in consciousness is either left unexplained or simply dogmatically asserted, no matter how complex or sensitive the investigation into the apparatuses of that uptake. The possibility that we simply ideally confer Nature onto nature is similarly excluded. It is not the contingency of the assertion of determinate existence that we are after, Schelling tells us; rather, it is the necessary expression and even realization of the laws of our spirit *in nature* that we are seeking.[77] The transcendental sources of cognition—its conditions of possibility—remain, for Schelling, contingent just so long as we cannot know their necessity as the expression of laws emanating from their source. This source can be neither mere willful positing nor a world imprinting a ready-made consciousness. In one sense, this line of argument extends Kant's epigenesis metaphor. Knowing that there is a guiding form that emerges from the real combination of heterogeneous substances (kinds of cognition) is not enough. In the absence of lawful cognition of the organizing principle of that combination and subsequent development, any unity we can confer on that development is always arbitrary, like the use of a tool.[78] That principle, as we shall see, is an expanded sense of intuition as the nonconceptual contribution to cognition. Intuition is divided for Schelling into different stages (or "epochs") of knowledge and its construction. The substrate of cognition is thus a flexible "spiritual organ"[79] which defines human being, knowing, making: intuition on a scale from intellectual to most sensible.

This organ became, I am going to argue here, the emblem of Schelling's specific version of a Critical metaphysics or organology, a series of systems marked by perhaps the greatest sympathy with the risk of dogmatism. We can see this already in the dictum that Nature should be conceived as realizing the necessary laws of cognition. Schelling's position within the development of Idealism—the *problem* of Schelling—is simply this, that he is the most insistent that the Critical philosophy cannot be completed scientifically until it offers metaphysical knowledge as knowledge of being qua Nature. His willingness to confront the categorial systems of Kant and Fichte with that Nature in philosophical experiment is unparalleled among his contemporaries.[80] That risk is expressed as early as 1795, in the *Letters on Dogmatism and Criticism*. This work considers both systems valuable, but ultimately leaves no doubt about which possesses the more salutary methodology:

> Nothing appears to me to prove more strikingly how little most have grasped the *spirit* of the critique of pure reason, as that almost universal belief that the critique of pure reason merely belongs to a system, since clearly the specificity of a critique of reason must be this, that it favors exclusively *its* system, but instead rather either presents or at least prepares the canon for *all* systems. But there admittedly belongs to a canon of *all* systems as a necessary part a general methodology: but sadly nothing can counter such a work except if one takes the methodology which it presents for *all* systems for the system itself.[81]

This notion of canonicity for the system of Critique renames *not* the canon of the understanding, but the *discipline* that Kant attributed to his methodology. As we have seen, Kant turned that discipline into a canonical pedagogy, "disciplining" systems after 1785 that did not respect the central Critical insight. Here Schelling imagines something similar, but marks a certain space for less polemical interactions between systems. Criticism provides the measure of (all) other systems, but only methodologically. The process of amalgamating the various systems with the Critical methodology would be more than merely identification and exclusion. Indeed, what we have seen in the *Naturphilosophie*—the careful reading for philosophical content of a broad spectrum of natural scientists, including Kielmeyer and Blumenbach prominently—is something like an elaboration of this program.

If, in the passage quoted above, there is nevertheless some commitment to the "canon of the understanding" argument in Kant—the delimitation of claims to correspondence not with putative objects but with the faculties which first make those objects possible—this Kantian strain in Schelling is shifted backward, as it were, into the metaphorological register of the *organon* by 1800, in the *System*. And that recuperation of the *organon* in a canonical post-Kantian context combined with the organs of the *Naturphilosophie* to allow for a second articulation, after Hölderlin's, of organology. Organology made retroactive use of natural science in the service of a new metaphysics. Through an exploration of intellectual intuition and the nonfit between world and microworld of the artwork, that metaphysics opened onto a doctrine of subtle intervention in Nature, and ultimately onto a metapolitics of the pure deed (on which see chapter 7).

The introduction to the *System* informs us that Schelling is looking for an *organ* of transcendental philosophy. The task of the *System* (on which more in a moment) is in need of an instrument. In the final account, Schelling will state clearly that this *organ* is also itself in need of an organon, a word he might well have used in the introductory materials to any system. Instead, however, he allows the biological overtone to hover over his considerations of consciousness—we will see that this is hardly an accident. What, then, is this organ, and what is its *organon*?

The *System* begins from the premise that knowledge is the combination of a subjective and an objective element. Knowledge (*Wissen*) must be *of* something (its object); knowledge implies a knower (its subject). The mechanism that causes the correspondence—or sometimes "combination"—between the two elements is the famous *third* (as I indicated above) which, in its various systemic formulations, had been the leading theoretical question after Kant. Reinhold and Fichte had reduced this *third* to a single apparatus (representation; absolute self-identity), trying to solve the Kantian epigenesis problem by deducing experience from a single principle. As Manfred Frank has shown, by the time Fichte had arrived in Jena in 1794, a critique of this single-principle deductive form for philosophy was already well under way.[82] Schelling's contribution—on view in various ways and emerging piecemeal through his systems—was a combination of this foundational effort with a profound respect for the open texture of Criticism.

If there was to be a "single third," it would have to be available in some cognitive form. Schelling agreed with Fichte that this principle could not be a mere element of empirical consciousness, but also adhered to Kant's distinction between the "transcendent" (exceeding the bounds of cognition) and the "transcendental" (providing the conditions of that cognition).

He thus writes that "philosophy [is a] continuing history of self-consciousness, for which that which precipitates in experience merely serves as it were as a document and monument [*Denkmal*]."[83] The latter point is emphasized as a reformulation of Kant's search for that within experience which does not stem from experience.[84] The former point—that the system of idealism is a history of self-consciousness—opens already a distinctive Schellingian starting point in this constellation. True, Fichte had tried to provide a genetic account of self-consciousness, one which narrates the emergence of logic from the ground of the ability to identify the self ("A = A" is derived from "I = I"). And as we shall see, the figure of that ground was anything but simple identity. Yet Schelling's suggestion that a "running history" of self-consciousness is the generic basis of transcendental philosophy suggests both that the structure of self-identification is at least partly alterable or in development, and that the *System* itself will be the fulcrum of that alteration. Self-consciousness might take different forms, and its true determination might be given first in the activity of the philosopher (or of his reader). This conceptual attitude (which differs, I think, from Fichte's, at least up to 1800)[85] is textually spread over a series of organs.

A single but historical third, as the mechanism for producing correspondence between subjective and objective, is needed. The duality (subject and object) in each judgment allows, however, for two approaches to every cognitive act. If we start with the object—as we have seen above—we have *Naturphilosophie*, the goal of which is to prove idealism, to demonstrate consciousness in Nature. The perspective that starts with the subjective element of the judgment is that of transcendental philosophy. Its task will have to be the proof of the reality of consciousness. But since all knowledge—including self-knowledge—is divided into subject and object, transcendental philosophy will be that knowledge where the subject becomes its own object. Thus the proper realm of transcendental philosophy is only the space between judgments and our reflection on those judgments. All content must be considered only from within this perspective—for transcendental philosophy, the *I* is methodologically absolute. Following an invitation from Fichte, Schelling brackets any possible contribution from the objects of knowledge to inquire, initially, into knowledge as object.[86] For this investigation, there can be no question of "external" objects. There can be only the conscious activity of object-oriented reflection (whether the object is a "thing" or consciousness itself) and, Schelling now continues, an unconscious activity that constructs and construes the object-world. Since, within the frame of judgmental knowledge, there is only the subject and the object—and since these are merely the representing and the represented—the two activities cannot be attributed to an "outside" and an "inside," but only to differing levels of awareness. Thus the apparent externality of the object-world is, for transcendental philosophy, merely the unconscious content of the represented. Two activities are postulated, corresponding to two apparent worlds—the conscious and the unconscious, the subject and the object. The *System*'s perspective is absolutely subjective, but is open to its own completion in the absolutely objective response from *Naturphilosophie*. The object of the system is thus the subject, and its organ "the inner sense." For Kant, the inner sense is merely that which occurs only in time, not in both time and space. It is the locus of self-reflection, but is not otherwise privileged. The empirical self is presented there to the understanding, but this makes no difference to the latter's rules (its canon). For Schelling, following Fichte, there is more to it than that. The inner sense is the locus of coincidence of subject and object. There are still two selves, or more precisely, there is the self-intuiting-itself (subject) and the self-intuited-as-itself (subject as object). This structure is called *I* or self-consciousness, and its re-production (and ultimately objectification) is the task of the *System*. Schelling compares the inner sense—as the organ of philosophy—to the inner sense of Kant's mathematician. For Kant, mathematical objects are constructed fully (but according to the rules of the *a priori* forms of intuition)—we make, but not arbitrarily, the triangle with which we will perform our geometrical proofs. As Kant had it, we cannot construct the object-world (we can only make it possible in its form). Schelling points in the opposite direction: the object of the transcendental philosopher (the subject) is already the constructive activity itself. How far down that construction goes is left an open question for philosophy, but absolute for the *System*. The unconscious construction of phenomenal objects is in turn the object of the *System*—the organ of philosophy is the intuited activity, on a scale of consciousness from zero (object) to complete (self-consciousness). Schelling can thus write (and what follows in the present section will be devoted to an analysis of this central statement): "The objective world is merely the original, still unconscious poetry of the spirit: the general organon of philosophy ... *the philosophy of art.*"⁸⁷ The statement is very precise, despite its poeticizing sheen. I will summarize the argument that Schelling will make over the course of his exposition, and then treat it in detail. What produces the *organon* is a specific approach to the philosophy of art that houses deep consequences for the metaphysics suggested by the very phrase "general *organon* of philosophy." Philosophy itself will turn out to need an *organ*, one which, in its operation, is the very "poetry of the spirit" in turn serving as the *organon*. The organ of philosophy—which will shortly be named "intellectual intuition"—is thus made concrete only in the exposition of the system. The organ can serve as *organon* only in the guise of self-objectification in an aesthetic product. Art is, for Schelling, the objectification of infinitely productive rules. In one sense, these rules must be re-produced (mimesis). In another, they focus our attention on the moment where there can be no mimesis (the absolute contradiction of the subject as conscious and as unconscious). The reality of the poetry of the spirit must be proven in aesthetic cognition, crystallized in an art object, and exhibited as an invitation to the experience of another subject's internal cognitive contradiction. This exhibition then serves as the *organon* for all philosophy, as a general and unique way of operationalizing its organ (intellectual intuition). It also suggests, as we will now see, that the self-objectification offers a flexibility in the mimetic order that must border on ontological freedom, or the ability to alter the world metaphysically.

The complex self-intuiting-itself/self-as-intuited circumscribes the entirety of transcendental idealism. This is why Schelling writes that "intellectual intuition is the organ of all transcendental thinking."⁸⁸ The self's self-grasping cannot be of the order of an empirical judgment—it does not combine an empirical intuition and a concept. In fact, one of the activities it self-perceives is just that: judgment. The ability to attend to the emergence of judgments cannot itself be simply one more judgment. This line of thought, which is taken from Fichte, is explained particularly clearly by the latter. For every object of consciousness, there must be a subject. If this holds, however, for self-consciousness, we confront an infinite regress of self-subjects and self-objects, none of which can be called *I* in the sense of self-identification as both conscious and self-conscious. Because this requirement—that object-reference and subject-reflection should occur simultaneously in order to make the notion of *I* legible at all—cannot be grounded by the endless separations of subjects from subject-objects, it must take on recursive immediacy. The self must know itself to know objects and itself simultaneously,[89] and it must do that in an immediate sense.[90] Immediate knowledge, Fichte tells us, is called "intuition." Since this intuition is clearly not sensible, it must be intellectual intuition (a decision clearly not as innocent as its terminological derivation).

This structure remains essentially unchanged for the Schelling of the *System*. Indeed, as he dubs this structure the organ of philosophy, he reiterates the analogy to Kant's geometer. Intellectual intuition is the condition of possibility for transcendental philosophy, as the *a priori* forms of intuition are for the mathematician.[91] And just as the coordinates of ideal space are a postulate for the Euclidean geometer, intellectual intuition is a postulate for transcendental philosophy. It fulfills the condition—Schelling is simply interpreting Fichte—that the point of origin of speculation must be the point of indifference or coemergence of form and content, of logic and object.[92]

If Schelling differs from Fichte here, then it is in subtle textual inflections that allow his conclusion to raise this intellectual intuition to consciousness—Fichte nowhere implies that this is really possible—in aesthetic intuition. Those inflections are terminological borrowings from the *Naturphilosophie* ("product," "produce," "activity"). Thus, when Schelling writes that "the I is nothing other than *a producing become an object for itself*, that is, an intellectual intuiting," the stakes of the terminological choices are very high. For Schelling, the activity implied by *becoming*—self-objectification through systematic philosophizing—is just as essential as the recursive immediacy at the basis of the conception. That immediacy is anything but certain on the basis of Fichte's argument. The organ of philosophy must be operationalized, it must be recognized as a *Produzieren*. All that remains is to find the proper cognitive medium for the realization of that recursive immediacy, or of the conscious intuition of the self.

That medium is, of course, the philosophy of art. It is of the utmost importance not to confuse art and the philosophy of art. Just as with Hölderlin, we had to be careful to distinguish theoretical and generic discourses from the activities of artistic production and reception, so with Schelling we must be careful not to commit the error of making "art" the organ of philosophy *tout court*. Not every human must be a genius in order to be human, for Schelling; but every human must and can participate in the ontological self-realization of the *poetry of the spirit*. Note that *Poesie* is strictly separated from *Kunst*, the mere mechanical execution of construction. We might translate *Kunst* as *techne*, and note that Schelling's argument prepares the way for philosophies of technology which do not share the assumption that the critical contradiction present in "poetry" is not present in the products of the mechanical arts.

Intellectual intuition is the organ of philosophy; aesthetic intuition is intellectual intuition become objective; and the philosophy of art is therefore "the only true and eternal organon at the same time a document of philosophy [*das einzig wahre und ewige Organon zugleich und Dokument der Philosophie*]."[93] This famous statement does more than make aesthetic judgments constitutive. As Dieter Jähnig has shown, it makes aesthetics constitutive for philosophy.[94] It makes beauty a factor—or rather a keystone—in system building. We should not hurry to conclude that this is an "aestheticization" of philosophy—it certainly is not that in terms of exposition, as any reader of the *System* can attest. It is not that systematically, either: the *philosophy* of art—the reflective medium of art-reception—is the *organon*, not artworks or art-production. Art is the object of that organon, that is, it is the source of the rule book for metaphysics, not the book itself. Artistic production and artworks allow the *System* to link self-consciousness and object-consciousness into fully conscious, recursively immediate self-recognition in the philosophy of art. The organon operationalizes the organ—art makes the structure of the *I* manifest, objective, in short: cognitively available.

But look at the phrase again: "organ at the same time as a document of philosophy." The documentation of intellectual intuition, of human activity as such, is the world of artifacts, the objectifications of rational human endeavor in the physical world. Freedom and necessity are there *joined* in concrete terms, just as they are in the organs of *Naturphilosophie*. The "necessity" end of the antinomy of the antinomy is left partly with nature, with which workers in art and in the factory have to struggle. Schelling's philosophy tries to undermine that very necessity by displacing the pair necessity/freedom onto a world where they are already combined. That is the world we would call *techne*, and that is the world, for Schelling, in which we live. The possibility presented by the productive contradictions of art adumbrate the problems that arise when the products of the mechanical arts (technology) seem to take on their own contradictory life. Marx will inherit this problem from the possibility of organology, as we shall see.

If we return to the claim that the history of self-consciousness makes use of the precipitate of experience as a document, we can see, in a first step, why Schelling adds the mysterious ellipsis "at the same time a document [*zugleich und Dokument*]" to the famous phrase above. The philosophy of art will be the *organon* of metaphysics, concretizing (because making operate) the self's intellectual intuition. In doing this, it will serve not merely as the most general rule for philosophy, but also—as it had been for the pre-Critical Kant—as the singular, objective evidence, the documentation *a posteriori* (but only for the *I*) of that very foundation. The organon—art *for the I*, or philosophy as reflective of art as cognitive and material process—is the rule; the document is the preserved material evidence. Once the human makes art (really, once *techne* is introduced into the order of things), the world is fundamentally human—organological.

The organ is intellectual intuition, but is not concrete until it is objectivized (the goal of the entire system) in aesthetic intuition. Schelling reiterates his judgment that an absolute identity of subject and object is an intellectual intuition, and the "organ of all philosophy."[95] This organ is the means by which the first four parts of the *System* (principle and deduction of idealism, system of theoretical philosophy, system of practical philosophy, and teleology) are established. The conscious and unconscious activities of the *I* are first deduced from the principle of self-consciousness, and then opposed in theory (which examines the unconscious activity as its explicit content) and practice (which takes the conscious activity as its infinite task). Repeating the framework of the *Critique of the Power of Judgment* in the space of about thirty pages, Schelling turns to teleology as the first possibility of their conscious unification. The opposing tendencies—unconscious, object-constituting activity as the establishment of brute externality, and conscious, meaning-making activity as conferring conceptual *teloi* on that matter—combine to imply an awareness of the unity of the two activities. The production of Nature can be shown to be the result of unity the *I*'s opposed activities. But it can only be *shown* to be this—it cannot be immediately so known, or intuited.[96]

For that we must have recourse to art. The desideratum is "*the identity of the conscious and the consciousless in the I, and consciousness of this identity.*"[97] The world—*the consciousless*—had been the poetry of spirit. Schelling now separates out *Poesie* from *art*. The general activity of material production does not always contain the regulated blending of form and content that specifies art. Indeed, not any such combination can count for Schelling as art. Art must depict an infinite conflict (real opposition) finitely. The character of the work of art is the finite presentation of infinite contradiction. *Poesie* is the formation of a (micro) world, an integration of the general and the particular revelatory of the rule of their harmony (which here is infinite conflict or real opposition). The most basic contradiction, however, is that between conscious and unconscious activity in the *I*. Thus, art is the objectification of the self at its most primitive. It is the possibility of consciousness of the unity of consciousness and unconsciousness. The world is populated—haunted—by infinite reconciliation, spiritual satisfaction in material form. This makes the genius for aesthetics what the *I* is for philosophy.[98] That means nothing less, of course, than that the genius is the organ of the philosophy of art, itself the *organon* of philosophy. The microworld—seemingly mimetic—that the genius fashions poetically is actually revelatory of the infinite contradictory unity of the *I*. But it is so revelatory *only through the organon*. The instruments of philosophy—the *I* and the genius, constituting the world and the microworld qua work of art—are united only *for the I*, that is, in the mutual support of transcendental philosophy and aesthetics. The *organon* for metaphysics unites the general world-activity of the *I* with the particular world freely created by the genius. The *organon* is the link not merely between general and particular, but between general and particular organs. Because those organs make worlds, the *organon* is a candidate for the metaphysical task as such: the insight into the relation between encapsulated orders of rules and their things.

To return to the problem of the *phusis-techne* analogy: is it certain that, for Schelling, there is a fit between those general and particular worlds? I argue that there is not, but that this consequence is left partly unexplained in the *System*.

Schelling rejects the notion of an imitation of nature in the following terms: "Thus it is self-evidently clear what we should hold about the imitation of nature as a principle of art, since, so far from the merely accidentally beautiful nature giving art the rule, instead what art produces in its perfection is the principle and norm for the judgment of natural beauty."[99] The unities of Nature—so perfect as they were in the *Naturphilosophie*—cannot serve, for the transcendental philosopher in collusion with the philosopher of art, as the rule for artistic beauty. That beauty must make the infinite conflict not between Nature and its organs but that between self and world, between conscious and unconscious, explicit. That beauty is thus not a reflection of the rules of any nature. No order of things legislates art. Indeed, it is even suggested that the observation of nature (at least from an aesthetic standpoint) is ruled over by the all-too-human definition of beauty given here. It is not far from that determination to the possibility that the microworld of the work of art could have ontologically innovative capabilities.

If Schelling has swept away the *techne-physis* analogy, ridding art of mimesis and founding it in a higher unity based in a deeper contradiction, the result of his work up to 1800 is that *techne* and *physis* are always combined. The world is constituted as artifacts—organs and simultaneously *documents*—of the contradiction that humans are for nature. This essential synthesis, and the notion that we live in a world with contradictions that run as deep as the ones we are capable of producing, gives Schelling's philosophy the mark of relevance. What is distinctive about his philosophy is not the minor difference from other Idealists, that he is interested in the concrete, the material, in the dogmatic and the real. Manfred Frank has argued that this prereflexive moment, this excess of the real, is what links Schelling to Marx.[100] But this reading—at least for the young Schelling—does not tell the whole story. What appears as the reappearance of mystical elements, dark powers, the madness of genius, in these early writings, is instead an anticipation of the confrontation with an uncontrollable real that emerges and takes shape *as* the synthesis of freedom and necessity. Organs, Schelling's metaphysics suggests, might take on their own lives. And when they do, as we shall shortly see, they run roughshod over the revolutionary purposes for which they were invented, cancelling the distinction between poetry and technology. Schelling's version of organology provides a critical insight into the first moment of this historically essential separation. If we live in a world made of synthetic organs, the "real" is always a series of syntheses, and it is only in the vicissitudes and apparent finalities of these syntheses that we encounter history and nature. To confront the totality of those syntheses at any given moment is what Schelling and his Romantic cohort call "mythology," as chapters 6 and 7 demonstrate.

SIX

Universal Organs: Novalis's Romantic Organology

> Everything can become an experiment—everything an *organ*.
> —NOVALIS, *Notes for a Romantic Encyclopedia*

Everything can be made an organ.[1] The eighteenth century had produced the metaphor of the mind as animal body, developing according to the laws of embryogenesis, its parts to be used by the will. The implication, drawn by Hölderlin and then most provocatively by Novalis, was that the parts of the mind, its capacities, were organs. The analogy is strange on its surface: after all, there are no minds (yet) that are not *in* animal bodies. Nevertheless, the problem of embryogenesis has proven a fertile metaphorical ground in the history of epistemology. The story can be traced back to Aristotle. But it was at the end of Aristotelianism, when the new scholastics of the eighteenth century were urgently seeking the metaphysical rug Newton had pulled out from under them, that the odd metaphor, *the human mind is an animal body,* arose. Because Newton had provided a way to conceive of the heavens and earth according to quantifiable force-measurements, his method was adopted across the European continent. But it could not be adopted in medicine and the life sciences because "quantifiable" did not apply to the phenomena of the living. Two models arose for the explanation of embryogenesis: preformation, which held that a microscopic but already structured ("preformed") animal existed in the parents' seminal fluid; and epigenesis, which held that *un*quantifiable but real forces traced the pattern from anorganic seminal matter to the formed organism. This difference—the difference between mechanists and vitalists, in other words—was transferred into epistemological terms starting with Leibniz, who explicitly claimed both that tiny preformed animals existed in spermatic fluid, and that the rules for thinking preexisted actual thought in structure of the mind. The consistency of this metaphor—which, I repeat, is a tenuous one, since the metaphorical source is also the literal body of the metaphorical target—strikes the reader of eighteenth-century philosophy. By the end of the century, Kant and his former student Herder both stood on the side of epigenesis: somehow, the *mind developed* according to unquantifiable forces. It was neither preformed nor was it entirely "empty"—Kant would dismiss Locke's notion of the *tabula rasa* with a further biological analogy: an empty mind could never generate a thought, much as inanimate matter could not generate an animal.

Friedrich von Hardenberg, or Novalis, borrowed terminology from these debates in order to fundamentally overthrow the premises of thought that enable these analogies. In other words, I want to show that Novalis's use of the word "organ" to found Romantic metaphysics, or what I call "Romantic organology," is intended to dispense with models altogether. Romanticism is typically cast as siding with the "animal body" analogy, as having a deeply organic model of art, epistemology, science—in short, an organic metaphysics. The scholarship, and not only on Novalis, abounds in claims that the Romantics are "organicist" thinkers, that they use qualities of putatively organic process (like autonomy and holism) to construe other processes, especially that of art-production. Indeed, I think it's fair to say that the scholarship is undecided about whether the Romantics biologized aesthetics or aestheticized biology. Either way, though: they did something like that to everything else (literally). I disagree. The Romantics were certainly interested in the life sciences, but they wanted more than a model from them. Indeed, as I'm going to argue on the etymological and philological force of the single word "organ," their thinking—or at least that of Novalis—was not driven by models but by the problem of function.

Take the model often proposed as an indirect source of Novalis's (and Romanticism's) "organicism," that of John Brown's "revolution" in medicine.[2] Brown (1735–1788) was a Scottish physician whose legacy consists in the decisive break with humoral pathology, with its limited range of therapies—purging, enemas, bloodletting. This break came with the introduction of a new model of life. For Brown, life was a balance of forces, and disease was merely disequilibrium. Too great an excitation of the nerves led to "sthenic" diseases (smallpox, measles) involving a surfeit of force, too little to "asthenic" conditions (typhus fever, plague). Brown construed life as a balance between excitement and excitability (which became, in German, *Erregbarkeit*), establishing a scale of eighty points for each. As the one rose, the other fell, and the zero/eighty mark with no excitability, where the body cannot respond to any stimuli, was death.[3] The animal was mesodynamic, defined as the middling state or balance of excitative forces. As the historian of medicine Nelly Tsouyopoulos has argued, it is probable that this model of life gained recognition in the German states because it relied on a model that seemed analogous to Fichte's framing of consciousness.[4] Brown's popularizer in Germany was named Andreas Röschlaub (1768–1835), and his success went partly by way of convincing Schelling to adopt the system.[5] The wild popularity of Schelling's early writings had a massive and still historiographically undertreated influence on an entire generation of scientists. Röschlaub figured the body as a unit that was fundamentally passive, receptive of impulses that, insofar as they were actually received, became elements of an active system. The body received electrical impulses (or fluid, as it was still considered) that were irreducibly foreign in both form and material to the body. But this merely altered the active system that was the whole body itself. The excitability or receptivity of the body was housed in its unity as a sensing (and thus voluntary) complex of organs. Brown's quantity-based scale was perceived as reducing life to mechanism, but he had included a category of illness that made the point about organization clear enough. What he called "indirect debility (asthenia)" were diseases that rendered the patient insensible through too much excitement, and they represented the hardest task for the physician, who would have to increase excitability while decreasing excitement temporarily. Whether construed as mechanical or organic forces, this type of disease made the dividing line between body and environment clear, and pointed up their interaction as independent entities as a therapeutic problem. Thomas Broman writes that this made pathology "a branch of physiology,"[6] making diagnosis dependent on the understanding of the energetic balance in the body. This balance was no longer humoral,[7] and as we have seen with Schelling, it fit nicely into the emergent fascination with electrical phenomena in physiology. Excitation had to be measured for its general effects on the body to be understood, and that body's ability to maintain its balance was the revolutionary topic of the new medicine.

Tsouyopoulos's point is well taken: this model is precisely that proposed by Fichte for consciousness, which separates itself from external stimuli only to create the pure passivity needed to make pure activity meaningful—or, in Fichte's proprietary vocabulary: "The I as well as the Not-I are both posited through and in the I *as limiting each other mutually*, i.e. such that the reality of the one cancels the reality of the other, and vice versa."[8] Fichte does not here propose the source of the self-constituting differentiation of the I and the Not-I, but instead states only how consciousness is structured as the division between itself and another—as transcendental cognitive capacity. It is crucial to see that here no cause of consciousness is given, just as Brown had obviated the call to find a cause for life. The state of tension—"life is a forced state,"[9] Brown had written—is both what is and what must be studied, biologically and metaphysically. The living being incorporates whatever might excite it, calibrating the external shock and the internal dynamic state; consciousness constructs and re-presents whatever might be available to it, construing it according to a cascade of internal capacities, establishing cognitive homeostasis. Tsouyopoulos's suggestion is powerful: the two models are nearly identical. Add to this that the Romantic generation inherited a century of analogy between the animal body and the human mind, and the model of Romantic thought seems anticipated. Given Novalis's readings of Schelling, his extended notebooks on Fichte's philosophy, and his reserved enthusiasm for Brown's work,[10] we might expect this discursive matrix to be the source of Novalis's conception of the organ.

But it is not. The extraordinary fact is that neither Fichte nor Brown uses the term "organ" in any systematic way. The analogy between the animal body and the human mind, which Hölderlin had exploited for the purpose of a revolutionary tragic theory, and Schelling had sought to find in neutral electric states, came to Novalis as the analogy between two models, models which both suggested but did not explore their enabling conceptual mechanism, the organ. Novalis would isolate this element and universalize it, typifying the new approach to speculation called organology.

Friedrich von Hardenberg, who adopted the nom de plume Novalis ("he who cultivates new land") for the publication of his fragments, prose fiction, and poetry, wrote to Caroline Schlegel just two years before his early death: "Now I live entirely in the technical [*in der Technik*], because my years of apprenticeship are ending, and bourgeois life, with its many demands, comes ever closer."[11] Known even among the Romantics as a sentimental spirit,[12] Novalis had lost his beloved Sophie von Kühn two years before, and wrote often of *philo-sophie*, "love of Sophie." The picture that emerged was that of a frail and transcendentally focused poet. This image, however, is contradicted by Novalis's next biographical move: he took up serious study of the natural sciences in preparation for work as an overseer of mines, eventually of all mining in Turingia before his untimely death. The period, however, of "bourgeois life" was marked by a poetic and philosophical frenzy that produced two novel fragments, countless collections of fragments, and an attempt to refashion the "encyclopedia" in a Romantic vein.

Novalis brings the tension in the picture of his life to a point in the term *Technik*, which we often translate today as "technology." This meaning emerged in German around 1900; around 1800, *Technik* was not the artifacts that surround us, the devices that aid us, but instead the technical elements of any discipline, the "technique" that was necessary for any type of production. We might almost read *Technologie* for *Technik*, since the former was a discipline founded in 1777 by Johann Beckmann, a discipline that would allow the state to regulate the production sphere of society, or what we would now call the production economy. This was, after all, what Novalis was doing: educating himself about science in order to take on a regulatory role in the mines, to function as part of the state's apparatus for regulating the economic part of its society. That would be a "bourgeois life" indeed.

The passage, however, contains another determination: the years of apprenticeship result in a technical life; indeed, *Technik* emerges as the result of "years of apprenticeship." The causal flow of the the second clause retroactively determines *Technik* as the result of education, specifically formation (*Bildung*), the term commonly applied to Goethe's 1796 novel *Wilhelm Meister's Years of Apprenticeship*. Novalis's allusion is an obvious one for his reader, Caroline Schlegel. He stands at the very point the protagonist Wilhelm occupies at the beginning of book 8: his education is complete, and it is only now beginning. Goethe writes: "Wilhelm saw the nature as if through a new organ."[13] The metaphorical organ that allows Wilhelm to bridge the gap between the felt indeterminacy of his apprenticeship and his new orientation in life is recast by Novalis as *Technik*. His life is institutionalized—the Freiberg Mining Academy plays the role of the Tower Society for Wilhelm, showing him how he has arrived at this point and granting him the tools to proceed.[14] This blend of novelistic self-construal, regulatory and technical understanding, and interest in statecraft hardly seems to make up the profile of the most mystical of the Romantics. It does, however, provide the terminological and formal basis on which Novalis fashioned his last writings. These were attempts to make metaphysics and discipline, the universal and the concrete, combine to produce new forms, new literature, new media. Novalis understood himself as a practitioner of new forms of writing that might generate new forms of thought, nature, and government. Romantic organology became technical, indeed "technological" in his writings, as he sought to demonstrate the metaphysical depth and disciplinary breadth of the lexeme *organ*. His program remained fragmentary, but its impulse—the plastic synthesis of subjects and other objects in historical development at natural, social, and governmental levels—was the most complete statement of the organological program. Organology, as Novalis defined and practiced it, was not "organic" in the modern sense but *technical*, functional—it provided a wide range of tools (methodological, metaphysical, literary, regulatory) to revolutionize and administer the historical and natural worlds.

Where Hölderlin had inaugurated Romantic organology, introducing it as a new metaphysics of judgment with an ethics of tragedy, Schelling had given it classical form, retaining its physiological and logical provenances as he made natural science, aesthetics, and theology the instruments of philosophy. It was Novalis who would give the doctrine its name, as well as its most robust—and most deeply political—form. He would do this through similar discursive matrices to those of his predecessors. He would agree that there must be a new—as he put it—plastic metaphysics, true to Kant's Critical vision and capable of remaining open to the contingency of phenomena and their history. He would insist, with his friend Friedrich Schlegel, that this new metaphysics should be expressed in the form of a mythology, a new totality of beautiful propositions (indeed: a bible) that could wield political force. Thus organology retained its dual mission: the progressive unification of the universal and the particular (in judgment's reflexive and open system) and the potential alteration of the world itself, both as Nature and as history. Where Hölderlin had envisioned an ethical organ, and Schelling a metaphysically moral organ, Novalis would be most insistent on a political organ. His version of organology is, in this sense, more thoroughly expressive of its attachment to the post-Revolutionary circumstance than those of his peers. Novalis agreed with them and in some ways went further than they did on the point that a flexible metaphysics was the logical outcome of Kant's impulse to include analysis of the knower in all analyses of the known. He also agreed that the result of that inclusion was the possibility of organs of knowledge that coincided, partially and actively, with the known. Metaphysics, in order to become modern after the Copernican and French revolutions, had to become technical or *Technik*, had to gain efficacy in the epistemological and political welter of the waning eighteenth century. For Novalis, this efficacy meant that organology could not remain on the political sidelines. It would have to provide what he improbably termed an "intellectual intuition of the political," the very substantial link between knowledge and action implied by the combined etymologies of "organ." It would be able to do that only through the new medium Novalis would propose: the Romantic encyclopedia.

The Early Romantic project of a "new mythology" insisted on the possibility of radically new forms of cultural and political unity, forms that would be derived from the content of reason. This project thus had to be culturally totalizing, and its production would need to include revolutions in all aspects of that culture. For Novalis, working in the Jena atmosphere of Schelling's suggestions and Friedrich Schlegel's aesthetic program, several subprojects attached to the new mythology emerged in organological terms. First, in dialogue with Friedrich Schleiermacher, a new cosmology would have to be written. This cosmology makes the distance between Herder's quasi-organology and the Romantic doctrine most clear: the continuous and self-sufficient teleology of cosmological enjoyment proposed in the *Ideas* and even more explicitly in *God* was replaced with a world now defined as incomplete. Schleiermacher's revisionist anthropological theology served as a basis for Novalis to introduce a kind of hole into the graph of the universe, a discontinuity built into any attempt to spatialize knowledge of the world. Novalis often conceives of human knowledge as mathematical—organs are sometimes described as both differentials and integrals—or cartographic. These two types of spatialization of lawful cognition of the world present a flat space that the subject observes from the other side. A hole in the graph (my terminology—Schleiermacher speaks of a "hole" as a "negative revelation of the universe") serves to point to what Novalis conceives of as a constitutive inclusion of that observation in the construction of the graph. Like a curve with a single point missing, the universe is graphed or mapped by an included but really contradictory moment—as we shall see, the Critical organ. Cognition thus both includes the subject in the pictured world produced by that cognition, while marking that inclusion as contradictory. The organ is what separates subject from object, even as it serves to remind that there is a real interaction between them. That real interaction can only be rendered incompletely, because its object (cognition as form and content) is itself incomplete. From the pictured content of the world to its representation to the representer there is is a discontinuity or a hole. That hole is the human organ, the total possibility of human assimilation and alteration of nature. In keeping with the Aristotelian sense of *organon*, that possibility had to remain a possibility for this quasi-cosmology to obtain. The universe is interrupted by the human, and each functions as an incomplete cipher of the other. The metaphysics of the organ is one of permanent theoretical incompletion with paradoxically systematic pretensions.

That incomplete quality finds its philosophical home in Novalis's organological revision of the Critical project, which expresses the second subproject of the mythological impulse. By replacing Kant's term *Vermögen* (faculty) with "organ," Novalis provides the most explicit basis for a Criticism stripped of the finality of intuitions. This philosophy calls for a reflexive analysis of the organs of knowledge, rather than "faculties," in all cognitive endeavors. This allows Novalis to claim lawful specificity and potential innovation in the Critical investigation of any cognition, indeed any form of cognition.

"Organ" thus comes to designate both a cognitive act and its content (more strictly, its sphere of applicability). The term serves to help Novalis follow a suggestion from Lessing and Hemsterhuis to the effect that new organs could arise. These Critical organs—lawful specificities of knowledge in development—became the basis for the twin subprojects of the new mythology for which Novalis is perhaps best known: Romantic encyclopedics and the writing of a "new Bible." In both these cases, Novalis is in dialogue with *The Conflict of the Faculties*, insisting that these books will be organs of reconciliation between the metaphysically isolated Kantian university faculties. These books were thus meant to be *organa* of the new mythology, the human universe of metaphysical organs in developmental revision of their self-critical cognitive production. The generic constraint of encyclopedic writing functions as tragedy had for Hölderlin: it is the organ of intellectual intuition, the discipline of all disciplines, or the medium of metaphysics for the instability of the modern world.

This multifaceted project was political. Drawing on a combination of Fichte's and Hemsterhuis's moral writings, Novalis followed his peers in noting the necessary ethical ramifications of organology. If metaphysics were reestablished, it would have to be tied to doctrines of norms and action, as it had always been. Novalis insisted that this consequence could not remain innocent of an explicit politics. His writings on the "moral organ" for the legitimation of the political thus most explicitly tie organology to its social-historical context. Novalis was almost ostracized from the Jena group because of his controversial presentation of these views, but his vision of the politics of organology is the most daring. It makes clear that both laws and institutions are organs, and that their developmental status means, in a post-Revolutionary world, that they must be made to conform to the reason at the basis of the new mythology. The actual, for Novalis, is organological. In this proposition, Novalis anticipates the Young Hegelian interpretation of Hegel's dictum that the "actual is the rational"—the actual must be made to conform to the rational. Here we can see why it makes sense to call Romantic politics "radical": it insists on a change from the roots up. And yet organs are not only in the roots—they are at every level of the world and its history. Novalis's organology is, in this sense, "German Ideology": it makes critical demands on the present from the perspective of an ideal future. And yet its very terms—its very organ—are a mythological expression of the multitemporal predicament of a "world" caught between imminent institutional dissolution and the reifying syncopation of restorations. An organ thus infused with productive energies is the basis of modern metaphysical ideological construction.

Novalis's blend of novelistic self-construal, regulatory and technical understanding, and interest in statecraft hardly seems to make up the profile of the most mystical of the Romantics. It does, however, provide the terminological and formal basis on which Novalis fashioned his last writings. These were attempts to make metaphysics and discipline, the universal and the concrete, combine to produce new forms, new literature, new media. Novalis understood himself as a practitioner of new forms of writing that might generate new forms of thought, nature, and government. His program remained fragmentary, but its impulse—the plastic synthesis of subjects and objects in historical development at natural, social, and governmental levels—was the most complete statement of the organological program. Organology, as Novalis defined and practiced it, was not "organic" in the modern sense but *technical*, functional—it provided a wide range of tools (methodological, metaphysical, literary, regulatory) to revolutionize and administer the historical and natural worlds.

Mathematics, according to Novalis, is "perhaps nothing more than the *exotericized* force of the soul we call the understanding, made into an *object* and an *organ*." Mathematics is objectivized in the instruments of production. In the same fragment, he continues: "The *system of the sciences* should become a *symbolic body* (organ-system) of our innder side—our spirit should become a sensibly perceptible machine.... [T]he world is a *sensibly perceptible* imagination become a machine."[15] The machine on which the mind is modeled can only be a book—a comprehensive account of human faculties: an encyclopedia. Novalis's so-called *Romantic Encyclopedia*—the system of the sciences—has been characterized in almost as many ways as the bizarrerie of views it contains. Scholars have called it religious, semiotic, mystical, medial, whimsical. The passage I just quoted, however, provides entry to its basic problematic: the coincidence of function and material, or the *organ*.[16] The encyclopedia is a "system of organs" displaying not an animal's body but our mind; this mind, in turn, should become—through math, for example—a visible machine; and the world itself is one of our organs in this objectified state. In other words: the encyclopedia seems modeled on an animal body that is an analogy for a human mind that constitutes the world itself as a machine. Untangling this web is a typical kind of hermeneutic task for the interpreter of Novalis.

A Hole in the Graph of the Universe: Cosmological Organs

Friedrich Schlegel's "new mythology" was based on the claim that a religious feeling linking the subject to the universe, and poetry to philosophy, is also rooted in the organ of the imagination, with explicit reference to Friedrich Schleiermacher: "The understanding, says the author of the speeches on religion, knows only of the universe; where imagination reigns, there you have a God. Exactly right: imagination is the organ of the human for the divine."[17] Schleiermacher—and especially his theologically radical *Speeches on Religion for the Educated Amongst Its Despisers* (1799)—is the model for Schlegel's aesthetically focused version of organological mythology (see chapter 7 below). And where new *organs* of art, even *artificial organs* (*Kunstorgane*) were brought into view by Schlegel, Schleiermacher provided the theological organs for the "new religion" that would operate as part of the new mythology. Novalis would bind all these factors together into the fullest statement of Romantic organology.

For Schleiermacher, religion is a sense.[18] The positive content of philosophy, the sciences, and morals is subtended by the total human who binds them together in a more primitive, even more negative imaginative acitivity: the intuition of the universe.[19] This intuition, which makes up humanity's humanity, becomes the principle of communication between those of different religions, and indeed the basic principle of communication in general. The fundamental intuition of the universe is shaped by the organ, which is brought to life by an excitation (*Reiz*) from the universe—the play of the positive and the negative in the religious organ makes the basis for a true communication between believers of different faiths.[20] Since the introduction to religious intuition is different for each person, mutual communication and yet substantial (perhaps we could say organological) tolerance is possible: "But you would not be able to refuse respect and admiration to him whose organs are open to the universe and who, far from all conflict and contrast, is raised high above all striving, flooded with the influences of this universe and become one with it, if you observe him in this delightful moment of human existence, reflecting the divine rays unchanged back onto you."[21] This sense cannot be taught, just as art cannot, but its organ can be opened. And openness is the essential *desideratum* and even definition of that organ's activity, which is a mediator (*Mittler*) in the sense in which a priest is.[22] It produces the religious sense in general and—because it is objective—the positive doctrine of particular religions. And yet it also maintains a dialectic of positivity and negativity internal to its operation. It is passive, general, negative: it opens onto the universe. Yet it is also concrete, active, and positive: it is the substrate of the cosmos itself. Novalis would both adopt the term "organ" and elaborate its partly negative essence in debate with Schlegel about a "new religion."

Schleiermacher's religious organology has surprising Kantian roots, and surprising cosmological consequences—the latter will complete the picture of the discursive context into which Novalis placed his organology. In his early *Short Depiction* (undated, but between 1793 and 1799), written after reading Jacobi's book on Spinoza, Schleiermacher calls into question Kant's account of the infinite in the *Critique of Pure Reason*. His line of critique is a refinement of the notion that the thing-in-itself is a problem for Kant. Schleiermacher argues that the infinity within the faculties (which Kant locates as the basic formal characteristic of reason) cannot be claimed to be immanent and transcendent simultaneously: if it is in the faculties, Kant is a Spinozist; if not, he falls into his own category of dogmatism. The result, for Schleiermacher as for the others in the Jena circle, was that the question would have to be bracketed in favor of a continued application of a critique of the various areas of human cognitive endeavors. And Schleiermacher is thus led to religion as the object of a critical approach to the "infinite." This approach is characterized by the "essence" of religion: the interaction between the faculties of intuition and feeling.

Schleiermacher glosses one of the basic doctrines of the *Critique of Pure Reason* ("thoughts without content are empty, intuitions without concepts are blind")[23] as follows: "Intuition without feeling is nothing and can have neither the right origin nor the right force, feeling without intuition is also nothing: both are only then and therefore anything, when and because they are originally one and unseparated."[24] Schleiermacher, in insisting that the organ of the infinite is the interaction of intuition and feeling, may, however, be referring proximately to Fichte's own revision of Kant in the *Science of Knowledge*: "Intuition *sees*, but is *empty*; feeling *relates itself to reality*, but is *blind*."[25] Kant is simply stating the doctrine that all cognition takes the form of judgments. The proposition of a given judgment is successful when it unites the content of some intuition with the generality of some concept. Fichte's take on this passage comes in the practical philosophy of the *Science of Knowledge*. In striving infinitely to realize its true, absolute nature, the *I* builds upward from the apparent passivity of sensation (*Empfinding*). But in the realm of feeling, there is only blank relation to "reality"—feeling is blind but connected. Intuition, on the other hand, "sees." Fichte appears to be contradicting Kant, but he is really building in two directions simultaneously. Intuition sees because it is constructed in a descending categorical activity on the part of the theoretical ego, and marks a kind of threshold of conscious cognition as the practical ego emerges in the other direction. For Schleiermacher, the source of cognition is feeling—it connects us to reality, as it does for Fichte. Intuition, on the other hand, is not empty, but—as Novalis would say—plastic. If the two are to become one, they must do so by way of the infinitization of intuition. The name for that process, which underlies the religious "sense," is "organ."[26]

The religious organ is balanced between the consumptive and resignative drives: it is the field of potential poised between action and passion. The peace and poise that come with the operation of that organ are counterindicated by the speed and industriousness (not to speak of industrialization) of the contemporary world.[27] And yet sociability (*Geselligkeit*) proceeds precisely from this balance:[28] the play of the positive and the negative in the critical religious organs forms the basis for a truly social society. This is indeed, then, organology: the sense for the infinite turns into a theological social doctrine, a kind of post-Revolutionary metapolitics. But before I turn to Novalis's reception of this cocktail of mythology, religion, and aesthetics, I want to note the cosmological revisionism of the *Speeches*. This revision, which introduced a kind of incompleteness into the cosmos, forms the entry point into Novalis's organology.

The full radicality of Schleiermacher's vision emerges only when he defines the relationship between God and the universe. A religion "without God" can be better than one with, he writes controversially. God is, in one sense, simply the feeling, the manner of intuition that comes with the proper relationship of the human to the universe.[29] And this higher intuition intimates that the universe is like a work of art, harmonious internally, unified according to a law not expressed inside it, and therefore, in one sense, partial.[30] This partiality is reflected in our faculties as the divide between possible and actual lawfulness. A possible relation, in this sense, is a "negative revelation": "And even if you can think connections that you don't see, so also is this hole a *negative revelation of the universe*, an indication that in the required degree in the present temperature of the world this mixture is not possible, and your imagination is a view out past the current limits of humanity, a true divine inspiration, an unintentional and unconscious prophecy of what will be in the future."[31]

The universe is the totality of being and thought—it is complete, includes possibility. This possibility, as we will see emphasized in Novalis, includes the future. The current "temperature" of the world, however, is another matter. The world is only the set of the actual, only that part that is explored in the sense for the particular. Part of that actuality, however, is the totality of the human and of humanity. And because the human is in the world, but has a sense (a religious sense) for the universe, the world is incomplete. It is a discrete totality, or a nontotal entirety. The religious person is simply aware of this condition, and her feelings and even thoughts emerge from the social and intuitional organs of that negative space.[32]

The imagination, playing freely between the understanding and intuition, is made constitutive not of the world, but of the relation between the nonwhole of the world and the universe: of God. Schleiermacher clarifies:

> In religion the universe is viewed [*angeschaut*], it is posited as originally acting upon the human. And if your imagination depends on the consciousness of your freedom such that it cannot think that which it originally thinks as acting upon it in any other form than that of a free being; good, then it will personify the universe and you will have a God; if your imagination depends on the understanding, such that it always stands clear before your eyes that freedom only makes sense in the individual and for the individual [*Einzelne*]; good, then you will have a world and no God. I hope you will not take it as blasphemy, that belief in God depends on the orientation of the imagination; you will know that imagination is the highest and the most orginal part of the human, and other than it there is only reflection about it; you will know that your

imagination is what created the world for you, and that you cannot have a God without world.[33]

The human religious organ is dual, active and passive, positive and negative. It treats the world, which is a precipitate of its interaction with the universe, as a holy but subordinate construction. God is not in that world but is immanent to the religious sense for the incompleteness of that world. God, intuition, the religious organ: these represent a hole in the graph of the universe.

Novalis worked with this problem as he developed his cosmology and Bible project, both before and in dialogue with Schleiermacher. Novalis claims that "thinking is, like the *blossom*, certainly nothing but the finest *evolution* of the plastic forces.... [T]he organs of thought are the world-generating ... nature's genitals."[34] The universe is interrupted, for Novalis as for Schleiermacher, by a constitutive partiality. Novalis—more than Schleiermacher—articulates the part-object "world" as the precipitate of the organs of thought:[35] "This higher formation [*Bildung*] ... uses the world, which is a world just because it does not completely and *totally* determine itself—and thus still remains in many ways and directions determinable.... To the world belongs thus that which does not abs[olutely] completely determine itself ... and through this is disturbed and changed in its essence."[36] *Formation* becomes mutual—yet constitutively incomplete—through its organs. Those organs, in stark contrast to those of the Herderian system, are not placeholders for enjoyment. They are not merely relata, but determining actors in the interrupted universe. Where the *Ideas* had insisted on the complete teleological self-enjoyment of God's universe, the dialogue *God* was even more emphatic. In the final discussion, the revised Spinozan universe is gathered into a series of laws. Organic unities subsist and persist; they amalgamate themselves with similarities and resist the dissimilar; and they "imprint" themselves internally and on other beings through a process called "similarization" (*Verähnlichung*).[37] As the rules unfold, the properties of opposition are included in a plane of immanent organic unfolding. The *plenum* persists. The universe is infinite, includes God and the self-naming animal (the human), and is even temporal.[38] But it suffers no break.

Kant had, in his inaugural dissertation of 1770, defined a world as the result of synthesis: "In composito substantiali, quemadmodum analysis non terminator nisi parte quae non est totum, h.e. SIMPLICI, ita synthesis nonnisi toto quod non est pars, i.e. MUNDO."[39] Here we have another example of what I have called the despecified Critical system. Novalis strips away the specifics of the faculties, not to return to the noumenal metaphysics of the inaugural dissertation, but in the service of a construction that maintains its organological status on the edge of reality and ideality. It is literally in process, chiasmically extending to both total object (world) and total subject (spirit), but based in the systemic attitude that system is not representationally complete. It is only in such a system, for Novalis, that efficacy can come to be.[40] Subject and object are the graph of a single mathematical function.[41] The organs of creation are holes in the graph.[42]

The systemic incompleteness of the universe is reflected in the Romantic *organ*.[43] Novalis writes that organ into his fragments, generalizing the doctrine formally:

> Tools arm the human. One can indeed say, that the human understands how to produce a world, he merely lack the suitable apparatus, the relevant armature of the tools of his senses.The beginning is there. Thus the principle of a battleship lies in the idea of the ship's architect, who is able to embody this thought through masses of humans and suitable tools and materials, by as it were through all this making himself into a monumental machine [*zu einer ungeheuren Maschine*]. And thus the idea often suddenly demanded monumental organs, monumental masses of matters, and the human is therefore, where not *actu*, in any case *potentia* a creator.[44]

The martial organs of this anthropological textual shard include literal tools, the senses, and the possibility of new organs (on which more below) out of the terrible grandeur of material. The human is *potentia* creator, and as we shall see, this state of *potentia* is anything but a mere adjective. The form of the fragment reflects the organological program—the compact sentences filled with strained, hyperbolic vocabulary describe the doctrine of construction (the engineer's ship). Just as construction needs a principle that could make it more than the *techne* of machine building, that could make it potentially active in the real metaphysical universe, so the fragment needs an idea which is not stated in it. This fragment—like Hölderlin's *Empedocles*—is an invitation to construction of its historical and philosophical content. The form of the fragment, combined here with its organological message, invites the reader to adopt the heuristics of the Romantic viewpoint, to consider Romanticism as the attempt at and simultaneously the object of a historical construction that might exceed its discursive bounds. If the result of the construction is unanticipated, this is because the real here is intentionally left in reserve, precisely because it is identified with the possibility that produces it. Novalis places us at the point of production of the real where it can still become, where it is not determined but determinable. The fragment is the formal consequence of the Romantic revision of Kantian Criticism—that is, organology—because it concretizes forms of cognition in a formal container open to revision at both the formal and content levels.[45] It is the writing of the history of the *a priori*, of the organs of metaphysical and ultimately political possibility.

The same view is expressed in cartographic terms by the third traveler in perhaps the most organological of all aesthetic productions, *The Apprentices at Sais*.[46] Beginning with an admiring plaint of an alienated apprentice in a hermetic school of nature, the explicit goal of which is to "lift the veil of Isis," the narrative becomes a chaos of voices. The second section, entitled laconically "Nature," begins with a depiction of human relations to nature, not unlike that of Gotthilf Heinrich Schubert's lectures *Views from the Night-Side of Natural Science*, or Schleiermacher's myths about the origins of religion. This devolves into a discussion between the apprentices on the uses of and attitudes toward nature, which is followed by the arrival of a "playmate" who recounts the new mythological fairy tale of Hyazinth and Rosenblütchen. This tale repeats the structure of the entire narrative's desire. It begins with naïve original relation to love, continues with fascination and alienation with another (intellectual development, presented by an old man who passes through their hometown), and after long searching, finishes with the lifting of the veil of Isis to reveal—Rosenblütchen. (In fragments, sometimes it is the self rather than Rosenblütchen who is revealed.)[47] The apprentices leave the hall where they are gathered, and a conversation between the forces of nature they have conjured ensues. The hall is then occupied by a group of travelers, who carry on yet another conversation about the nature of nature. They are then included in the group of apprentices for a final speech by the master—yet another voice in the seeming infinity of nature myths (*the human goes multiple ways* . . .).

Nearly every voice in the fragment includes a consideration of instrumentality. The third of the travelers points to the arbitrariness of any "net" that is thrown into the sea of nature. The nonfit between system and experience is the condition for investigation. The voice of third traveler then switches to the cartographic metaphor:

> Do you not think that it will be precisely the well-executed systems from which the future geographer of nature will take the data for his great map of nature? He will compare them, and this comparison will first let us get to know the strange country. But the knowledge of nature will still be infinitely different from its interpretation. The actual decoder will maybe be able to get so far as to set in motion many forces of nature for the production of great and useful appearances, he will be able to fantasize on nature as if on a huge instrument, and nevertheless he will not understand nature.[48]

The experimental method makes nature an instrument, and recognition goes systemically missing in its mapping endeavors. And yet the preparation of instrumental data returns, in the second speech by the third traveler, in a more positive light. The map becomes an artifact of the retrospectively real, ontological work of the cartographer.[49] His work is rooted in that of the artist, whose perspective is that of the organologist: "The artist should best emphasize activity, for his essence is doing and producing with knowledge and will, and his art is to use his tool for everything, to be able to create an image of that world, and for this reason the principle of his world is activity, and his world is his art."[50] The universe is a map, or perhaps more precisely a graph,[51] that functions and is known systematically to be nonwhole. This critical differentiation—the infinite yet discrete curve—is where the *organ* finds its home. The nameless voice (the narrator) of the beginning of the second part might be identified most strongly with the cosmology of this critical graphing. He states, perhaps most authoritatively:

> Even earlier one finds, instead of scientific explanations, fairy tales and poems full of strange imagistic characteristics, humans, gods, and animals as communal master workman, and hears the emergence of the world described in the most natural way. One hears at least the certainty of an accidental, *tool-like* [*werkzeuglich*] origin of the same, and even for the despiser of the unregulated products of the imagination this representation is meaningful enough.[52]

The world comes to be as a tool. The origin is, then, the organ, real and ideal, that makes this instrumentalization possible. To do that, the world and the subject must be produced by an incompleteness in the universe. The cosmos is organological.

The *Romantic Encyclopedia* relates, under the heading "COSMOL[OGY]," that the organ is the integral and simultaneous differential of the opposed tendencies of the inner and the outer. It homogenizes and separates, isolates and causes interaction. For these two opposed techniques in constructing a graph of the cosmos, it is "its general *function*."[53] The organ thus makes a zero-point—its apparent unification is always negative, because its essence is contradictory. "*Mediatedness*," writes Novalis, "*strengthens immediacy* on the other side . . . as organs touch, souls harmonize."[54] The organs of the universe's incompleteness are also the organs of human souls—in fact, the soul's organ, as we will see, is the very contradictory function that seeds the universe with incompleteness. It is the critical organ, the systemically necessary split between knower and known. Novalis will not give up on this partially Kantian point, but as we have just seen, he will literally construct disciplines from its negativity, turning the canon of the understanding into the organ of a revised Criticism.

Much of Novalis's writing can be understood from this vantage point. His projects of a Romantic encyclopedia and a new Bible are—as we shall see—both elaborations of this critical moment. Ultimately, the "moral organ" he borrowed and developed from Hemsterhuis also rests on this basis, as do the political organs which alienated him from the Jena circle. The encyclopedic project intersected with the biblical project in just the moment of Novalis's explicit rejection of the specific version of Criticism that Kant had defended in the 1780s. In other words, the basis for Novalis's organology was a critical organ.

From the Senses to the Bible: Critical and Encyclopedic Organs

Kant had rejected "organs of reason," and had even gone so far as to construe the organs of the professional faculties at the university as mere guides for practice, with no metaphysical ramifications. The Bible, the rule book for physicians, and the legal code were "organs" of a different sort: mere tools to orient practitioners. And when it came to the medical faculty, Kant chose to write not of a metaphysically contentious conflict but of the area of legitimate sharing of knowledge that could occur between philosophers and physicians. He framed the final chapter of *The Conflict of the Faculties* as a response to a work by Christoph Hufeland, the most influential opponent of the Brownian revolution. Kant might have chosen his response to Soemmerring (by the time of final publication in 1798) as an example of the medical faculty overstepping its bounds. But instead, he chose to focus on the morals of physical care for the body, of the way in which the mind is influenced by the use of that body, and on the prevention of the waning of life-force that is called aging. He gave a short disquisition, in other words, on dietetics. The very decision to do so shows the difference Novalis would work out in his reading of Kant on the backdrop of the shift in medical discourse. Hufeland, sometimes called an "eclectic" or a "historical" practitioner, insisted on the bedside wisdom of the old methods, even as Brown's doctrines exploded in popularity. In 1798, he wrote a scathing attack on the Brownian system, defending the recommendations he had given on the lengthening of life through the protection of life-force the year before. Kant yoked this program to his own sense of navigating the aging body, providing a set of guiding notions that could, in principle, be collected as an "organon" for living in a body—perhaps he had his *Anthropology from a Pragmatic Standpoint* of the same year in mind. In any case, no organ of reason could exceed the pragmatic guidances offered here. Novalis would pick precisely this bone with Kant, marshaling both the "organ of reason" precluded in the *Critique of Pure Reason* and the "organs" of practice in the *Conflict*. And if Novalis was interested in the medical and epistemological advances made by Brown and Fichte, he also had his own speculative program, one limited neither by the finitude of the understanding nor the pitfalls of models. In the last four years of his life, Novalis produced a sustained meditation on the term "organ," and what emerged was an entirely new type of speculation, the full-fledged version of Romantic organology.

The narration—which is not ascribed to a single voice—of the beginning of the second part of the *Apprentices* contains an anticipation of Gotthilf Heinrich Schubert's idea that the observations of primordial humanity are an essential part of the history of nature itself, and closely resembles the opening of Schlegel's *Dialogue on Poetry*.[55] Indeed, the organological origin of the world—its necessary production of tools as its very nature—makes humans in turn the instruments of perception: "Thus we can consider the thoughts of our forefathers about the things in the world as a necessary product, of a self-representation [*Selbstabbildung*] of the then-current state of earthly nature, and we can read in them, as the *most suitable tools* [*den schicklichsten Werkzeugen*] for the observation of the universe, the main state of the same, the then-current relation to its *inhabitants*, and of its inhabitants to it."[56] For Kant, our cognitive tools were too deeply hidden and ultimately incapable of providing rational insight into the ground of cognition; for Schelling, the "hidden" organ was not too deep, but lay in the incomplete but complementary methods of *Naturphilosophie* and transcendental idealism. Novalis targets not the hiddenness of the organ but its capacity, its technical failure for Kant. The organs of those primordial observers, themselves tools of observation, stand in for a technical capacity of the organ which Novalis would develop in a vast discourse of the organ. This term runs throughout his work, as we have already begun to see. In the last years of his life, it became the explicit topic of conceptual and mythological concern—indeed, it demonstrates the link between Novalis's philosophical and cultural-technological projects. The hole in the universe makes the Critical-encyclopedic project possible, and makes its characterization as a Bible necessary. Those projects are supported by a "Critique of organs" in which organs are fungible, historical, new. The possibility of such organs—the possibility of possibility, as we will see—makes radical politics in an organological vein necessary.

The entirety of this project is sketched in a series of extraordinary letters between Novalis and Friedrich Schlegel from the fall of 1798 to the spring of 1799. The letters are the best surviving example of what the friends called *symphilosophy*. As such, they attest to the breadth of the project, which included Schlegel's novel *Lucinde*, Novalis's encyclopedic and biblical efforts, Schlegel's own notion of a Bible—all on the basis of a new philosophy.

The central statement is Novalis's response to Schlegel's mention of a reiteration in the footsteps of Muhammad and Luther, literature as Bible:

You write of your Bible project and I too have fallen on the idea *of the Bible* in my study of science in general and its body, the *book*—the Bible—the *ideal of any possible book*. The theory of the Bible develops, gives the theory of authorship or wordsmithing in general—which simultaneously yields the symbolic, indirect doctrine of construction of the creating spirit [*Constructionslehre des schaffenden Geistes*]. You will see from the letter to the sister-in-law [Caroline Schlegel, married at the time of writing to Friedrich's brother August Wilhelm], that a comprehensive work is occupying me, absorbing all my activity this winter. This should become nothing other than a bible project—an attempt of a universal method of biblicizing—the introduction to a true encyclopedics. I am thinking here to generate truths and *ideas in the grand sense—genial* thoughts—to produce a living, scientific organon—and through this syncritical politics of intelligence to break ground for myself to *true praxis*—the veritable process of reunion.[57]

The Book[58] is a function, and Novalis wants to offer its Critique. The introduction or propadeutic to the New Bible would be the creation of a universal method: an *organon* that would offer—perhaps surprisingly—two only slightly distinct objects, an encyclopedia and a Bible.[59] The encyclopedia adopts the Critical function of the Bible; what would be a "new bible" (and perhaps persists in the "spirituals songs") becomes a medium for any possible speculation, the generic constraint on all metaphysics—as we shall see. The quoted passage demonstrates the compact presentation of the entirety of the project. The proposed project moves extensively toward a new religion, even as it seeks to deepen its approach on the basis of a doctrine of spiritual construction tied to organological metaphysics.[60] This project then becomes the basis for a "syncritical politics of intelligence" (*Intelligenz*—which seems to mean both "intelligentsia" and "intelligence"). The connection, in the context of the encyclopedia, to syncretism is clear. "Syncriticism" would then be a collection of disciplines—this is how Novalis imagines the encyclopedia—connecting concrete organs to metaphysics and critique. He writes of this collective discipline as a "groundbreaking to true praxis . . . a perfecting of *intellectual tools*."[61] The Bible and encyclopedia projects seem to run together, each containing a generalized Critique followed by content. In each case, too, the content will be treated substantially from the perspective of the Critical organ. This is why Novalis takes issue with the demotion of organon in the *Conflict*. For him, the critical container will be the Bible, indeed the Bible will be an organ of Critique or an *organon* for metaphysics. Novalis takes direct aim at Kant: "Kant's Conflict of the Faculties is a nice specimen of lawyering—a web of fine chicanery. Kant has become, as you fault Leibniz for, *juridical*— ... the phil[osophical] faculty, like the most severe sinner, is to be defended. The phil[osophical] depiction of this conflict would have been the most beautiful defense of the phil[osophical] faculty. Kant is, with respect to the Bible, not *à la hauteur*."[62] In the intention to build an encyclopedia, Novalis marks Kantian Discipline as his point of reference. Kant misunderstands something about the Bible, he cuts through its possibilities too quickly, indeed legalistically. Novalis goes so far as to accuse Kant's treatment of the university faculties of being nonphilosophical.

The disagreement is profound. Novalis and Schlegel are both working with Lessing's *Education of the Human Race*, which had claimed that a new "gospel of Reason" must be written. Indeed, Lessing had gone so far in that work as to say that revelation was not rational, but that the revealed was given to humans in order to be made rational. That formula of intentional secular reduction—preserving, as it does, the now disputed content of dogma—is the precursor to Schleiermacher's play of negative and positive as much as of the New Bible project.[63] Schlegel's nostalgia for Lessing—if only he were still alive, much of the project would not be necessary—makes up the one axis of this complex. The other is Novalis's engagement with Kant, which, in spite of his uncharacteristically negative designations, is terminologically deep.

We have seen the development of disciplinary organs in Kant's *Conflict of the Faculties*. Novalis singles out the demotion not of the theological faculty but of the organa of the faculties in general. Indeed, the accusation of legalism is tied to just this terminological shift. Novalis's dual project—encyclopedia and Bible—is an *organon* in the sense Kant had abandoned in the 1770s. This designation is not accidental—it is a reaction to a close reading of the *Conflict*.

The specific conflict between the theological and the philosophical faculty—the "legal" version of this conflict[64]—is that between church-faith and rational faith.[65] This division is in the Bible itself, according to Kant, but only in the way in which it is read by the different faculties. The philosophical reading, which insists on the rational kernel of moral teachings, is the canon of religion in general. The higher faculty's reading, which is based on revelation, and focuses on the content of the stories, is the *organon* of religion. The *Conflict* had demoted *organon* to an "aid" or "help" for practice in each higher faculty. The Bible is precisely this demoted tool, insofar as it is read by the higher faculty. It is the source of dogma, which finds its disciplinary home in the pulpit. Kant glosses *organon* with the even more deflationary *Vehikel*, which is repeated throughout the text.[66] The legalism that Novalis detects is not in the separation of the faculties, but in the disciplinary rejection of their substantial and political disagreement. Novalis agreed with Schleiermacher that substantial disagreement could be both politically and religiously salutary.

"Much must be remembered against Kant['s] Conflict o[f] t[he] Fac[ulties]" (fragment 782)—for Novalis, as a practitioner of the new mythology trying to create a new Book, the restrictions the *Conflict* terminologically placed on communication between the faculties could not be normatively accepted. And, in fact, the hermeneutic and disciplinary problems presented in this debate went, for him, to the very heart of the Critical project. With Schlegel and Schleiermacher in dialogue, Novalis specifies his complaint against the *Conflict*:

> Religious doctrine is scient[ific] poesie. Poesie is among t[he] feelings [*Empfindungen*]—what phil[osophy] is with respect to thoughts.
> (Self-thought—*self-feeling*.)
> Religon is synth[esis] of feeling and thought or knowledge
> Rel[igious] doctrine is thus a synthesis of
> Poetics and philosophics.
> Here true *dogmata* emerge—true principles of experience, i.e. from princ[iples] of reason (direct)—philosophemes—and from principles of belief (indirect)—poemes [*Poëmen*]—truly composed—not mut[ually] limited, but instead rather mut[ually] strengthened and extended principles.
> (Reason is directly a poet [Poët]—directly productive imag[ination]–*belief* is indirectly poet indirectly productive imag[ination].)[67]

The poet writes *dogma* from the productive imagination of reason (which constitutes the world) and the indirect production of belief. The "self-thought" of reason must remain in contact with the "self-sensation" of feeling, in which religious feeling emerges. In one sense, religion unites poetry in philosophy; in another, poetry is the container of actual *dogmata*. Derived from *dokein* ("to seem [to one]," but also "to think"), Kant's excluded word becomes the temporary crystallization of experience, the expressive moment between the self-reflexive systems of reason and of feeling. It seems that *dogmata* are simply propositions that combine these two: in short, synthetic judgments *a priori*. The metaphysics of everyday judgment is built into the heart of the new religion.

The producer of *dogmata* will turn out to be the organ, nestled in the definitional core of the *organon*, Bible and encyclopedia. Novalis writes, just before the above quotation: "Does mysticism kill reason?—Kant means dogmatism—dogmatism is *relation-cancelling* etc. activity or inactivity" [Dogmatism ist *Verhältnißaufhebende* etc. Thätigkeit oder Unthätigkeit].[68] This unusual definition of dogmatism after Kant contains the whole vision of the organon for metaphysics. The larger difference between Kant and Novalis on the score of mysticism highlights the difference in their Bible teachings; the central adjective "relation-canceling" the difference in their conceptions of Critique. What relation would be cancelled in dogmatic thought?

As we have seen, Kant's Bible is hermeneutically divided between faculties, and their communication is a "legal" battle over political pragmatics. The Bible is a mere ecclesiastical vehicle. It is, however, the document of the canon, too: it contains the truths of reason and of reason's belief, which the philosophical faculty must unearth. Novalis's Bible is a function, and an object. It is a book, containing a record of real dogmata, a history of organological Nature. It thus both contains the historical truth and is irrelevant otherwise. Its own history, as Book, makes up its true worth. Not the content—not even the rational kernel we might and should discover by reading the Bible—but the collocation and cultural force of the Bible should interest, for Novalis. The Bible is simply the total container of teachings—*dogmata*, results—and must therefore be reproduced. The assumption is that *true dogmata*, philosophy itself, will change over time. The Bible must reflect the dual histories of self-sensation and self-thought.

If these histories are synthetic judgments *a priori*, why is "dogmatism" characterized as "relation-cancelling"? It seems to me that "relation-cancelling" applies adjectivally only to "activity," making "inactivity" equal to "relation-cancelling activity"—dogma is the reduction of cognitive activity, not merely a polemical word but a sense of which parts of the organ of reason are at rest, and which in active function. We must produce dogma, but this cannot fully cancel the relational activity of the organ of reason.

But what relation is meant? Dogmatism is, according to Kant, the doctrines of reason without a previous critique of the faculties. The relation cancelled in dogmatic thought is thus that between Critique and philosophy. Philosophy by itself is the production of *dogmata* without the built-in assurance of their source and location in the system of knowledge. Novalis's reading of this problem agrees with the basic gesture: activity that cancels the self-reflexive examination of cognition in its very acts is actually not activity. *True dogmata* are those which are produced with critical energy, with simultaneous attention to form (thought) and content (feeling, likely here by Schleiermacher's lights)—*or* they are those teachings which are subjected retroactively to such critique with sensitivity to the historical circumstance of both parts of cognition. That is, the Bible is the totality of propositions produced by a certain doctrine of judgment itself, a certain moment in the hylomorphy of human cognition. Their truth is in their witnessing, as the *Apprentices* has it on my reading above, of the condition of the constructed world, natural as well as social, at a certain time. The critique of the Bible is thus an *organon* for reading itself—this is why it overlaps with the encyclopedic project—or for the evaluation of judgments as producing the relation between self-knowledge and knowledge. If that relation also produces the nonwhole world (and we shall see that it does), then the Book is an *organon* for metaphysics.

The bearer of such historically critical *dogmata* cannot be the proposition, which, by Kant's own admission, is not in itself critical. The name for that bearer is *organ*. Novalis transcribes from the *Critique of Pure Reason* in his *Kant-Studies*: "Dogmatism on the other hand is the dogmatic procedure of pure reas[on] without a *previous critique of its organ*."[69] Either Novalis has here confused this with later passages,[70] or "organ" simply makes more sense to him as a characterization of the parts of intelligence. In any case, Kant had actually written that "dogmatism is thus the dogmatic procedure of pure reason *without a previous critique of its own faculty [Vermögen]*."[71] The shift from *faculty* to *organ* reflects the program of organology. By making Critique a question of organs, fundamentally in development as judgmental combinations of the physical and the spiritual, Novalis insists on the history of possibility, indeed on the possibility of possibility. He comments on his transcription:

> /*Critique* is education for the organ of reason, through directing attention to this organ—through critique, one acquires for oneself a *secure sense* for the organ of reason [*Vernunftorgan*], so that one learns to use it and to differentiate strictly between its functions and all others./[72]

Critique is an experiment, and makes pure reason into an organ. Indeed, it involves itself in the development of the actual organs, and in so doing, means to provide the substantial link between the general (metaphysics) and the particular (the individual disciplines of the encyclopedic efforts).[73] Organ is no longer taken only as the physical sense-organ, but in the breadth of its etymological possiblities. The Bible as *organon* of the encyclopedia makes the relation between the general and the particular substantial. We have, no longer disciplinary organs, but disciplines *as organs*. Critique, in order to avoid dogmatism, must particularize its object. The faculties under Kant's investigation become the generalized organs under Novalis's. These organs can be literally anything (see the chapter epigraph). Thus, the *organon* of the encyclopedia, which is also the New Bible, is a discipline Novalis dubs *organology*:

> ORGANOLOGY. The tool as such cannot be thought at leisure. An organ is, according to its very concept, in movement and thus in connection with its excitation [*Reitz*] partly immediate, partly mediated through the product. The dead body, thought as dead, will give us no information about the force and its connection with the body. Observe [instead] the living organ and the tool in movement.[74]

Tool here is constrasted with *organ*, but only to point out a higher concept of the organ, which applies to both.[75] That higher concept grounds Novalis's replacement for Kantian Critique, organology. Indeed, the concept of the organ here comes to replace the metaphorically organic judgment I located in Kant's doctrine of judgment in chapter 1 above. Organs are the forms of judgment, and they make what Novalis calls "magical idealism"—and what I will call "transcendental realism" to demystify the doctrine (see below)—possible. They are both called organology because the organs, physical and cognitive, that they attend to and eventually manipulate, cover the expanse of being in general and in particular.

If organs are the forms of judgment, then we must ask, what is the difference between an organ and a category? For Kant, the categories were unities of judgment under which synthesis (of intuitions and concepts) occur. "Organ" for Novalis is a designation of the form of that synthesis as it occurs. "Higher realism" is attention to the categories that holds open the possibility of their own development. Novalis writes: "Higher physics or higher mathematics or a mixture of both has always been understood under phil[osophy]. One always sought through philosophy to make something *workable* [*werckstellig*]—one sought an all-capable organ [*allvermögendes Organ*] in philosophy. *Magical idealism*."[76] This description of the historical search for an *organon* intentionally conflates that term with the organs at issue in "magical idealism," the distinctive doctrine that, over time, humans should make their involuntary organs fully voluntary. Its classical statement is: "The active use of the organs is nothing other than *magical, thaumaturgic* thinking, or the *intentional* use of the world of bodies [or the "body's world"—*Körperwelt*]—for willing is nothing other than magical, *forceful* capacity of thought."[77] "Magic" functions for Novalis as a way of talking about real interactions between seemingly opposed entities, like body and soul.[78] Novalis thus reads the history of metaphysics as the search for an organ of intervention in the world, the first site of which will be our own bodies. This is because they are the site of our specific, historical organs, and the users of tools more generally. The sense themselves become historical, their respective fields of potential a matter of potentialities (*Potenzen*) or possibility.[79] This solves a preliminary difficulty in considering organology a revision of Critique. There can be no Kantian critique of organs—indeed, for Kant, organs fall into the realm of physiology (thus the medical faculty) because they are empirical. The "physiology of reason" is only a metaphor. Novalis responds with magical idealism. "Organ" is more than the completion of the metaphor of the mind as animal body. It makes the organ the literal locus of bodily, cognitive, cosmological, and social-political force, the site of any occurrence, that which allows it to be seen in a temporal and phenomenal chain of events, integrated into the world as a set of rules expressed only in these sites. It is thus the locus of any scientific or historical work to be done.

Novalis bases this potential work on a specific doctrine of the senses, freeing them from the limitations of their fields of potential by identifying *sense* with *tool*: "Sense is a tool—a means. An absolute sense would be means and end simultaneously."[80]

The sentence makes up the distance between magical idealism and transcendental idealism, by making the field of possibility of any given sense into an instrument itself. Novalis elsewhere calls this *organibility*.[81] Between the specifics of supposedly material and the generalities of the supposedly rational lies the previously experienced mixture of the two through an open-ended formation called "organ." Each sense has a field of perception attached to it, and the sense is the locus of the field, the point that makes it possible for us, the transcendental point for each such field. Novalis writes: "For the eye the visible world is *a priori*—for the ear the audible world is *a priori*—for the more organ the more world *ap*[*riori*]—for the organ of thought the thinkable world *a*[*priori*] and so on. All these worlds are only different expressions of different tools of a single spirit and his world."[82] Organs are the points of origin for worlds of senses, which are transcendental conditions of those worlds and vice versa. The mutual transcendence of world and sense is gathered in the point of the organ, where the rules of the sense-field are located and malleable, as we shall see. Fields of potential, precipitated out of historical formations of this very site of the formation of categories,[83] which are in development, changing. Their activity is the activity of the universe. They are the organs of transcendental realism.

Transcendental realism is my name for this doctrine, the most robust form of Romantic organology. I use this name because "magic" denotes the real for Novalis, while idealism refers to the conditions of possibility that Kant calls "transcendental."[84] The organs of that philosophy, however, are in development. This means that they are not merely developmental or evolutionary factors in biological mass, but instead describe the history of possibility. Possibility is not, as Kant had had it, the formal coincidence of categorical contribution on our part with time as a form of intuition. Instead, it is the very hole in the graph of the universe. The divide between subject and object, which is also the condition of possibility for their reunification in each judgment, takes on different forms according to the whole complex of human natural, social, and indeed cultural-political perception and action. The only word that could characterize that form in general was *organ*.

The "organ" is supposed to allow critique to become "applied" to any possible contents of cognition: "Synth[ethic] judgments are ingenious—not antinomial, one-sided judgments—Idealism grasps one sort of one-sided judgments—Realism the other. Critique [Kriticismus] grasps synth[etic] judgments. Method of *judging synth[etically]*—system of synth[etic] judgments. Common—higher Critique. Applied Critique [*Kriticism*]."[85] Critique is the method by which the organ potentializes any field of possible cognition, making the set of its rules—the way in which it can pass from possibility to actuality—alterable. Such a Critique sets the rules for the power and legibility of its own action. The organ is both the possibility of the field in which it intervenes and the alternate possibility of that field's transformation. "Applied Critique" brings the function and history of the organ as cognition and capacity full circle with the reality of its mediation. It operates in the reality of mediation in individual fields; yet its own possibility ranges across those fields, and therefore reserves for itself the general applicability of its capacity. Mediation is synthesis *a priori*, its philosophy *Crititque*, its very possibility the organ.

Novalis shares the approach that Fichte had given to Schelling, that of methodological stringency, the deferral of the question of the thing until the form-content complexes of judgment began to respond to this question independently. Schelling presented this approach in latency, always incomplete, always anticipatory; it is made a complete program by Novalis. The statement of transcendental realism is that sense is organ, where "organ" denotes both the physical locus of the sense and its instrumentalization—*organibility*. The distance from the physical organ (which is defined literally and traditionally as the locus of a field of possibility) to the organ as the *use of the senses* is the distance from transcendental idealism (even in its stringent forms in Fichte and to an extent in Schelling) to transcendental realism, which is magical idealism. Organs of sense are tied to spectra of possible cognition; the sense *as organ* becomes the means to alter the nature of cognition. This is why "we must bring the body as the soul under our power. The body is the tool for the formation and modification of the world—must then make our body into the *all-capacious [allfähigen]* organ. The modification of our tool is modification of the world."[86] To potentiate a sense—to make a new field of perception available—is to make a *novum* in the world, to alter or modify the world through the organ that presents the intersection of possiblity and actuality at all points. Organology construes the world as this constant intersection, its rules potentially alterable by the very organs that bear them out.

By conceiving of the senses as loci of possibility precipitated out of an instrumentally emergent "world," Novalis literalizes "organ" across the divide of object and subject. The effect for metaphysics and politics is that the world is then fully potentially constructible, but only through a kind of absolute devotion to its precipitated form. Analysis of the made world is meant to free the fixity of organs and to allow participation in the making. Novalis writes of *"series of tools.* Chain o[f] senses—that *supplement* and *fortify* each other."[87] This passage might be taken from the Simondon tradition of organology, and it deepens the sense of this philosophy. Tools and organs are extensions of each other because they operate the same way, as the material and functional locations of worlds or fields of sense or action. Because they are extensions, however, they can be used for different ends, changed in their use, redirected. This is obvious of the tool, and in a sense, of the organ too. But this means that the worlds and the fields of perception attached to them can be modified too. And that modification, that construction of the very rules that constitute the phenomenal, is organological action. The system of organs through which this action occurs is the organ of reason—as we shall see shortly, the Romantic encyclopedia.

We should then not only have epigenetically developing categories, but the suitable tools to work on their development.[88] The categories as Kant speaks of them are the universal forms of the world as phenomenon. The Copernican revolution, traditionally understood, is just this reversal: the world's very ability to appear is formally subsumed under the categories, while what appears is a matter of contingency (from a categorial or conceptual standpoint). The categories would be the organs of the mind, as we have seen. Novalis makes this literal, and by doing so, he makes the form of appearances *possibly different*, or in a word, historical. Why should the organs of the mind develop epigenetically and then congeal in their maturity? This question, on which Novalis wavers, even as he pursues its theoretical possibility, forms the basis of organology as a member of the family of Critique. The Copernican revolution made the lawfulness of Newton's universe reflect in the lawfulness of the subject, freeing the flow of phenomena from rational anticipation and the rational will from empirical determination. Novalis responded by making the subject's very lawfulness a matter of development, half of the phenomena and half of their very constitution—*the formation of the organ of reason*. The phenomenal freedom of the world in Kant is transferred to the transcendental subject, which is thereby not simply reified but made real in Kant's sense, in principle open to new discoveries, new and higher and more rational unities. So far from trapping the world in the net of the subject, this makes the world, for Novalis, first cognizable, first the property of cognition with its palette of categories, intuitions, feelings, sensations. In short, perception imposes an open framework on its resulting objects, but neither the object nor the subject ultimately precedes that interaction. The program of systemic imposition thus has to proceed by painstaking attention to the appearance *qua* organ. Thus Novalis's method is in a third position with respect to Kant and Fichte, whom he characterizes as "lower and higher natural history."[89] Kant has removed the one-sidedness of scientific construction according to a "single criterion" (the sun), and Fichte has uncovered the laws of the multiple-criterion system. Fichte is thus the "Neuton" or "2n[d] Copernicus," the inventor of laws for the interior world-system.[90] Novalis occupies a third position, which is based on a dynamic relation between the pre-Kantian systems and the interior constructions that follow on their critique. "Organ" designates the use—theoretical and practical—of the changing interaction between the pre-Kantian and Fichtean positions. As we will see below, for Novalis, Kant is the "lower organ" and Fichte the higher. In this context, this means that the method proposed by the *Critique of Pure Reason* takes the position of a concrete organ, the designator of a field of epistemological possibility crucial to the contemporary intelligentsia. The goal, then, must be to use this "lower" organ to influence the world in such a way that that organ itself is raised to a higher power. Metaphysics must realize itself in a politics that alters the very constitution of that metaphysics retroactively. Critique is transcendentally real: it makes up the historical formation of cognition that leads to a better world, which will in turn change the organ of its production. The constant is "organ" as the general term for that mutually instrumental development. Novalis is a third Copernicus: following the construction of systems of law for the cosmos and the subject across the Critical divide, Novalis allows contingent flow to attach to both systems. The division between the two—their nonfit at the level of representation—constitutes the subject, which thus becomes the locus of organological capacity (the organ), or the possibility (as subject) of possibilities (of individual fields of potential, or organs).

The name "organ" thus also denotes appearance's subjective/objective contingency or more precisely possibility, and thus demarcates the field of possibility of all possibilities. It allows construction of not the world (which is incomplete, constructed but present) but of the universe. That construction is no longer merely the pure creative act of an intellectual intuition (whether Kant's or Fichte's), but instead the production of possibilities. The organs of that production are necessarily mythological, biblical, encyclopedic.[91] Reason as organ is thus not free from empirical determination, but free to determine empirically. For that, it needs fields of possibility in the plural, and it must have an active relation to them, that is, they must become its instruments. The retroactive constructions of natural history are thus metaphysically innovative. They leave the *physis-techne* analogy behind, and they do so with the immediate task of revising the moral, social, and political orders.

Encyclopedics is itself the organ of its completion.[92] Among the more than seventy passages devoted to the term Novalis seems to have coined, we find: "The highest elementary science is that which simply treats no *determinate* obj[ect]—but instead a pure *N*. Thus also with art [*Kunst*]. Making with hands is also a specialized, applied making. N *making* with the N organ is the obj[ect] of this gener[al] doctrine of art and of art."[93] The function of the organ becomes general—it is "potentiated" (*heraufpotenziert*) to "N." *Kunst* unites the high calling of a "system of poetry" with the new genre of all scientific endeavor: the Romantic encyclopedia. By subjecting science to art, Novalis seems to carry out the encyclopedic work of negating the content of all specific sciences in favor of the general form of the lexicon. The encyclopedia is not only its contents, as Novalis notes, but also its form, which is a medium. But Novalis rejects the arbitrary alphabetics of Diderot and d'Alembert, opting for a different methodological center for the encyclopedia. Friedrich Kittler has suggested that this center was the processing of data in the "night" of the negative. Encyclopedics passes, for Kittler, from the book technology of the mid-eighteenth century to the medium of spirit, with its source in pure motherly language and its goal in the real philosophical language would reference. But Kittler's "encyclopedia" is Hegel's *Phenomenology of Spirit*; Novalis's encyclopedia goes unmentioned.[94] This is because the *writing* of the encyclopedia is not the Hegelian concern, and so this genre—which Novalis develops into a medium—goes missing from Kittler's grammatology. The Romantic encyclopedia is meant to fulfill the task of organology as Schlegel had defined it, to be both real and ideal, both written and speculative. It is typical of Novalis to turn Schlegel's notion of irony into a matter of sheer philosophical ambition and utmost seriousness. And this is precisely what he does with the encyclopedia, which becomes the third genre—after the fragment and the novel—of Romanticism. It becomes the medium where *Poesie* but also *Technik* combine with metaphysics to contain the possibility of possibility. It becomes the location for the creation of new capacities.

The central statement on encyclopaedistics runs as follows:

> ENC[YCLOPEDICS] *universal poetic* [*Poëtik*] and compr[ehensive] system of poesie. A science is complete *1.* when it is applied to everything—*2.* when everything is applied to it.—*3.* when it, as abs[olute] totality, considered as a universe—is subordinated *to itself* as absolute individual with all other sc[iences] and a[rts], as relative individuals.[95]

Paratactically moving from poetry to science, Novalis repeats Johann Heinrich Lambert's determination—see pages 78–79 of this study—that the *organon* should be applicable to anything (1). But Novalis's *organon* retains a general receptivity (adopted from Franz Hemsterhuis's organ—more on this presently): everything can be applied to *it* (2). The crucial third term for this new genre, however, is that its practice involves its simultaneous universalization and subordination to its own principle. Looked at internally, an individual science is a totality, a universe from which the other sciences are also visible—from physics we can see life, but we cannnot explain it within the bounds of the methodology of physics. *Each science* relates to all the others in this way, as Novalis argues throughout. But only "encyclopedics" subordinates itself to the general principle of scientificity as such, because its content is the sum of all the other sciences. It does not do the work of those sciences, but *writes them*, produces them culturally. The encyclopedia is the genre of "applied critique." It is perched between speculation and poetry, it does not grasp but contains all things, and it works indirectly on them through "poetry." It is a generic container that provides mediation between the general and the particular, both *practically*—the book is meant for use, after all—and *theoretically*. It is the organ of representational content and strategy, the point at which new fields of cognition and reality might emerge. The Romantic encyclopedia is the genre of Romantic organology. More: it is the *medium* of poetry and philosophy, the vanishing point where speculation and materiality meet. It is neither negation in general nor poetry's symbolic transformations specifically. It is the genre of genres, the organ or organs that connects the order of poetry to itself and to the other orders of rational action, of *techne*. It is the very *Technik* that his own *years of apprenticeship* result in, and it is the common medium—the organ—that Schlegel called for. Romanticism calls for precisely this: speculation that is grounded in the methodologies of specific sciences, and that adopts the technological or medial shifts occurring around it in order to intervene in the flow of history and the seeming brute objecthood of nature. That is a call to arms—*tools arm the human*—that retains obvious relevance in our contemporary world. Novalis, having grounded his doctrine in the practice of encyclopedic writing, would extend it to the moral, social, and political spheres.

The Absolute Annihiliation of the Present: Moral, Social, and Political Organs
MORAL ORGANS

The senses, for the transcendental realist or magical idealist, are historical formations of possibility. The historical condtions of possibility called organs form the possibility of that possibility because of their dialectical subject-object structure. For Novalis, this means also that organs have a future. And that future is political. Organs are tasked with the political: radical instauration of institutions from the theoretical nothingness which its capacity to abstract has reduced them. That pure possibility is subtended by attention to their nevertheless actual, contemporary forms. Thus a link is created between system and history, and this philosophical gesture is at the basis of German radicalism going forward into the nineteenth century.

Future organs had been a topic of discussion in at least two previous authors with whom Novalis was familiar. In a fragment by Lessing entitled "*That There Could be More than Five Senses for the Human*," the Enlightenment stalwart defines the senses as areas of order and mass (*Ordnung und Maß*) within the realm of representations. Noting the gap between these fields of possible perception and the totality of possibly perceived things (different types of matter are posited as the limits of individual senses), Lessing reasons that the discovery of other senses is possible. The obvious candidates are electricity and magnetism, since we know these phenomena indirectly—what in principle could prevent a direct means of knowledge in these areas?[96]

As Fritz Mauthner showed long ago, Lessing may well have influenced Novalis's beloved moral philosopher Franz Hemsterhuis. The latter's conception of a moral organ was determinative for Novalis's own larger cultural projects.[97] Much has been written on Novalis and his relationship to Hemsterhuis, and I do not intend to repeat that excellent work here.[98] Instead, I offer a short summary of the Novalis-Hemsterhuis relationship, followed by a suggestion about the moral sense as organological conscience in the second part of *Heinrich of Ofterdingen*. I then press on to the political organs.

The Dutch philosopher Franz Hemsterhuis wrote against sensualism and materialism throughout his career. His project was, in fact, a new mythology in the most literal sense. Works like *Alexis or the Golden Age* (1787) make clear that an *organe morale* is meant to restore the classical paradise on earth. Stadler has suggested that this project was first taken to be a kind of guiding fiction, and then later became a poetic fact that should be realized.[99] Mähl has emphasized Hemsterhuis's pessimistic attitude—ultimately Pauline—toward organs: they exist as mediators because we are not face to face with the beautiful world, and we are not in the fullness of time because we have organs at all.[100] In the *Letter on the Human and His Relations* (1772), this ambivalent attitude is built into the definition of the organ. The organs are literal for Hemsterhuis, and they form the simultaneously separative and synthetic containers for the perception of objects. Indeed, they are the locus where signs are separated from objects—where criteria of recognition are abstracted from aggregates of material—and thus become the basis of the semiotics of memory—that is, the ability to remember an isolated sign is the ability to think of an object by linking that sign to it.[101] "Organ" is defined as passive, but in this very gesture, also as receptive of activity. Thus its stasis—the duration of its modification by an affecting object—is its essence, but its ability to be moved is just as essential.[102] This definition is extended to a putative moral sense, which casts the very perception of the world in moral terms.[103] The central ambivalence is about the persistence of organs in the return of the Golden Age. In one mood, Hemsterhuis imagines an umediated, pure moral perception. In another, he says that this condition will make us "tout *organe*."[104]

For Novalis, this condition of total organicity was not ambivalent. There is, perhaps surprisingly, even in the moral sense which organology provides, no yearning for the unmediated. Yearning (*Sehnsucht*) fulfills a function, but is not determinative. Instead, the activity of the organ—its ability to potentialize the actual, make it an organ—provides the basis for Novalis's adoption of the moral organ from Hemsterhuis, because it provides a theoretical basis on which to make "possible" the moral coloration of perception.[105] The epistemological question was set in the notes on Hemsterhuis:

> /Seeds of future organs—perfectibility of organs. How does something admit of being made into an organ?[106]

The connection between the perceptive organs and the moral organ is thus not trivial. The potentializing activity I have identified above must play a role in the creation of a morally perceiving organ. The understanding and reason "express the organs or faculties for relations."[107] As we have seen, this means that the organological faculties—crystallizations of cognition in development—are in Critical or substantial relation to their putative content. If they are cancelled, we fall into dogmatism. They can affect that content, all the way down to its moral coloring. This means that "every finite being is a tool."[108] Organology should allow the world to be rebuilt morally, relying on absolute abstraction on the one side, and absolute attention to historical formations of knowledge and its constructed world on the other. In principle, for Novalis, this realism about the historical conditions of possibility should make the world alterable all the way down, through the ontological activity of poetry:

> And if philosophy makes *perfected* poesie possible through formation of the external whole or through *legislation*, then poesie is as it were the purpose [*Zweck*] of philosophy, through which end it first obtains meanings and graceful life—which end poesie forms the *beautiful* society, or *the inner whole*—the world family—the beautiful economy [*Haushaltung*] of the universe—as philosophy through system and state pairs and reinforces *the forces of*

the individual with the *forces* of the universe and the rest of humanity—makes the whole into the organ of t[he] individual and the individual into an organ of the whole—so poesie—with *respect to enjoyment*—the whole is the object of individual enjoyment, and the individual the object of total enjoyment. Through *poesie* the highest sympathy and coactivity—the most inward, greatest community—becomes actual. /Possible thr[ough] philosophy.[109]

Philosophy is organology, the organification of the whole-part relation. It establishes the mutuality—categorical community—at the basis of actual community, which must be based on poetry's ability to make organs the objects of enjoyment. This enjoyment retains the form of the organological universe, however: it actualizes—both in the sense of making actual and in the borrowed sense of making contemporary—what philosophy has made possible. Philosophy after Kant is legislation (*Gesezgebung [sic]*), but its potential syntheses are infinite. Where the human's tools are unsuitable, Novalis writes, philosophy's task is to make them serve. Every finite being is caught in this potentially endless circulation of ends and means. Philosophy makes the beautiful relation of whole and part possible by making that relationship mutual; poetry makes it real by expressing it as mutual enjoyment. The new mythology and its religion emerge from the negativity of the organological universe into the positive historical realm of discourse and politics.

This structure is most clear in the unfinished second part of the novel-fragment *Heinrich of Ofterdingen*, which is ironically titled "Fulfillment." Heinrich, having finished his journey of education, finds himself in a deserted landscape where he encounters a monk named Sylvester, who once taught his father. The father's recognition and rejection of—but simultaneous nostalgia about—poetic fulfillment in the blue flower had the tone of the novel as the paradigmatic Romantic epistemological space, split and unified by the discursive distance from fulfillment and the bitter sweetness of desire for that completion. Heinrich wants to find what his father could not, and, as in *Wilhelm Meister's Years of Apprenticeship*, the apparent end of the apprenticeship with Klingsohr is emphatically not the end. Heinrich asks Sylvester when the lack in the world will come to an end. Sylvester responds: "can music be explained to the deaf?" Dismayed, Heinrich responds: "then sense would be merely a share in the new world opened by the sense itself? One could only understand the thing if one possessed it?"[110] The "only" force in the world, according to

Sylvester, is conscience, and it is connected to the senses by structural analogy and by a substantial link:

> The universe [*Weltall*] breaks down into infinite world always in the grip of larger worlds. All senses are in the end a single sense. One sense leads like a single world gradually to all worlds. But everything has its time and its manner. Only the person of the universe is able to understand the condition of our world [*das Verhältnis unserer Welt einzusehen*]. It is difficult to say if we can really multiply our senses with more senses, our world with more worlds, inside the sensible limits of our body, or whether every addition to our cognition, every newly-won capacity should merely be counted part of the training [*Ausbildung*] of our current world-sense.[111]

The extension or addition of cognition tools may remain in a total order we do not yet perceive all of, or it may be that we can build beyond that world. In either case, however, the incompleteness of each world is essential to the training, the *Ausbildung* of extended capacities. Being is beyond those capacities, but only appears because of them. This is the pathos of Romantic ontology, but it appears only as one narrative voice, as a problem that does not determine transcendental activity, but must remain an open question in the context of fiction production, specifically that part of narrative that one can call the *years of the journeyman*, the empty and unstructured space beyond individual *Bildung*, the years of institutional and perhaps political work. The "fable" is, for Heinrich, "the total tool [*Gesamtwerkzeug*] of my present world": "Even conscience, this power to generate sense and worlds, this seed of all personality, seems to me like the spirit of the world-poem, like the accident of the eternal romantic reunion [*Zusammenkunft*] of the infinitely changeable total life [*Gesamtleben*]."[112] *Everything can be made into an organ*: the sense of this dictum is ultimately a constitutive morality in the world—whatever the extent of that world—the historical ability to introduce, slowly and in connection with the most concrete of appearances (critically), difference into the categorical structure of things. Sylvester confirms the poetic, organological innovation suggested by Heinrich,[113] and Heinrich responds with the final word on the new theology: for that doctrine to come forward as science, it will need the moral organ that is conscience. That organ gathers the historical senses and makes their ultimate possibility possible: the personalization of the world.[114] Sylvester confirms: the general mediator is conscience itself,[115] now conceived as the active sense, the sense par excellence as organ. The novel "ends" with the promise of a politicization of organs in the secularizing program of a new religion.

That program was explained to Schlegel as the "negativity of Christianity," the total annihilation of the present as the condition for the production of the past (antiquity) and the genuinely new future. This is an organological expansion of the play of negativity and positivity in Schleiermacher. Organs bear the functional conceptual task of the reproduction of the world, and that task is not only moral, but also political. The "new Christianity" is paradoxically more secular than Kant's biblical organa. Kant's moralization of belief (which redirects content to a different function) finds an even more secularizing tendency in Novalis's new mythology, which alters the function only by replacing its content. What seems to be a new religion is a new set of tools for a function designated religious. But then, those tools do not leave the function innocent: they make it an organ of political production.

SOCIAL ORGANS

The political is, for Novalis, the critical joining of sense and will, and, as such, it is derived from dialogue with Fichte.[116] Among Novalis's various opinions on Fichte, his adoption of and revision of Fichte's doctrine of organs from *The Foundation of Natural Right* (1796) is most political. The notion of connecting sense and will was one of the primary tasks of the Critical method, one primarily associated with Fichte's metaphysics.[117] In this context, Novalis casts the incunabula of Criticism in organological terms: "Fichte is the re-drafter of Kantian critique—the 2nd Kant—the higher organ where Kant is the lower organ ... he sets the readers down where Kant brought them up to."[118]

Kant's work is the basis of communication, the self-imposed passivity of the organ; Fichte's the active organ, and ultimately recognitive communication itself. Novalis takes his terms here from Fichte's *Foundation of Natural Right*, an application of the *Science of Knowledge* to the political realm.[119] This work marks out what we might call the ethics of reason, the specific doctrine of action (and ultimately law) that results from the *Science of*

Knowledge's determination of the capacities of reason. The *I* sets itself as absolutely free, but can only do so in setting limitations for itself. These limitations take the form of a fungible sensible world—the entirety of the parallel deductions of the theoretical self and the practical self. For the rational being to inscribe itself into a social order, however, it will have to have a conception of limitations that it does not give itself, but which come from elsewhere. These are the so-called "impetus" (*Anstoß*), the beginning of embodiment in the recognition of other rational beings.[120] This relativizes the notion of the individual, who can only exist as one among many. The set of legal relations is established by this mutual appearance of external limitation, which is actually based on the recognition of others as rational beings (and thus an act). In order to have this relation, the individual must posit himself as an embodied being. Indeed, the central paradox is that, while the body itself falls under the constructive freedom of the absolute *I*, the appearance of the restriction of freedom must occur, without metaphysical restrictions thereby being placed (from where?) on the *I*. Fichte thus posits two organs, a higher and a lower, a concrete organ of rational communication and recognition, and a lower, bodily organ (sometimes the body itself, as complex). Ultimately, these two organs are identical in the will, but they are separated through recognition of others, so that other can influence the *I* not only in manipulation of the body, but in communication with reason. Fichte writes:

> Evidently, a double manner of determining the body's articulation, which for now might even be called a double articulation, or a double organ, the two sides of which relate to each other in the following way: the first organ (within which the person produces the canceled movement and which we shall call the higher organ) can be modified by the will without thereby becoming the other (which we shall call the lower *organ*). To this extent, the higher and lower organs are distinguished from one another. But furthermore: if the modification in the higher organ is not to lead automatically to a modification in the lower, then the person must also restrain his will from thereby modifying the lower organ: thus the higher and lower organs can also be unified through the will they are one and the same organ.[121]

The lower organs are those of the senses—completely heteronomous with respect to the influence of others, but completely autonomous with respect to the I itself. The higher organ is a sort of concrete schematic capacity—its basic schema is the human face—within reason. It is thus completely autonomous with respect to others, but is also what we might call *heautonomous*—necessarily free with respect to itself. Two types of material correspond to these two organs—*rough material* and *subtler material*. The subtle material is modified by the will itself, and makes communication rational in the sense that it is both transmitted concretely and leaves the higher organ under the power of the possessor's will.[122] This entire structure must be attributed to another being for the individual to exist at all. Thus there is an interaction, or *Wechselwirkung*, between both the higher and lower *organs* and between individuals communicating through them. The form of that interaction is organic, its perception intuitively understanding. Its science is called anthropology,[123] its method is intuitive understanding, and the immediacy of that recognition makes its fundamental principle that "human form is necessarily holy to the human."[124]

Thus, in spite of his "limitations," Fichte presents Novalis with a higher organ of rational human commonality—a social organ—that he himself defined.[125] Fragment 452 of the *Romantic Encyclopedia* uses these organs to discuss the body-soul problem. Lower organ is excitation [*Reiz*], and higher is reflection in the soul. The complex, clearly taken from Fichte, applies the logic of the critique of organs to the historical conditions of cognition. The direct excitement of the lower organ is indirectly (nonrepresentationally) reflected in the higher organ. The result is feeling, which is thus harmonious or nonharmonious. No element of this system offers a real solution to the classical *commercium*-problem. This description lacks the activity of real critique (the making-possible or making-organ), functioning rather to prepare the imagination of social organs.

The surprising location of actual reception of the Fichtean doctrine is in the conversation between the travelers in the *Apprentices*. In one sense, the whole novel-fragment seems to take place in the "subtle material" of communicative reason, what the first traveler calls "the elastic medium."[126] For this traveler, the nature of attention and its complex social conditions must be investigated in relation to the body before we can hope to penetrate into the depths of nature. The "*Researcher* thinks he has perfected his craft if he, without disturbing that play, can simultaneously take up the regular business of the sense and sense and think at the same time. Through this, both perceptions win: the external world becomes transparent, and the inner world manifold and meaningful, and thus the human finds himself in an inwardly lively state between two worlds in the most perfect freedom and the most joyful feeling of power."[127] This traveler, who, implicitly leading the group as the first voice, will be among the community to receive the social-organological message of the master at the end of the fragment, ties the higher and the lower organs to the phenomenology of form-content connections which come to be called inner and outer organs.[128] If the novel is the subtle material of the higher organ, then its own higher organ is the voice of Fichte. One voice among many, he sets the social framework in its simultaneously subjective and objective traits. As the internal and higher organ of the constructivist novel,[129] he is thus the representative of social organology. Where Hemsterhuis had provided the basis on which to think the world as subjectively but constitutively moral (through the conscience), Fichte provided the basis on which to tie that moral world into a social sphere. The novel makes the humans in that sphere organs, the sensitive antennae of a possibly rational world. The concretization of that rationality is called politics.

THE POLITICS OF ORGANS

In the context I have established, it is easy to see why Novalis would call everything from a court to a church an organ.[130] With that discourse, he anticipates the social-political use of organ so familiar to our contemporary ears. What is less familiar—even alienating—is his talk of a "christian monarchy," of the love of the king and queen as the basis of the organological state.[131] The locale of these strange determinations—and indeed, the reason I have put politics last—reveals that they are the results of the other side of Novalis's "German ideology," the attempt to take the absolute abstractive drive (which is a reflection of the institutional dissolution occurring on the other side of the Rhein and the Atlantic) and join it with the concrete forms of contemporary institutional life. What Marx experienced as the international embarrassment of Germany, Novalis tried to pry open conceptually using his new system, organology.

Not only do the institutions of the feudal government become organs; the organs of metaphysics become political actors. Novalis articulated the final (and really only) statement of organological politics in a speech delivered to the Jena group in 1799, *Christendom or Europe*. The apparent mysticism and medieval nostalgia of the text alienated him from perhaps the only audience that could have appreciated his message.[132] After a vehement debate within the circle, Goethe was called in as an arbiter, and he saw nothing more in it than the others had. The text was not published in Novalis's lifetime.

Novalis's basic insistence is on the unification of state and culture—on a harmonious totality of society. And indeed, there is much that looks like nostalgia for medieval Catholicism on these grounds, right down to a bizarre retroactive momentary hope in the "secret society" par excellence, the Jesuits and their Inquisition. Novalis goes so far as to claim that the Reformation had simply destroyed Christianity, that the latter did not any longer exist after Luther's separatist activities.[133]

The theoretical heart of the essay is, however, the following statement:

> Now we turn to the political drama of our time. The old and new worlds are caught in battle, the lack and neediness of the institutions of state up the present have been revealed in fearsome phenomena. As if here as in the sciences a new and more varied connection and mixing of the European states was the immediate purpose of the war, if a new impulse on the part of Europe, sleeping until now, came into play, if Europe should awaken again, if a state of states, a political science of knowledge [*Wissenschaftslehre*] stood before us! Should perhaps hierarchy be the symmetrical basic figure of these states, the principle of an association of states as an intellectual intuition of the political?[134]

The present is the battle of the past and the future, and the state, now separated from its cultural and religious institutions (however partially), has become incapable of guiding its people through the passage of time. What is needed, then, is a political science of knowledge, a Fichtean doctrine (not Fichte's own) of political technology. The political *I* must be reflected in its institutions—the link between metaphysics and politics must become substantial. For that, it must be based on an intellectual intuition, the very instrument of Romantic organology. A nonrepresentational, innovative relation between the rational metaphysics of instrumentality and political representation must be created. This is a call to theoretical arms, or more precisely, tools. Novalis anticipates Hegel: the state must be saturated with reason.[135] But as philosophy could only capture its time in reason for Hegel, so could the state for Novalis only capture reason in its historical development. The state is a science of the possibly possible. Governance is organology.

Thus a profane connection between religion and the state must persist, at least during the battle for the present: "Do nations have everything from the human—just not his heart?—his holy organ?"[136] It is the connection between religion, the Critique of all knowledge in the Book, and the nation-state, that must take on the immediacy of the intellectual intuition. Statecraft is the final theater for the breaking of the bonds between *physis* and *techne*, and for its project, the actual tools of the present are needed. Monarchs, persons, love: the discourse of feudalism is the found object that must be judged in a single, immediate historical intuition dictating the form of a new, human state. Novalis writes: "The veil is for the virgin what the spirit is for the body, her indispensible organ whose folds are the letters of her sweet announcement; the infinite play of these folds is cipher-music, for the language is too wooden and too insolent for the virgin, her lips only open for song."[137] Note the reversal of terms: spirit is the organ of the body, and spirit's analogical folds are the infinite letters of a prophecy sung into the heart of the present. Organological politics would be the construction of a political order out of the nothingness it intentionally creates,[138] but in constant dialogue with the persistence of the historically existing world. This program remained a suggestion—Novalis died less than two years after his conflict with the Jena group. And yet the suggestion is, in a way, powerful enough. Whatever else the politics of organology should have been, it would always have been what I have termed transcendental realism, persistently attendant to the historical conditions of political possibility, even as it systematically negated them in the interest of of a plasticization of instututionality. This is the narrative space that Novalis could not fill, did not live long enough to spell out. I am uncertain that any of his contemporaries, perhaps even including Goethe, managed that—but the Romantics put that narrative and critical task on the table. Its utopianism was anything but a sentimental dream: it was meant to unite the abyss of alienated self-consciousness with the concrete ruination of the political order around it. To change the world, the Romantics thought, the modern subject would have to tarry in the night of interpretation.

SEVEN

Between Myth and Science: *Naturphilosphie* and the Ends of Organology

> Myth [here] is not simply inventing or fictionalizing but rather an organ of exposure of the historical world and the historical reality, that is, an organ of metaphysical awareness.
>
> —ERNST CASSIRER, *Towards a Metaphysics of Symbolic Forms*

The Emergence of Discipline between Myth and Science

Just after 1800, the disciplinarization of the sciences accelerated in Europe.[1] But the emergence of the narrow disciplines arose only slowly, and in a proliferation of methods, philosophical and polemical, that must appear strange to contemporary observers. Indeed, after positivism the methodological plurality became a problem for historians of science.[2] Ernst Cassirer wrote, about the period following Hegel's and Goethe's deaths in 1831 and 1832, that "there are now as many single forms of theory of knowledge as there are different scientific disciplines and interests."[3] Before experimental method gained predominance in the separation of the disciplines, and just as the modern research university came into being, what would eventually become the split between the humanities and the sciences was adumbrated in organological terms. What follows tracks the tension between myth and science in that split, and argues that Schelling contributed, in 1809, one last chapter to the book of organology.

The history of the term "organ" allows us to see an anticipation of the narrow disciplinarization that accelerated after the 1830s. Two striking uses show that myth and science needed conceptual instruments as they severed their connection. Gotthilf Heinrich Schubert (1780–1860) and Joseph Görres (1776–1848) adapted nature-philosophical meanings of "organ" to imagine a new cosmos. Organs connect the parts of a universe that is an extension of the sensory system of the human—the human heart is the "central sun of the microcosm." Görres's contribution to the ongoing Romantic effort to ground German culture in its own history had a scientific side, a side that borrowed Jena Romanticism's term "organ," stripping it of its methodological meaning and projecting it on a long-dead, quasi-astrological cosmos. Organs became the instruments of myth.

As early as 1798, my proposed counterpoint to Görres was already publishing the theses that would come to be called "phrenology." Franz Joseph Gall's (1758–1828) doctrine is everything but mythologizing—it is opposed to Romantic scientific efforts in their very roots. It rejects all metaphysics in science, and makes only apologetic efforts to respond to ethical concerns. The organs of the brain became instruments scientific and regulatory, meant to control behaviors through analysis of mental contents.

In this confrontation of organs—the end of organology—I single out two uses. The first is that of Friedrich Schlegel, an "aesthetic organology." Here, I look at the political ramifications of the term, arguing that it serves, as it did for Novalis, to make politics metaphysical. Schelling, on the other hand, contributed both his own notion of a "new mythology" and a last conception of the organ, making it the interface between action and cognition.

Mythological Organs

For the Schelling of the *System*, philosophy was born from poetry during the childhood of science, and, since "a system is completed when it is returned to the point where it began,"[4] the totality of the sciences now in their maturity must, to fulfill the organological vision, flow back into the ocean of poetry (aesthetic intuition as the organ of philosophy). The means for that return are historically evident:

> But what will be the tie [*Mittelglied*] that causes the return of science to poesie is not difficult to say in general, since such a tie existed in mythology before this (as it appears) undissolvable separation occurred. But how a new mythology which is not the invention of a single poet but instead a new generation presenting as a *single* poet, can emerge,—this is a problem whose solution can only be expected from the future destinies of the world and the further course of history.[5]

The mythological dimension of organology is necessarily tied to origin myths about science. As Schelling argues here, the unified cultural expression of a state of balance between poetry, science, and philosophy has always been mythology. The need for a new unity arises from the apparently irresolvable separation of these disciplines (for that is what they now are), and a national project—the unification of a nation as a single poet—comes into view.[6]

Schelling gave a definitive impulse for mythologization to the Romantic physiologists, the generation which followed and was largely inspired by him. He had extrapolated in the *System* from the organs/*organon* complex to a possible cultural container for the program of organology. This could not be philosophy in the old mode, nor could it be art without philosophy. It had to be presented as a higher unity of the two, and this, Schelling argued, was the very concept of mythology, its traditional role in the history of humans. As we will see, this impulse was to ground a series of cosmological physiologies.

Schelling's efforts would bear discursive fruit among his contemporaries. A series of Romantic attempts at scientific mythology emerged in the first decade of the nineteenth century. Two are treated here—those of Gotthilf Heinrich Schubert and Joseph Görres—because of their organological pretensions.

In building mythologies of organs and opposing them to the new, radical scientism they saw in their contemporaries, Schubert and Görres were drawing on a use of the term "organ" that was no longer as philosophically ambitious as it had been in Jena, and prepared the ground for the technical term that would arise in the next generation, that is, the biological term with which we are still familiar. This use came from the institutionalized form of *Naturphilosophie*, which loomed large on the European scene, in research agendas and textbooks, deep into the 1820s.

Its technical definitions of "organ," as we shall see in a moment, were rather traditional. Their context, however, was that of the construal of the universe in organic and even organological terms. Take this not unrepresentative passage from Lorenz Oken's descriptively titled *On the Universe as a Extension of the Sensory System* (1808): "The organ of the world must relate to the senses as the periphery to the center, as the skin to the brain, or as the sense-organ to the brain; the sense-organ is, however, just the lengthened, extended brain, thus the world-organ is just the extended sense-organ.... [T]he senses are the body of the brain, the world is the body of the senses, but both are one like body and brain."[7] The system is obviously inspired by that of Schelling. But where Schelling's organ is the methodological lynchpin in a system meant to speculate on the basis of science—where that organ binds and divides individual orders of nature and binds those orders together—Oken's organ is merely a factor, a represented individual, an object. It fulfills the same function—in fact, the same schema could be used to describe it. It unites and divides the human as human (in the difference and continuity between brain and organ), and it binds the orders of the human and the cosmos to each other—we will encounter this philosopheme again in Görres shortly. But it projects that organ onto a cosmos to which it claims humans are intimately connected. Oken has no space for the methodology that the Idealists maintained in their philosophies of science; it was ultimately changes in that methodology that produced positivism,[8] and not, as we might suspect, mere changes in the representational results of scientific experiment. Thus *Naturphilosphie* and its proto-positivist other present the loss of the philosophical methodology of organology, the vagaries of that ambitious discipline before its aftermath.

Oken agrees with his nemesis Goethe and their predecessor Albrecht von Haller that nature has "neither inner nor outer" but is instead an articulated continuum—organs all the way down; a cosmos of function. Oken claims there that the senses are extensions of the brain, and that all natural qualities perceived by the senses are extensions of those senses. He extends his cosmology into an epistemology, founded on the following dictum: "Sense is immediate consensus of the nerve-systems with the world ... in the nerve-system, sensing and moving are absolutely one."[9] It appears that Oken thinks of extensions of organs, articulated as functions from the brain to the world to the macrocosmic sun, as solving the Kantian question: whether and how what we know is meaningfully informative about the world. He shifts the Schellingian problem of the *identity of the knowing and the known* to the historical and physiological problem of historical knowledge, or the difference that drives the shifts in such knowledge, that relates it meaningfully to the world. This is the insight that Oken brings to Schelling's theory: the virtues of the discontinuity at the heart of the overarching cosmological force-continuity. For although the universe is an extension or "continuation" (*Fortsetzung*) of the sensory system, it is also a "continuance" (*Fort-Setzung*), a placing further or forward, of that system. It is all too easy to see only the continuity-thought, the remnants of Renaissance mysticism, here. But continuity is not, precisely, identity. The continuity between universe and sense is not the identity of subject and object, although both are necessary for life and cognition to exist in one being (the animal). Embodied cognition, however, is always an extensive reaching, an adaptation, a shift in nervous energies and ultimately, perhaps, a shift in the universe itself. Organs unite and divide that universe, and *Naturphilosophie* has not often been given credit for having emphasized the division, like that of a cell, historically infinite and producing new forms, continually new organs, of cognition.

The technical definition we still recognize today emerged in the early part of the nineteenth century, and was held by *Naturphilosophen* and their more conservative counterparts alike. Karl Friedrich Burdach would define term, for example, as follows:

> The products of life's formation are thus indeed what their name suggests: *organs*, i.e. tools through which something should be brought into being. Their activities are executions, i.e. performances of certain tasks that are configured and ordered such that they reach into each other to complete each other's tasks, and through their cooperation achieve the survival of the living body.[10]

Organs became the *tools of life*, the function-bearing, task-oriented parts that make up and preserve the "living body." The definition is familiar, and probably could have been accepted across the divide between *Naturphilosophie* and the emergent set of practices that would come to be known as positivism. As we will see, the tension between "the living body" and its arrangement of organs would be a major point of contention as in the period before Darwin. But the very association of "living body" and "organ" was late in coming, and it gave rise to the term "organism" only in the mid-nineteenth century.[11]

Gustav Carus defined the organism as identical to life, and both as unities that could create functional pluralities by developing organs: "But now it is clear that the concept of life and the concept of the organism are identical according to their essence; for we call that a unity which continously develops itself into a real manifold *in* and *out* of itself, insofar as it produces means to reveal itself, that is, in other words, creates for itself tools, *organs*, an *organism*."[12] The problem of the type—the question of what drives the organization—continues to haunt Carus, who, following Kant and the general trend of *Naturphilosophie*, claims that the unity of this plurality can only exist in the "world of ideas," since the organism is "conditioned . . . by an *image of its being prior to its being*." Carus uses Kant's definition of the organism here, suggesting that the idea that precedes the being—the purpose—is contained in the organ, with the Greek *tool* (*organon*) supplying the linguistic suggestion for that conclusion.[13] Organized life receives its empirical name, even as its conceptualization retreats into the transcendental sphere: Foucault's notion of the transcendental-empirical doublet (see the introduction to this study) is mirrored on the material and ideal sides of developmental biology in the mid-nineteenth century. *Naturphilosophie* confirms a transcendental term ("life") that materialist biologists simply assumed. The discipline was structured on this dilemma only once "organ" took on its modern meaning and allowed the empirical term "organism" to arise. This would remain the crux of the biological problem until Darwin proposed an immanent solution, empiricizing the concept of type.

Naturphilosophie exploited this relation and projected it onto the universe, intentionally poeticizing and mythologizing the infant discipline of biology. As late as 1854, Schubert could write:

> This life, which forms the colorful world of plants, which feels in the animal and effects movments, has in its service completely other bearers and gatherers (condensators) of the sidereally exciting forces of the world of bodies, as those

are, which the art of the human made for itself, we call those gatherers which reveal themselves in the most manifold forms of the inner structure and external parts: *organs* and thus enlivened bodies are all called *organic*.[14]

Organs have finally become the metaphorical bearers of life, the basis of organisms linguistically, even as they pass from scientific consideration for this pride of place. Schubert compares them to instruments humans make for themselves, but the implied purpose of the organs is watery. The meaning that has come down to the present has arrived: if organs are tools, then they are tools for whatever purpose the organism happens to use them for. Both the open-endedness and the flexibility of this conception became determinative around this time, but they determined what would largely be a metaphor.

Schubert's *Views from the Night-Side of Natural Science*, a series of lectures delivered in 1808, starts with the premise that only a full history of the human relation to nature can complete the scientific picture. A primordial unity of man and nature, an account of the split and the collateral epistemic effects of that split, and the principled possibility of a reunification are the necessary elements of that history.

Schubert clarifies the original separation in a gloss on Kant's "Copernican revolution": "In those days the spirit of the human did not give laws to the stars; rather they gave laws the very existence of the human, as to the motion of the earth, and the wisdom of the ancient world was: to do everything just as nature had taught it."[15] The Kantian revolution is literalized and inserted into a long evolution of human cognitive control over nature, the history of which is psychological as much as cosmological. It is then cast in organological terms: humans and their societies are the organs of nature and each other:

> At a quick glance the old ideal of the kings is seen in sublime brilliance, as they were a exemplum of the divine, mediators and preservers of the ancient harmony with nature. The law of nature and the higher influence were the first sovereigns of humans, and those had been elected as representative place-holders to whom that higher influence had communicated itself most inwardly. That time honored not the sovereign but the faithful organ of higher nature, and we still see in the oldest history of some peoples the noble king himself as the priest presiding in the service of nature, giving over his his grey head in

the observatory to the night, the consecrated eye preserving the ancient bond of the human with nature for his slumbering people.[16]

The king is twice organ, first as the truest knower of nature, and second as the legislator of human law in the image of this nature. The psycho-mythology of science's night is the story of the emergence of human cognitive royalty, the becoming-king or becoming-organ of each human in transition to democratic mutuality. The king becomes irrelevant politically precisely as humans become organs and then legislators of nature.[17]

This psychological new mythology found an explicitly physiological complement in the work of Joseph Görres, especially in his 1805 *Exposition of Physiology*, subtitled: *Organology*. A Jacobin[18] and dedicated Romantic physiologist,[19] Görres undertook to make Schelling's *Naturphilosophie* into what he called "*iatropoetics*"—a poetics of medicine or a physiological cosmology. He imagines a cosmos filled with organs as transmitters of forces between different orders (like Schelling's *Potenzen*). Every part of the universe is an organ. The cosmos has a central organ, which is analogized to a "central sun of the microcosm": the brain.[20] Görres seems to be quoting William Harvey here, who in his 1628 *On the Motion of the Heart* had spoken of his object of study as the "sun of the microcosm" [*microcosmi sol*]. But Görres calls it an organ where Harvey never did (although it is often translated that way). The universe was in fact a series of organs, mirroring the human body. Organs were not tools for philosophy but pictures of the harmony of the spheres. The step back from the philosophical history was noted by contemporaries: Goethe wrote in a letter that "There is evidence of a very good head in [the *Aphorisms on Organonomy*], though one is tempted to change its title to "organomania."[21] This is the system of extended sensory system, but all is organs. The orders of *Naturphilosophie* are connected exclusively through organ-relations, as though the elongated tubes of de Robinet had returned. *Organology* would be the study of function as an aesthetic doctrine, a mythology that blends empiricism and beauty.[22]

Although directly inspired by Schelling,[23] the resulting cosmology is Herderian in quality. An aether subtends the universe, and spreads itself out finely and thickly through the organs—individuals as the *referenda* of the totality of the universe—according to their composition, bridging divine and human, object and subject.[24] The Herderian picture (aether, organs as cosmic relations) is reinforced by Görres's reliance on analogy, which he licenses with his claim to be writing a scientific or medical poetics. The rhetoric of the work is Schellingian, but the method is not that of Romantic organology.[25] Görres offers a crystallization of organology.[26] "Organ" is not the locus of methodological struggle here, but the panacea covering the connections in the projected mythological universe. Görres's organology is firmly of the new mythological variety but represents a Romantic parallel to that of Gall: the premises do not need to be spoken, and "organ" becomes literal, functional, and cosmological again.

In his rather more famous editorial work on the Early Modern chapbooks, Görres makes the mythological dimensions of his project explicit. The immediate expressions of culture in the *folk epics* impress themselves upon the organ of the people.[27] Then—Görres analogizes to Homer for the Greeks—the *Volk* narrates its sense of self. This narration—the very chapbooks Görres is publishing—becomes the organ of a remembered second phase of nation building, the mediated unity which can be remembered for renewal, as it has lost its poetry and become the prose of the nation.[28]

Görres's editorial mythologizing presents the one side of organological aesthetic theory. In one sense, this part of the project is not *new* mythology, but the recuperation of old mythologies. Friedrich Schlegel presented organological aesthetics with a kind of *Kunstorgan*, an artistic but also artificial organ. His was the proper aesthetics of organology, mythological in pretension and metaphysical in conception. Where Novalis speaks of an aesthetic organ, the conversation with his friend Schlegel is never distant.

Friedrich Schlegel's version of the new mythology is perhaps the most political, and the most aesthetic. In the *Speech on Mythology*, part of Schlegel's mixed-genre treatise *Dialogue on Poesie*, Schlegel calls for a total unification of the cultural efforts of Romanticism. What he perceives already happening, especially in "higher physics" and philosophy, needs, however, a means for generic formation. As I pointed out in the introduction of this volume, Schlegel provides a kind of mission statement of organology in this context: "I too have been carrying the ideal of such a realism in myself for a long time, and if it has not yet come to be communicated, that is only because I am still searching for the organ for it [*weil ich das Organ noch dazu suche*]. Yet I know that I can find it only in poetry [*Poesie*], for realism will never be able to present itself [*nie wieder auftreten können*] in the form of a philosophy, not to speak of a system."[29] The means of mythological formation are named "organ," and poetry must be that organ. The shared goal of an "ideal realism" is tasked to poetic formation. Although Schlegel insists here that a system cannot help, his repeated use of Spinoza as example[30] suggests that the poetry he intends as philosophical organ will have a good measure of systemic unity. Indeed, the organ of philosophy here—we can now see—confers a higher unity on a whole discourse, one particularized in organology as Critical metaphysics, and generalized in cultural and scientific projects (such as the creation of a new [national] literature, a new "Bible," and a new politics). It is neither the putative reality of the imaginative synthesis of the system, nor the raw contradictory power of the fragment's rhetoric alone, that makes up that organ. The systemic unity that can be refounded with each flash of wit[31] is itself an organ, its energy originating in the ineluctable tension between its smallest part and its farthest reach.

The reason for a new mythology is contemporary poetry's lack of a "middle-point," a "sensuous center." The poetry of the ancients wove itself into a single large poem.[32] *On the Study of Greek Poetry* proposes a Classicism that resists the specific imitation of organs, but implies the need to develop new ones, thus placing the work in a transition from Classicism to Romanticism. The basic notion that the imagination is an organ.[33] But Schlegel is also searching for a new "legistlation" for modern poetry, which, because it is subjective (or broadly "sentimental" in the terms given by Schiller), is *interesting* but not *objective*. The term "organ" thus abounds here as a demand for the concrete laws of aesthetics which, in the *Dialogue*, are broadened to include the totality of culture—an organ is needed for a "full legislation," an aesthetic doctrine.[34] Perhaps most interesting in this organological revision of Classicist aesthetic doctrine is that Schlegel implies that new organs must be created by not imitating the local organs of individual Greeks, but instead their total culture. The statement is clearly a reworking of the Winckelmannian dictum that to imitate the Greeks, the Germans must become inimitable: "not *this* and *that*, not a single *favorite poet*, not the *local form* or the *individual organ* should be imitated: for an *individual "as such"* can never be a *universal norm*."[35] No imitation of organs: the Germans will have to build new ones, and for that, they will have to be Romantics. The formation of such organs is automatically revolutionary: "The great revolution will grip all sciences and all arts."[36] The organ of imagination, as the aesthetic organ of a general cultural revolution, allows the free movement of the poet between primordial chaos and its formation.[37] Artists themselves thus become the unifying factors in the sensuous world, and their cultural-political activity is organological: "Through artists humanity becomes an individual by tying the past world and the future together in the present. They are the higher organ of the soul [*Seelenorgan*], where the life-spirits of all of outer humanity meet and which inner humanity operates."[38] The collective soul needs the interface of inner and outer, and for this it needs an organ. That organ is the artist in imaginative activity, and his task is to illuminate the present from the perspective of the past and the future. The collective *sensorium* must range imaginatively over the expanse of time in order to act in the present.

The aesthetico-mythological organ is thus the means by which to make an arrangement of cultural disciplines real. On the one hand, there should be philosophy, the self-positing of reason, in its moral/practical guise, but also as intellectual intuition;[39] on the other hand, there is *Poesie*, which exists in the realm of "Chaos" and works more communicatively. It stands on the side of the "real," of *Realismus*, because it operates in the sensuous and creates therein. Religion stands between the two, the universal mediator and the mediator to the universe (more on this in a moment).[40]

As Manfred Frank has argued, the new mythology presents us with an early critique of bourgeois republicanism, not merely a defense of the new form of government. For Frank, the dichotomy organic/mechanical is used to separate between bourgeois laws (the laws of "mechanical," "Enlightenment" reason) and a possible form of organically synthesized community—a utopia.[41] Frank's point is both supported and metaphorically altered by an investigation of Schlegel's early essay (1796), "*On the Concept of Republicanism*." Schlegel argues, in response to Kant's "*On Perpetual Peace*," that there should be a "deduction" of the republican form of government. Indeed, he accuses Kant of insufficient Critical impulse with respect to government: institutions should be characterized according to their sources in the faculties.[42] And yet the talk of faculties is paralleled by talk of organs. Institutions of government can be organs precisely because they are the material expression of faculties: senses of the common body. For Schlegel, constitution and government differ as the permanent and alterable elements of the republican form, while represenationalism is the organ.[43] The constitution, however, might include insurrection (Schlegel clearly has the United States Constitution in mind). A constitution which rejects the possibility of revolution is simply blind, for Schlegel, since its power only extends as far as it actually constitutes. One which maintained the permanent possibility of insurrection, however, would "cancel itself" (*sich selbst aufheben*). And yet, the provision that revolution should follow failure of the constitution to maintain the republican ideal itself can be included: "That insurrection is thus *legal* which has as its motive the annihilation of a constitution whose government is merely a temporary organ and whose purpose is the organisation of republicanism."[44] Republican government—the only rational form of political constitution—may form for itself a provisional organ of self-correction when, in the course of organic events, reason fails to embody itself as constitution. Schlegel's new mythology includes a voice that might be characterized as the Jeffersonian strain of Romantic organology.

The Romantic Metaphysics of Morals, or, Schelling vs. Hegel, a View from Organology

By the time Schelling had moved to Munich in 1806, a significant new discourse of the organ had emerged. This was the so-called "organology" of Franz Joseph Gall, popularly known (then as now) under the name phrenology. The famous break between Schelling and Hegel, with Hegel's charge in the *Phenomenology of Spirit* that Schelling's absolute (which had also been his own) was a "night in which all cows are black," can be profitably read in this discursive light. From the position the young friends had held together (the absolute idealism of the "identity system"), and in common reaction to the esotericization of *Naturphilosophie*, the bitter rivalry between the "negative" and "positive" philosophies became a matter of the organ's extensive applicability. And where Hegel adumbrated his avoidance of organology (for which see chapter 8 of this volume), Schelling articulated in the *Freedom Essay* of 1809 a final organological position that united the electric and ideal, particular and general organs of his earlier career. I call his final organological synthesis the "Romantic metaphsyics of morals."

Even as Schelling was producing his *Naturphilosophie*, however, Gall was inventing phrenology.[45] In 1798, he published a mission statement of the program he would exhibit and demonstrate across Europe (especially in Vienna and Paris) in the coming decades. This writing was published in Wieland's periodical *Neuer teutscher Merkur*.[46] In a sense, Gall's "organology"[47] is everything but mythologizing—it is opposed to Romantic scientific efforts in their very roots. It rejects all metaphysics in science, and makes only apologetic efforts to respond to ethical concerns. If organs are the purely material loci of various capacities (*Fähigkeiten*) and tendencies (*Neigungen*), then the essay's title makes clear its interest: performance of function, *Verrichtung*. But where Schelling and the force-system-theorists had been interested in *Verrichtung* as a quasi-philosophical term, Gall employs it in a literal sense restricted to "execution." As Hagner points out, it is no longer the "soul" that has its seat in the brain—it is behavior. Organs are the material basis of observable capacities and tendencies. In the attempt to reduce the problem of the organ to brain functions, Gall nevertheless has to pay conceptual tribute to the Aristotelian complex. He writes that the eye does not see, the ear does not hear: the brain's organs and the brain as organ execute those functions:

> One errs if one thinks that the eye sees, the ear hears and so on. Every external tool of sense stands in connection with the brain through nerves, where at the beginning of the nerve an appropriate mass of the brain makes up the actual internal organ of this performance of the sense. Even if the ear, the eye, are quite healthy, then even the nerve of sight is undamaged; if the internal organ is sick or destroyed, then the ear and the sight-nerve are useless. Thus the external tools of sense also have their organs in the brain, and these external tools are the means through which the internal organs are set in connection or reaction with external objects.[48]

There is no "gland H" anymore, and this physiology is not quite as "fantastic" as that of Descartes, but the process is the same. Multiple organs are needed to filter the bruteness of objects into the finery of thought. But Gall is not interested in the problems of dualism—*something* allows perception and leads to behavior. Organs, for the localization theory that leads to phrenology, are internal to the brain, itself an organ. *Tool* denotes a mere instrument—the eye, the ear, the vehicle of transmission. The organs of the brain, as material as they are meant to be, must be internal and separated from the mere instruments of phenomenal collection. They thus perch in a sphere of potentiality, in true

Aristotelian style, mediating between brute perception (capacity) and behavior (tendency). Indeed, *capacity* and *tendency* describe just the potentialities the organs serve to communicate for any possible actuality. The system is less materialized—though it certainly includes a material moment, as do the other organological systems—than it is *behavioralized*, socialized.

The materialist moment of phrenology lay not in its organ-doctrine but in its inference from the brain's *organs* to the skull's form.[49] Organs might make up the potential spheres of perception as of action, but the social program of Gall's doctrine was not mistaken in developing a social and criminological hermeneutics from the skull. The identification of organs as possibilities of behavior does not imply, as Gall pleaded with the public to understand, the necessity of those behaviors.[50] Yet the doctrine of the skull's determination by the brain's organs—the skull as expression—was, for Hegel as for Schelling, philosophically unfounded. Their reactions to Gall—placed in the context of their own mythologizing efforts and those of their contemporaries—present us with a first separation of organology and experimental science. Gall's silence on philosophy is discursively different from the varying positions surrounding Kant on the conflict of the faculties. Silence can emerge only from a position of power, in Gall's case not the power of institutional and ideological backing, but that of a sense of methodological superiority. Organology and science began to miscommunicate when their premises are taken for granted. Success would be the result, not the philosophical grounding, of this science.

Gall visited Munich in 1807 to give a lecture at the Bavarian Academy of the Sciences.[51] Schelling responded with a short article, "Some Thoughts about the Doctrine of the Skull," where he questions the skull's meaning as fixed: "What Babylonian confusion could e.g. arise if the organ of the poet were first found in *Alringer* or the field-marshall-organ in General Mack, and then only were sought in *Schiller* and *Massena*; then one would have to look for totally other skull-raises."[52] The objection is not ethical, but epistemological: the Babel of organs which would result from their fixity would not merely contradict human freedom, but behaviors and identities would either lose connection to the system, or the skull would have to transform in real time with the vicissitudes of human behavior. By 1807, Schelling had little even polemical ink to spill on the systems he took himself to have so thoroughly refuted in the 1790s.

Between Myth and Science   265

Not so Hegel. Hegel used the platform of the *Phenomenology of Spirit* to attack and categorize systems of spirit-matter expression in his chapter on "observing reason" (*beobachtende Vernunft*). Consciousness (*Bewusstsein*) has constituted objects, and self-consciousness (*Selbstbewusstsein*) has emerged through the famous lordship-bondage relation. Reason has emerged as the recursive immediacy of object- and self-consciousnesses. Reason, for Hegel, will have to come to realize itself, but proceeds first along a detour of redetermining objects in the social world of self-consciousnesses. Science exists only in this complex, does its observational work at a specific level of self-realization of spirit, an uneven attempt to determine the world lawfully from the standpoint of reason's constitutive attempt to satisfy the foundational dissatisfaction of self-consciousness. Could there be an organ of reason for Hegel?

Observing reason conceives of an organ which does not occur elsewhere in the system. This organ is clearly borrowed from Gall, and yet Hegel contributes a systematic determination of it that anticipates his later objections to organology. Hegel considers two doctrines of expression in the *Phenomenology of Spirit*: Lavater's physiognomy and Gall's phrenology. Organ serves in the latter system to name the mechanism which potentializes a sphere of behavior, but Hegel correctly observes that organ should also be the term for the mechanism of expression itself. That which makes the skull's rises and valleys an expression of the brain's functions *should* be the organ. Hegel's use, borrowed as it is from Gall, thus relies textually on Romantic organology, using the latter to interrupt the flow of phrenological argument. Organs are here taken not merely as mechanism—this would reintroduce the Aristotelian problem Gall already has—but are projections from the specific standpoint of observing reason. Indeed, they are quasi-categorical—they are bastard categories.

The mistakes of physiognomy are apparent. Hegel applies "organ" (from Gall and Schelling) to Lavater's discipline, but then strips its use from his discussion of phrenology.[53] "Expression" is the wrong category to determine the relation between the material and the spiritual in both cases, but facial formation is potentially organ-based, while phrenology emphasizes the dead thing, the "bone" identified with spirit. Hegel does not reject "expression" because he adheres to a doctrine of preestablished harmony[54] ("expression" is, after all, a version of the *influxus*-claim), but because the

Phenomenology of Spirit tracks the self-constitution of spirit through stages of negativity. In the *Introduction*, Hegel writes his version of the "stringent Criticism" of Fichte and Schelling: thought can be taken as a medium or an instrument, but is correctly assessed to be in progressive mutual constitution with its putative objects. The problem with "expression" is that it takes two elements of representation and unifies them categorically, without including categorialization (synthetic conceptual unification) as part of the process. It is certainly possible to unite representations according to more or less rhetorically convincing strategies, but that possibility is arbitrary. It is opposed to the true possibilities of the concept as progressive synthetic constitution of objects.[55] Indeed, "observing reason" is the point in the narrative at which true categories begin to play a role:

> The unhappy self-consciousness emptied itself of its self-sufficiency and agonizingly rendered its being-for-itself into a thing. As a result, it returned from self-consciousness into consciousness, i.e., into that consciousness for which the object is a being, a thing.—However, this, the thing, is self-consciousness. The thing is thus the unity of the I and of being; it is the category. Since the object for consciousness is determined in that way, consciousness possesses reason. Consciousness as well as self-consciousness is genuinely in itself reason. However, it is only for consciousness, for which the object has been determined as the category, that one can say that it possesses reason.— But this is still distinct from the knowledge of what reason is.—The category, which is the immediate unity of being and what is its own, must pass through both forms, and observing consciousness is just the following. It is that to which the category exhibits itself in the form of being.[56]

The true goal of observing reason is the categorizing function that appears opposed to mere observation—categorizing reveals itself as the hidden work of that observation. This means that the true (conceptual) object of reason's observance is its activity—the object of this subject is the categorizing subject itself.[57] Category-work is synthetic in itself for Hegel, because it discovers lawfulness in an assuredly conceptual sphere. In order to recognize that this sphere is finally conceptual, the object will have to determine itself *as category*—as the immediate self-differentiating synthesis of I and being (the concept)—for reason. In fact, this is how reason will realize itself (in both senses). At the level of observation, the representation remains categorically underdetermined. This underdetermination means that observation grasps its objects—and in the case of physiognomy, itself as subject—as being. This asserted being can be juxtaposed with other asserted beings, but judgments that unify these beings remain arbitrarily possible. For conceptual possibility (real possibility) to come into view, reason must realize itself as self-realizing. But to do this, it will have to reach a nadir of self-objectifying assertion. It does this in phrenology.

The organ is that which makes the inner outer, the invisible visible, the *für sich* the *an sich*. It is essentially activity:

> To start with, it is only as an organ that this outer makes the inner visible, that is, into a being for others. This is so because the inner, insofar as it is in the organ, is the activity itself. The speaking mouth, the laboring hand, and, if one pleases, the legs too, are the organs of actualization and accomplishment that have the activity as activity, that is, the inner as such, in themselves. However, the externality which the inner achieves through the activity is the deed, in the sense of an actuality cut off from the individual. Language and labor are expressions in which the individual in himself no longer retains and possesses himself; rather, he lets the inner move wholly outside of him and he thus abandons it to the other.[58]

The term "organ" falls away, therefore, as Hegel follows first the physiognomist and then the phrenologist into a kind of navel-gazing observation. The activity suggested in the quoted passage is, for Hegel, the first hint of self-materializing observation, of the grasp of the inner in its passage to the outer, of the body possessed by reason. Thus, the term disappears in Hegel for two reasons. First, as we descend from organic physiognomy to the dead bone-matter of phrenology, observing reason stops uses "organic" terminology.[59] Second, even imported into physiognomy, the organ is literal for the mind construing its own body, but operates for that very mind as a locus of materialized reflection.[60] Insofar as that organ is objectified subjective activity, it remains tinged with the representational assertion Hegel is objecting to. This is why the term's use dwindles as the chapter progresses, and is not reintroduced later in the work.[61]

Phrenology lowers the already faulty conceptual content of physiognomy, judging spirit to be a bone.[62] Hegel calls this judgment "infinite"—it is a pure unification of opposites, and the most extreme form of judgment immanent to the perspective of respresentation. It is the ultimate false categorial assertion because it takes the two sides of representational assertion (the subject and the object) and identifies them.[63] It is a pure, unreflected category. It forces reason to grasp its own activity as such, because it points into categorial emptiness. For Hegel, the infinite judgment "spirit is a bone" is simply nothing other than reason's categorial assertion in itself. This is what makes it the entry into reason's self-realization: "Observing reason thus addresses itself to this wisdom, to spirit, to the concept existing as universality, that is, to the purpose existing as purpose, and henceforth its own essence is the object."[64] As observing reason prepares to become self-realizing reason, organ becomes a terminological artifact of the disappearing stage.[65] It was a terminological instrument of observing reason, and nothing more. Other than a dismissive—and intentionally vulgar—joke, Hegel has no more use for organology, Gall's or Schelling's:

> The depth from which spirit pushes out from its inwardness but which it manages to push up only to the level of representational consciousness and which lets it remain there—and the ignorance of this consciousness about what it says are the same kind of connection of higher and lower which, in the case of the living being, nature itself naively expresses in the combination of the organ of its highest fulfillment, the organ of generation—with the organ of pissing.—The infinite judgment as infinite would be the fulfillment of self-comprehending life, whereas the consciousness of the infinite judgment which remains trapped within representational thought conducts itself like pissing.[66]

Reason will realize itself in a play of immediacy and mediation that decreasingly requires materializing terminology. *Naturphilosophie* is itself brushed aside here, as the hermeneutics of spirit's self-becoming. Infinite judgment has a pale reflection in the contradictory coincidence of the organ of secretion and reproduction—nothing more.

If there is a critique of Romantic organology at work here, then it must dovetail with Hegel's critique of Schelling. That is a critique of the formerly shared "identity philosophy," perhaps Schelling's least organ-laden system. As I reconstruct Hegel's objection, it runs along parallel terminological and conceptual tracks. The terminological track is retroactively irrelevant, given the Schellingian objection to Gall's materialization arguments. The transcendentalization of the term "organ" in

Hölderlin as in Schelling shows that organology does not limit itself to the use legitimated for observing but not other reasons. If there is a conceptual objection, then it must be that Romantic organology is itself an example of the categorial emptiness of observing reason. This in fact syncs nicely with Hegel's basic critique of Schelling, to the effect that no principle of differentiability effectively links the real absolute with its emanating particulars. It is my contention that Schelling, if he is to have a response to this objection, will have had it only in the 1809 *Freedom Essay*, in a last but important formulation of organology that unites organological Nature with the organological I (the two poles of reason's observation).

The *Freedom Essay* is notoriously difficult, and represents Schelling's turn against Hegel and toward a "positive philosophy" in which a theological, irrational kernel resists the totalizing reason of his earlier systems. And yet, the essay explicitly unites *Naturphilosophie* with idealism by way of a reading of Kant's moral philosophy. Organology's final formulation in Schelling was tasked, from the beginning, with responding to Hegel. And although Schelling reached deep into the Christian tradition, orthodox as well as hermetic, to achieve his end, in a sense, theology becomes an *organ* here, before the late Schelling would return philosophy to its ancillary position with respect to theology. Because the problem was that of the activity of reason, it was natural that the context for such reflection should have been Kant's "practical reason," reason possessed of a rational will. Schelling adopted the project of Kant's rationalization of the will, but altered its source, making it a deed which, in the end, rationalized Nature, too.

The task of the *Freedom Essay* consists in a revision of Kant which sounds very much like the job of nature philosophy:

> It will always remain odd, however, that Kant, after having first distinguished things-in-themselves from appearances only negatively through their independence from time and later treating independence from time and freedom as correlate concepts in the metaphysical discussions of his Critique of Practical Reason, did not go further toward the thought of transferring this only possible positive concept of the in itself also to things; thereby he would immediately have raised himself to a higher standpoint of reflection and above the negativity that is the character of his theoretical philosophy.[67]

Andrew Bowie comments that "this is... actually another version of *Naturphilosophie*."[68] And indeed, the *Freedom Essay* seems to take both projects into its purview: to raise things to consciousness, now by conferring freedom upon them, and to objectify consciousness itself, by knowing—and ultimately by acting—it.

Schelling follows the passage cited above with the remark that "mere idealism" will not take him the step Kant did not go. In this context, however, it would be wrong to think that Schelling means "idealism" in general; rather, he means Kant's specific version, replete with its agnosticism about things-in-themselves and its (apparently sourceless) notion of freedom. Schelling goes on to call this concept of freedom "merely formal," and tries to replace it with "real" freedom. Kant thinks himself to be establishing a metaphysical doctrine of freedom based in morality—the synthetic *practical* judgment a priori (the categorical imperative) is, for Kant, a constitutive practical truth for us. Nothing supervenes the rule of this law, and the project of "transition" consists largely in "creating the kingdom of ends," as Christine Korsgaard has put it.[69] Key uncertainties remain within the Kantian model, especially what sort of role aesthetic, but crucially also ontology itself in the form of teleological judgments, can play in "helping" us to act freely.[70]

The *Freedom Essay* is notoriously hermetic in language, in outline, and in sources (the mystical tradition and above all Jakob Böhme).[71] Yet it is governed by a single figure—the relation between ground and consequent—interpreted according to the special concept of Reason Schelling was developing.[72] Indeed, the *Freedom Essay* begins with a reinterpretation of pantheism that proceeds by investigating this figure. Schelling rejects the possibility that "sameness" is intended in the pantheistic doctrine *Deus sive Natura*, defending his belief that judgments are characterized rather by active and differentiating synthesis than simple identity. Although the *Freedom Essay* quickly moves into a defense of the "real" concept of freedom, we should not underestimate the importance of this early distinction. Schelling here grounds his concept of the "indifference" of Reason: we have insight into the process we carry out, and we thus act and know simultaneously. Remarking on the dialectical movement between "A" and "A" in the judgment "A = A," Schelling writes:

Spinoza's most astringent expression is likely this here: The individual being is substance itself considered as one of its modifications, that is, consequences. Let's posit now that infinite substance = A, and the same considered in one of its consequences = A/a: thus the positive in A/a is still A; but on this basis it does not follow that A/a = A, that is, that infinite substance considered in its consequences is the same [*einerlei*] as infinite substance considered as such; or, in other words, it does not follow that A/a is not a particular individual substance (even though a consequence of A). . . . Therefore, even if substance dwelt only momentarily in its other consequences A/b, A/c . . . it would surely dwell in that consequence, in the human soul = a, eternally and, therefore, A/a would be divided from itself as A in an eternal and irreversible manner.[73]

This declared borrowing from Spinoza forms the main implement of the *Freedom Essay*, the strategy of which is to combine this tool with a Kantian metaphysical framework to generate a new Reason. Much of the above statement would, of course, not be accepted by Spinoza. Nevertheless, Schelling is dealing with a genuinely "Spinozan" problem, and it seems doubtful that, in the last account, this problem of predication was not also a question for Spinozism. In any case, Herder had laid the ground for such a reading of Spinoza by carefully raising the problem of different types of infinities in his *God: Some Conversations*.[74] The second of the *Conversations* is largely devoted to proving that Spinoza is not a pantheist after all, and this takes the form of denying the strict, predicative identity of things and God. Schelling's procedure is a different one: he reinterprets the sense of the copula, leaving Spinoza's identification intact but altered.[75] Nevertheless, as Manfred Durner writes, Herder and Schelling "emphasize the inner unity of nature and history."[76] The particularity of the seemingly contingent historical flow of things is tied, for both thinkers (and for both thinkers through Spinoza) to Nature itself. Rather than hedging however slightly on this unity, Schelling recasts the sense of unity as such along Romantic organological lines.

The moral action, for Schelling, is metaphysical, not because it meets requirements set up for another, constitutive metaphysics whose task it does not fulfill, but because the act here binds and opens to cognition in itself. Outside of time, it produces the ground of synthesis even as it occurs—and thus produces Reason. Reason is thus situated constitutively in the dynamic moment poised to become the future. It is as though

Schelling's romantic metaphysics of morals is simply Kant's embodied—constituting as it produces the temporal order it oversees. Freedom is simply the necessary engagement with this moving metaphysical ground. And freedom means making the self an organ.[77]

The most radical move of the *Freedom Essay* is, without a doubt, the identification of Will with Being (*Ursein*): the combination of Spinozism and Idealism literalizes and ontologizes the language of *action* and *deed* in Fichte and Hegel. *Wollen ist Ursein*—this metaphysics unites Idealism with *Naturphilosophie*, making will the source of theoretical insight. But for Idealism as a Schellingian project, this sentence places a potentially rational will in the position of providing the source of extension of reason, a "truly rational insight." Just as Schelling ultimately rejected Kant's restriction of metaphysical knowledge to the regularities of conceptual reasoning, so in ethics he rejected Kant's constitution of the experience of the act. The line between "hypothetical" and "categorical" in Kant is simply too cleanly cut for Schelling. Where Kant's system restricts rational willing to that which proceeds from the "standpoint" of constitutively practical reason, Schelling imagines a will that can make the external or the merely hypothetical categorical. This will cannot simply be "argued" for—it literally gets its (necessarily historical) information from its actions. Schelling was proposing a synthesis of deed with the possibility of moving intentionally but not arbitrarily into the future.

The act, for Schelling, binds as the judgment does for Kant. The act thus fills the copula, opening onto a future as yet undetermined but unfolding synthetically with our attention and intention. Reason therefore is not the ground but stands in a privileged, active relation to the ground, its movement being that of the ground in expression. We can say that Reason does not constitute the ground but is privileged to be constituted by it; and thus knowledge, paradigmatically of the act (since the ground is will) is conceived in Spinozan terms as immediate insight (*scientia intuitiva*) into the ground of Reason—knowledge of the cause literally through its effect (the vehicle of that knowledge itself). We can call this epistemological pantheism and a dynamically constitutive metaphysics of morals.[78] The moment of Reason we call the "future"—the movement in which synthesis occurs—is thus opened by Schelling in a plastic recasting of the Christian metaphysical tradition, and a synthesis of Kantian and Spinozan elements.

The history of Reason, history *qua* Reason, serves as the torque to pry open this new dimension of active predication, to *extend Reason* and extend its task. The interface between the rational will and its nonrational ground makes a theological organ for a metaphysical (organological) approach to politics. That interface tells us not how to act, but what an act *is*.

For Schelling, there must be a "rational insight" (*Vernunfteinsicht*) in the action itself. Schelling tells us, even as he quotes Lessing's *Education of the Human Race*, that the system of "true reason" would be "a system in which reason would actually recognize itself"[79] but that this recognition would not be enough. Schelling's future consists of the ur-movement provided for by the ontologization of the will, and the theoretico-practical indifference (the "indifference of Reason") created by the atemporal act.[80] He writes: "Only in personality is there life, and all personality rests on a dark ground that indeed must therefore be the ground of cognition as well. But it is only the understanding that develops what is hidden and contained in this ground merely *potentialiter* and raises it to actuality [*zum Aktus*]."[81] The twin concepts "life" and "personality" are thus reworked by Schelling on the basis of his new conception of pantheism. They are intentionally mythologized, and the personality of the human, immersed in hermetic emanation of life, becomes the *organ* of God.[82] In the key passage about Lessing in the *Freedom Essay*, Schelling points to the result of transformation of revelation into Reason: "We are of the opinion that a clear, rational view must be possible precisely from the highest concepts in so far as only in this way can they really be our own, accepted in ourselves and eternally grounded."[83] To conceive of the result of this process as human ownership of the truth is also to see that, for Schelling, Reason is capable of acting, is primarily *actu*, and as such, opens continuously onto a new dimension. This makes the relationship between two wills—general and individual—organological: "The general possibility of evil consists, as shown, in the fact that man, instead of making his selfhood into the basis, the organ, can strive to elevate it into the ruling and total will and, conversely, to make the spiritual within himself into a means."[84] Human selfhood is the necessary root of possible evil, but can be instrumentalized—made organ—not by God but by God's reliance on the very selfhood in question. The hermetic text is used to make the individual will the organ of the general, metaphysical will. But this makes the hermetic text the organ of a metaphysics of freedom, a metapolitical organology based on the progressive synthesis of practical judgments. It is not that we proceed *a posteriori*, that we "have examples" in this system. It is that the very acts that must serve as evidence for thought can be made exemplary. By following Kant's social rationalizing of the will, Schelling arrives at a metaphysical general will for which the individual will acts as an organ even as the philosopher of freedom makes theology his organ.

Just this line of thought would be reduced from its philosophical subtlety into the mysticism to which Schelling would later succumb. Schubert would use Schelling's thoughts to try to adapt the myth of the Golden Age to Bible belief, agreeing that selfhood is the "root of all our evil" and that it must be overcome through pious attitudes, to transform the self back into an "organ receptive to higher influence and directing the same."[85] In proposing the self as a lightning rod for divine influence, Schubert proposes a mythology no longer organological. The human would be the mere instrument of God, which is what the word here means. Organology evaporated, as Romanticism did, on the long itinerant journey that ended in Munich.

Not so in 1809 with Schelling. There, activity is not reified, as Hegel had worried. The organs of will are not those of mere self-objectifying representational projection, because they are—as is Being in the *Freedom Essay*—already self-differentiated. Reason grasps itself as will, as having started to act. It must trace the line backward to the origin of its act, only to realize that its apparent essence relies always on its status as expressed. Its actuality is actual activity, and organ is no longer the sign of an arbitrary hermeneutics of observing reason. It is, instead, a necessary term for the dynamic relations of subsumption and subordination in the will to know. Organs are meant to be the practical categories of a radical politics.

These organs are a synthesis of the early organological *Naturphilosophie* and the transcendental organology of the *System*. The electrical and ideal organs of those two systems are combined into moral or metapolitical organs. These organs—the ur-will as the conflictual source of insight and action—constitute both the source of insight into Nature as infinite productivity, and root the ideal organs of the transcendental philosophy in a source which is not foreign to them but not simply created by them.[86] This is, in fact, why the political part of organology is so important: it provides the possibility of intervention in nature and in history which is nonarbitrary but not preconceived. It tells us—now for the first time—that there can be something like a technological metaphysics. The politics of organology offers the temptation of a systematic philosophy open to the contingency of real phenomenal flow and yet always preparing to interrupt that contingency with human categories, both theoretical and practical. That politics thus bases itself on a revised metaphysical notion of *techne*: speculative activity.

PART III

After Organology

EIGHT

Technologies of Nature: Goethe's Hegelian Transformations

> In the forests of the Amazon River, as on the back of the high Andes I recognized how a single life, as if ensouled by a breath, from pole to pole, is poured out in stones, plants, and animals, and in the swelling breast of the human. Everywhere I was flooded with the feeling of how powerfully those Jena relations affected me, how I, raised through Goethe's views on nature, had been as it were armed with new organs.
>
> —ALEXANDER VON HUMBOLDT to *Karoline von Wolzogen, 14 May 1806*

Prologue: Eruption in the Academy

(Probably) 2 August 1830: Frédéric Soret hurries to Goethe's residence in Weimar, the news of the July Revolution in Paris fresh in his mind. Goethe responds to the visit with apparent sympathetic immediacy: what does Soret think of this "volcanic eruption," with everything in flames, and closed doors opening to reveal the true proceedings? Soret's response is ambivalent yet excited: the events are "terrible" (*furchtbar*), yet where else could the story have ended? The royal family had to be driven out. But this guarded optimism, a sort of liberal kernel cloaked in a calculated, conservative rhetoric, met with Goethe's dismissal. The event of the time—the epoch shift truly in the air—was not in the Revolution at all, but instead in the open skirmish in the Paris Academy of Sciences between the anatomist Georges Cuvier (1769–1832) and the zoologist Étienne Geoffroy de St.-Hilaire (1772–1844). Soret reports "several minutes" of complete stasis in his thoughts as a reaction to this strange, sudden shift in topic and enthusiasm.[1] After all, it was a war of the nontriumphant. To be sure, Geoffroy de St.-Hilaire's version of transcendental morphology, which compared the development of single organs across different species, ultimately construing zoology as the study of a single animal differentiated into different species. Cuvier may have won the discursive war, ending such speculative talk with his comparative studies or the *arrangement* of organs to classify different species—he created the schema of the four *embranchements* vertebrata, articulata, radiata, mollusca—but his opposition to any form of evolution put a definite limit on his contribution to biology. He was always polemical, attacking Lamarck posthumously for reversing the relationship between capacity and organ: "Other wants and desires, produced by circumstances, will lead to other efforts, which will produce other organs: for, according to a hypothesis inseparable from the rest, it is not the organs, that is to say, the nature and the form of the parts, which give rise to habits and faculties; but it is the latter which in process of time give birth to the organs."[2] Cuvier would displace organs from their pride of place in biology, just as much as the cell theory did on the other side of the Rhein. His comparative method, however, lacked all temporal elements. The combination of misfires—Cuvier's move away from organs and transcendental morphology's insistence on open-ended development—would clear enough categorial space for the evolutionary hypothesis to emerge, as we shall see. Goethe was right.

Nevertheless, Goethe's reply seems reactionary. Dismissing the news (which he clearly understood) about the potential breakdown of the Metternich consensus, Goethe seems to play his own Wagner here, withdrawing into the laboratory and away from life, idealistically setting his hopes for a new epoch not in institutional change but in abstruse shifts in zoological methodology in Paris.[3] Indeed, as he makes clear in the subsequent conversation (the minutes of Soret's stupefaction), what is at stake is a triumph of "spirit" (*Geist*) over matter in the sciences, a triumph which is German (that of *Naturphilosophie*) in an increasingly French-dominated institution;[4] Geoffroy has heralded the entrance of the "synthetic" method in the natural sciences into the leading institution of the day. This seems a far cry from the libertine and cultural-revolutionary enthusiasms of the author of *Werther*, let alone of *Torquato Tasso* or the *Roman Elegies*. If one thinks of the collectivist spiritualism of the French radicals of the time, or the soon-to-emerge German "Ideologists" in Berlin, Goethe strikes the figure of intellectual senility, missing the epoch-making mark of his last years in favor of personal obsessions making up the least of his cultural legacy.[5]

None of these impressions holds water. By examining Goethe's adoption and adaptation of Romantic organology, I will show that this first ending to my terminological history is one of philosophical synthesis with political possibilities. So far from isolating himself in the apolitical and the scientistic, Goethe sought, cautiously and painstakingly, to found a practice of observation that could not only tenderly attend to the phenomena in their generality and concreteness, but that could change this phenomenal world categorially, in short, to found a system of experience capable of altering the world, a technological metaphysics.[6] He developed this transformative system in oblique dialogue with the mature Hegel,[7] and since this system seeks a rational agency *in* the historical world during its present constitution, we may be justified in calling Goethe the first Young Hegelian—or in saying that he anticipated the fundamental gesture of that radical movement he appeared to foreclose on in August of 1830.

GOETHE'S ORGANS

Perhaps the best-known words in *Faust*, which Goethe was still completing when he turned his attentions to Paris, run:

> Two souls, alas, are dwelling in my breast,
> And one is striving to forsake its brother.
> Unto the world in grossly loving zest,
> With clinging organs, one adheres;
> The other rises forcibly in quest
> Of rarefied ancestral spheres.[8]

Faust describes not his mind but his animal soul as so many lustful organs, clinging to and keeping him in the world. This passage is mirrored in the essay "Polarity," where Goethe refers both to "two souls" *and* to the the distinction "spirit/matter," adding to the polar pairs "two body-halves." As Astrida Tantillo notes, the direction of the cut across the body is ambiguous. Perhaps it is precisely between the rational head and the sex organs, thus further mirroring the distinction between souls across the spirit/matter axis in the body itself.[9] Organs, however, play a greater role in Goethe's system than this casual mention might lead us to believe. Indeed, the formal role occupied by the term here is representative of a more general function I will examine in this chapter. In Faust's speculative-experiential monologue, the organs of the lower soul are the sexual organs as such—yet then again, they are, in the soul, merely the desire of attachment to the world. And then again they are both (lower) spirit and their functioning bodily counterparts, polarized and unified across the spirit/matter divide. This emphatically dialectical terminological gesture already points in the direction of Goethe's revision of the organological doctrine: uniting function and field across dual divisions—especially spirit/matter—the organ becomes the *organon* of an intentionally quasi-philosophical approach to experience and science which ultimately also serves as the foundation for the alteration of the object addressed by that experience and that science: the world. Goethe's organology is a technological metaphysics, one drawn from the same etymological sources as Romantic organology, and perhaps conceived, late in Goethe's life, partly through the prism of that Romantic impulse.

CANON AND ORGANON: HEGEL'S CRITIQUE OF KANT

As we have seen, finding a way back to the organological impulse in metaphysics was a primary factor in philosophical Romanticism in Germany. Maintaining the critical edge of Kant's thought remained central even as the Romantics painstakingly removed from the Critical philosophy what I have called Kant's "intuitionism" (his reliance on a "material" condition of possibility for the legitimate rules of reason). And while this methodological intervention allowed for the "*organon*" of reason to ground this new metaphysics, the term was enriched by the legacy of that other term, "organ," which provided the field (dynamic, developmental nature) for the *organon*'s (rational) function. The new metaphysics was, as Frederick Beiser has put it, rational and organicist simultaneously. But in this etymological conflation, it was more: it was meant to be interventionary, to make possible rational alteration of the world in a nonarbitrary manner. It was *technologia transcendentalis*, an attempt not to describe the world but to grasp it in its subject-implicating development. The task of grasping was only partially representational, and its other part was oriented toward changing it. The broad course of organology, into which Hegel and Goethe entered in conversation in the 1820s, was thus neither a return to "content" nor a bland dialectical formalism,[10] but instead a conception of the very content of being as the very forms-in-development of the human organ(on) itself, which thus, in keeping with the challenge first presented in these terms by Hölderlin to unite practical and theoretical philosophy, was capable of rational intervention in that content itself, the world in history.

Hegel's entrance into this discursive field is marked not by his relationship to the Romantics, but by his critiques of Kant and Goethe, respectively. Indeed, it was a sort of oblique conversation between Goethe and Hegel that brought a first finishing episode to the organological adventure. And that episode was always entangled in both men's relationships to Kant.

Hegel everywhere critiques Kant. In his most famous—and most hostile—assessment of Kant's philosophy, he presents the Critical system as the obverse of vulgar empiricism, which ignores the rational binding activity it uses for its putatively formally innocent observations. Hegel cites *Faust*:

> Chemistry calls it encheiresin naturae,
> Mocking itself and not knowing how,
> Has the parts in its hand,
> Is missing only the spiritual band [*Band*].[11]

As elsewhere,[12] Hegel uses Faust's indictment of the ideology of empiricism to mark his objection: that thoughtful analysis occurs in the allegedly naïve observation of the empiria. The obverse of empiricism's error, then, is overcommitment to analysis. And indeed, Kant's error is to cling too closely to the forms of thought (the categories) as fixed forms. The categories are, as we have seen, tied to the empirical forms of space-time intuition, but this is not Hegel's objection.[13] Nor, as it is easy to think, is Hegel's objection that Kant retains a "thing-in-itself" beyond thought. Or rather, it is not to the "thing-in-itself" as such that Hegel objects, but rather to the way it emerges within Kant's thought. For Hegel, the critical impulse remains incomplete in Kant's work, because an obdurate "objectivity" remains external to the rational work the categories do, not outside consciousness but within it. That is: Hegel is relatively unconcerned about the problematic "thing" because he is already concerned with the establishment of internalities and externalities inside consciousness in Kant's thought. The legitimate work done by the canon is fixed in its form, not due to the "material" of perception (the influence of which must remain technically problematic for Kant), but because the positivity of its laws can express only one kind of objectivity, one Kant describes as "constitutive," and which conforms, ultimately, to the mechanical-quantitative world-image of Newtonian physics and the qualitative-intensive scales of perception. The world so constituted is not or not only prey to a final exterior, but (also, and more importantly) to an internal splitting into an interior and an exterior, an impassive "material" element (ultimately related by Kant to apprehension) and a therefore necessarily incompletely constructive formal element (categorical synthesis).[14] Where the empiricist ignores the rational work he is unconsciously transporting into his observation, Kant misses the "empirical" element in his supposedly transcendental analysis of the forms of judgment. To repeat: it is not that Kant treats of genuine empiria in his analysis. Hegel's reading is subtler than that. The problem is that Kant maintains the problem of the empirical—its essence—in the categorical system. In trying to analyze what we do when we synthesize *a priori* (when we think nonarbitrarily but also informatively about the world), Kant imports the contradiction of thought and being into thought itself. The characteristic of being which does not allow of penetration by thought—does not allow of true understanding—is simply reproduced as the "material" of that thought in the judgmental form. The categories are fixed by a permanent contradiction, now not between thought and being, but between thought's form and thought's content. Hegel's critical question is: why should we regard this contradiction as permanent?

Thus where the empiricist eternalizes the world (in contradiction with thought), Kant eternalizes the forms of thought, in internal contradiction. Because of this determination of thought's form, only an analytical canon was possible. Hegel writes:

> *Determinateness* remains something *external* even at the extreme . . . for *thinking*; it remains simply *abstract thinking*, which here also is called *reason*. The latter—this is our result—delivers nothing other than *the formal unity* for simplification and systematization of experiences, is a *canon*, not an *organon* of truth, is not able to deliver a *doctrine* of the infinite, but instead only a *critique* of cognition.[15]

A canon serves only to regulate and simplify judgments, not to extend them—in Hegel's terms, to establish valid abstract statements, but not to extend knowledge to an objective, or truthful, form. That form, according to Hegel, could not be fixed, but would have to be in movement. Neither the world nor its knowledge is in a fixed form: their interaction is in constant revolution, mutual informing activity, or what is usually called the dialectical relation.

Kant had divided the Rationalists' distinction between *metaphysica generalis* (ontology) and *metaphysica specialis* (rational psychology, cosmology, and rational theology) into a canon of the understanding and a "regulative" system. For the form of judgment possible under the canon, the latter disciplines presented problems of a kind that could not be synthesized constitutively, since propositions about the immortality of the soul, the infinity of the world, or the omnipotence of God could not correspond to any intuition. Hegel's tack here is not to claim that we know those antinomic propositions constitutively, but that our constitutive knowledge is *also antinomical*.[16] The contradiction in thought between its objectivity and its form is affirmed, but as a part and foundation of thought's dialectical development. The instrument of that antinomical knowledge-process is the concept.

Perhaps the first (and certainly the most famous) articulation of this thought-in-movement is in the introduction to the *Phenomenology of Spirit*. Here Hegel levels critiques against earlier philosophers for working with a false notion of what thought itself is. Wavering between the poles of "passive medium" and "tool," philosophers have missed the methodological point: thought certainly mediates, but in doing so is not merely formal, but is the formal presentation of any possible content. Hegel's point here—at the beginning of an introduction to a book in turn intended as the introduction to his system—is not that thought and being are dialectically mutually informative (this is the result of the system), but that they appear within thought to be so. Taking a cue from Kant's antidogmatism, Hegel establishes a methodological baseline: what we know is presented in the form of thought, for us there is no "outside" to this form of content-presentation, and thus philosophy must take its impulse from this very form-content admixture, including its internal contradiction. The beginning of philosophy is where we take note of the concrete forms of thought, what Hegel came to call "thought-determinations" (*Denkbestimmungen*).

Note that Hegel rejects the vulgar organological notion of thought as a "tool." We should not be fooled into thinking that this leads him away from the more general organological path, for the sketch of the dialectic of the concept that follows is named, precisely in reaction to Kant, an *"organon."* The concept—as representation and as process—is the organon of metaphysics.[17]

This is not the place to explain the entirety of Hegel's concept-logic, but the basic procedure should be clear.[18] A representation of any sort has two sides: on the one hand, it is as itself a unit; on the other, it is related to something (to which it putatively refers), its object. The object is the intended content of the representation, but the critical method intervenes between intention and conclusion, and demands that we treat the object as provisionally nonindependent from its representational form. Thus Kant's thought-internal contradiction reappears, but now without its finalistic determination. The representation as unity is as representation a referring function, the object of which is uncertain already for the representation—the uncertainty of the reference leads us to imagine the object and the representation as external to each other. This process occurs, however, conceptually: we produce the concept of the object and the concept of its concept. This division is then seen as a "lack" in the representation, and drives us to create a new unity: the concept must be corrected, and applied to a new, more appropriate object. But this process of internal division is not limited by an external object: it is the nature of conceptuality as such. To approach philosophy from the perspective of the concept is to name this entire process "concept."

It now becomes possible to see why the concept can be a "synthetic organon" in Kant's sense. Because it works through the concrete interaction of form and matter without ever treating the matter as permanently resistant or finally "external," conceptuality provides its own index of concreteness. The categorical function—synthesis or the creation of types of unity out of given manifolds—is carried out by the concept, which is not only an element in the process but the methodological basis (or perspective) which allows the very analysis of that synthesis. Thought-determinations are really the concrete aspect of thought itself.

The analysis of conceptuality as process reveals that the "material" of judgment is concrete, not finally resistant to but dialectically implied in that very process. The concept as "organ" (in the Romantic sense) thus provides its own extensive field for its function, delimiting the extreme possibility of that field in concrete functional instances. Thus the concept is "objective": "logic becomes one with metaphysics, the science of things grasped in thought, which in turn were meant to express *the essentiality of the things*."[19] The "tools" themselves are deneutralized, and their putative objects are leveled into their field's functionality. Content is included in the form of conceptuality, treated from the perspective called the "concept," which *as this perspective* is the organon of logic and metaphysics simultaneously. The categories become the dynamic organs of the concrete, the world as conceptual process.

ORGANS OF MEDIATION: HEGEL'S CRITIQUE OF GOETHE

Hegel accuses Goethe of doing science without instruments.[20] The concreteness of the concept, allowing for synthetic knowledge of the truth of objects in their mutuality with the concept itself, is missing from Goethe's method, even as his experience reveals the schema of the truth:

> In experience everything depends on which sense one approaches reality with. A great sense has great experiences and glimpses in the colorful play of appearances the thing on which everything depends. The idea is present and actual, not something over there and behind. The great sense, like e.g. that of a Goethe, which looks into nature or history, has great experiences, glimpses the rational and pronounces it. To go further, one must be able to recognize the true even in reflection and determine it through relations of thought. The true in and for itself is thus in both these ways not yet present in its proper form. The most complete manner of knowing is that in the pure form of thought.[21]

Hegel dubs this type of rational recognition "*sensible* intuition" (*sinnvolle* Anschauung—literally "sense-filled").[22] Playing on the dual sense of the word *sense* (*Sinn*), Hegel accuses Goethe of leaving the term in this conflated state, where it deserves to be separated and rejoined to make up "the generality of matter [*das Allgemeine der Sache*]." Hegel writes that

> "Sense" is namely that wonderful word which itself is used in two opposed meanings. Once, it signifies the organ of immediate apprehension, then however we call sense the meaning, the thought, the general part of a matter.

And thus sense relates on the one hand to the immediate external part of existence, on the other hand to the inner essence of it. A sensible [*sinnvolle*] observation does not e.g. separate the two sides, but instead contains in the one direction also the opposing one and apprehends in sensuous immediate intuiting the essence and the concept simultaneously.[23]

Here as elsewhere, Hegel toys with the organ as not merely the part-expression of the whole, but also the two-sided expressive point of the whole's development.[24] Implicitly, he accuses Goethe of lacking this two-sidedness of function and expression in his presentation of affairs. But then, Hegel himself, as we have seen, never exploits the sense of organ systematically. Goethe's "sense" divines this general truth but remains without an organ for the determination of its insight: "with great sense he went naively with sensuous consideration to the object and had at the same time the full *presentiment of a concept-like coherence (die volle Ahnung ihres begriffsgemäßen Zusammenhangs).*"[25] The conclusion of the *Doctrine of Colors* is correct, but its method is lacking: Goethe's much-vaunted "intuitive" approach to nature lacks an internally insightful organ that could make it a candidate for inclusion in the metaphysical innovations of the early nineteenth century.

Hegel's objection is to a lack of rational accounting for the conclusion of the observation, a kind of meta-conceptual doubling (in his system, concept as representation and as process). The lacking function was called "organ" by the Romantics, yet Hegel's accusation is ambivalent with respect to Goethe: only one step is missing in Goethe's method (where the Romantics, for Hegel, misstep from the very outset). In the *Encyclopedia*, the Romantics are accused of having the opposite of Goethe's problem: they have a "conceptless instrument"[26] where Goethe is missing only the organ.

Goethe's insistence on method in his own work speaks directly against Hegel's critique, most pointedly in a letter from 7 October 1820 to Hegel himself: "Here we are not talking about an opinion to convince anyone of, but instead of a method to be communicated, one which anyone can avail himself of, according to his manner, as a tool."[27] Yet an anomaly in Hegel's own presentation of Goethe may already point us in the direction of Goethe's response to this critique. As pointed out above, Hegel uses Faust's *encheiresin naturae* to reject empiricism's claim to conceptual innocence.[28] In doing so, he implicitly recognizes that Goethe's method is not *empirically* naïve: Goethe sees "reason" (*das Vernünftige*) in nature. And yet Goethean methodology, while not committing the error of Romantic *Naturphilosophie*, retains a naïvety about just that judgmental analytical division of nature. Further, the passage makes clear that Goethe stands principally on the same side as Hegel in terms of the relationship of metaphysics to physics. Physics unknowingly defends a metaphysics, projecting a "world" which is ruled by a body of positive laws not visible but only calculable. The relation between general and particular is, for Hegel as for Goethe, unsatisfactory. (Newtonian) physics produces an unhappy parallel but not synthesis between the general-quantitative and the qualitative-particular, not truly obviating metaphysics but producing a metaphysical image not grasped or graspable by experience.[29] Hegel characterizes this as an overemphasis on unity: a formula is produced but is exterior to any actual individual or event in the physical world, and nature is split in two. An implicit metaphysics arises, a split world parallel to the split thought-world of Kantian methodology.

What is missing is a categorical shift.[30] In Goethe's case, it seems, this would simply be the activation of these categories for the constructive participation in the conceptual world-process. With clear reference to Goethe, he writes:

> The determinations are not indifferent for philosophical universality; for it is the self-fulfilling universality that simultaneously contains difference in its diamantine identity. The truly infinite is the unity of itself and the finite; and this is now the category of philosophy and thus also of *Naturphilosophie*. If the species and forces are the inside of nature [*das Innere der Natur*] and with this universal the external and singular is diappearing, then one still demands a third step, the inner of the inner [*das Innere des Innern*], which would be the unity of the universal and particular according to the foregoing.[31]

Goethe's method leaves a remnant of the division of inner and outer nature intact, because the inner and the outer are not doubly reflected as the dialectical core of identity and nonidentity which is the concept, the true organ or category of (nature) philosophy. Goethe observes but does not participate in the internal rationality of nature:

> The naïve spirit, if he views [*anschaut*] nature in a lively way, as we often find this validated by Goethe in a sensible way, then he feels life and the universal coherence [*Zusammenhang*] in it: he divines the universe as an organic whole

and rational totality, even as he feels an inner unity in himself; but even if we bring together all the ingredients of the flower, no flower will emerge. So one has called back intuition back into *Naturphilosophie* and set it above reflection; but this is the wrong path, for one cannot philosophize from intuition.³²

This passage is immediately followed by a citation of Albrecht von Haller's famous words, "No created spirit penetrates to the inside of nature / happy he, to whom she only shows her exterior shell." Rejecting this sentiment as an eternalization of the pseudo-metaphysical split in nature, Hegel quotes Goethe's rejection of Haller's poem, itself a poem entitled *"To the Physicist"* (Newton):

> "Into the inside of nature–"
> Oh you philistine!
> "Penetrates no created spirit."
> Only may you not remind
> Me and my kind of this word:
> We think: place for place
> We are inside.
> "Happy he, to whom she
> Only shows her exterior shell."
> . . . .
> Nature has neither kernel nor shell,
> She is all at once.
> You should above all test yourself,
> If you are kernel or shell.³³

Goethe thus also polemicizes against a putatively "exterior" nature and its impenetrability by reason. Hegel concludes his section on *Naturphilosophie* with the correct version of the generalizing (or "theoretical") and particularizing (or "practical") attitudes toward the reflective negation of nature in observation. He thus leaves an ambiguous challenge to Goethe. Goethe's words have stood in for the (correct) critique of empiricism (twice), a one-sided approach to nature's true Reason, and the norm to which that Reason should tend in correct nature-philosophical methodology. Even in Hegel's ambivalent assessment, we begin to suspect that Goethe may have a response, an "organ" with which to go about his thought-work.

FROM ORGANS TO CATEGORIES

Taxonomy and Idealism

Goethe's work forms a direct contribution to the family of philosophies I have been treating in this study, an alternate or response to the Romantic[34] organological project. That contribution emerges slowly and piecemeal in Goethe's works, forming less a "doctrine" than a conceptual attitude, and as such a robust response to Hegel's ambivalent organological critique.[35]

Goethe sometimes addresses his relationship to philosophy directly, especially in his scientific works. Yet his characteristic attitude in this regard is—I think—found in a literary text, the epistolary "novel"/aesthetic treatise *The Collector and His Circle*, published in 1799 in the *Propyläen*, a periodical meant to contain the heart of the classicist Weimar program. The collector himself (and it is hard not to see Goethe's primary identification with this figure)[36] delegates the taxonomic efforts to Julie and maintains a distance from the abstract brilliance of the "philosopher" in the group, insisting on a sort of cognitive innocence and tradition (his collection was started by his grandfather) in his relation to art. And yet, the novel narrates his reconciliation to the necessity of human cognitive intervention in both the production and the judgment of art. This reconciliation is clearly signified by the central debate between the philosopher and the *guest*. The stranger presents the reader with an explicit ideology of the attitudinal innocence of the collector: he constantly brushes aside the "metaphysical," foundation-searching thought-style of the philosopher, and maintains that Classical art had room for the ugly and the monstrous. It is at this crucial moment that a Romantically ironic note is sounded with a reflexive reference to the very organ (the *Propyläen*) in which the novel has been published.[37] Before withdrawing from the conversation and giving the pen to the philosopher (!), the collector attempts to convince his interlocutor with an example he cannot disagree with: the *Laokoön* statues, and the tradition of their Classicist interpretation. Both Lessing and Winckelmann are cited: surely no raw ugliness or horrifying thing (*Entsetzliches*) is included in this kind of art, but only that which is recuperated by a greater formal beauty.[38] The collector's horror at the *guest*'s response ("no") is then turned into a genuine affection for the philosopher, who steps in to defend reason and its—organological, as we shall see—role in the production and judgment of art. The novel thus doubles back on to Goethe's own history: having struggled with philosophy and its deductive stylings, he had, in 1798, begun a serious study of and with Schelling (on which more below). He had not and would not aspire to write philosophy: he left the defense of reason to the "philosophers." Yet he included reason, and its discipline (metaphysics) in his intellectual itinerary. This inclusion was not incidental: his oblique contribution to organology, as we shall see, was indeed about progressive inclusions of the intuited and developing world. Resigning the ideological pen to the philosopher, Goethe diagonally supported, included, and engaged in philosophy's characteristic task. From this engagement emerged a quasi-philosophical episode in the organological adventure.

Goethe's first significant use of the term "organ" (in his metamorphosis-writings)[39] and the first dislocation of that use into his aesthetic writings in the late 1790s seem to fall easily prey to Hegel's accusation that Goethe shows us "reason's exterior" without its internal "organ."

The *"Metamorphosis of Plants"* (1790) contains two central doctrinal points in his natural-scientific thought. The first is "type-theory":[40] the notion that development occurs according to a specific underlying form, one divinable from a correct perception of nature's individuals. An underlying form metamorphizes into the various parts of the plant. Rooted in the same teaching is the second point: the name for this form is "organ," the part expressing both its own function and its relation to the whole. Because this more general name is given to each separate form, the second doctrine is what Goethe named as a discipline: morphology. Development occurs through processes of concentration and expansion, forming "new" organs out of the functions of old. The two doctrines are related but not identical, and even seem to beg a question about form: what *is* the "organ" which, generating the model, schema, plan for the organic individual, is also encapsulated terminologically and ontologically in each "part"? That tension is at the heart of Herder's proto-organology, and it was from a Herderian setting that Goethe seems to have taken the term.[41]

Recall that the terms of Herder's abortive "organology" were the following: an organically developing Being qua force which, as God/World stood in mutual, expressive interaction with all of its parts, thought, material, or otherwise. Those "parts" were called "organs": functions with internal *teloi* expressing the necessary course of nature and culminating in humanity itself, the ultimate organ. The question seems cast in Herder's terms. Herder's pullulating organs populated a single plane, a mere extension, an exterior of teleological entities with no breaks. The methodology he had defended was based in analogy, in structural similarities between various organs proliferating in the organic unity of an undulating "Being."[42]

It is easy to place Goethe's early use of the term "organ" in this Herderian context. Take the more significant of those uses in the *Metamorphose* itself:

> Hence we may observe that the plant is capable of taking this sort of backward step, reversing the order of growth. This makes us all the more aware of nature's regular course; we will familiarize ourselves with the laws of metamorphosis by which nature produces one part through another, creating a great variety of forms through the modification of a single organ. . . . Researchers have been generally aware for some time that there is a hidden relationship among various external parts of the plant that develop one after the other and, as it were, one out of the other (for example, leaves, calyx, corolla, and stamens); they have even investigated the details. The process by which one and the same organ appears in a variety of forms has been called the metamorphosis of plants.[43]

> Thus we have sought to follow as carefully as possible in the footsteps of nature. We have accompanied the outer form of the plant through all its transformations, from the seed to the formation of a new seed; we have investigated the outer expression of the forces by which the plant gradually transforms one and the same organ, but without any pretense of uncovering the basic impulses behind the natural phenomena.[44]

We note here, in paragraphs that open and close Goethe's reflections, that terms for "exterior" occur multiple times, that "exteriorization" or "expression (*Äußerung*) is central, and that organic forces are everywhere at play. Goethe has successfully applied the Herderian monistic matrix to natural-scientific observation, leaving the "concept" of the plant announced in his 1787 letter to Herder somewhere in the methodology itself, allowing it to emerge in the narrative flow of his written observations. A failed organological metaphysics became the methodology of successful empirical observation. I would like to suggest here, however, that the philosophical impact of that methodology did not truly emerge until the inclusion of this writing in the "Morphological Notebooks" in the second decade of the nineteenth century, when it in fact became included in the history of organology.

The migration of the term from the Herderian context to that of the late 1790s, from biology to aesthetics and beyond, is marked by Goethe's engagement with Karl Phillip Moritz, who uses the term in the same framework, but for aesthetic purposes:

> Thus every higher organization, according to its nature, grips the organizations subordinated to it and tranfers it into its being: the plant unorganized matter through mere becoming and growing; the animal the plant through becoming, growing and enjoyment; the human transforms not only animal and plant through becoming, growing and enjoyment into its inner being, rather it also takes everything that is subordinate to its organization, through the brighest of all of them and smooth relecting suface of its being up into the reach of its being and puts it back outside itself, beautified, when its organ formatively completes itself in itself [*wenn sein Organ sich bildend in sich selbst vollendet*].[45]

Goethe's reflections on Moritz's essay *On the Formative Imitation of Beauty* (1788) come from conversations between the two men during Goethe's Italian journey, less than a year after his famous letter to Herder. The text slides analogically from one form and its organs to the next, and the creativity of the artist is attributed to the distinctively "formative" aspect of the human "organ." "Webs of organization" intertwine in an *"active force"* (which is greater than but related to "thought-force" [*Denkkraft*]—Herder's term in *God*). And two epigenetic forces—formative drive and sensible force (*Empfindungskraft*)—mirror (*abspiegeln*) the whole in its organization. The former, organic/ontological forms a perfect microcosmos, but the latter has a "single point" in the whole missing. The organic genius is thus again mirrored by the faculty of taste; a sense for the beautiful does not require genius. This relation is that of man to woman as well.[46] Thus the organs of the plant (for example) become those of the imitator of nature: ideal organs, as in Herder, are attributed to the figure of the genius. These organs confer on him the ability to extend the ontological hall of organic mirroring into the human world. Being is one and differentiated, in art as in nature.

This type of mimesis changes the frame of reference for the term, but makes no change to Herder's cosmology of organs. At best, this system differentiates thought-force into a finer taxonomy. From a Hegelian perspective, however, it is just that: an extended taxonomy. If Goethe was going to defend a kind of mimesis, it was going to be a different one.

In the 1790s, Goethe started to use the term "organ" with some frequency again, but now with a generic difference. This difference is, I think, the first clue to the emergence of a particularly Goethean idealism, and the harbinger of his mature organology.

Goethe's theoretical writings are marked precisely by a strong taxonomic bent. Thus the essay "Simple Imitation of Nature" (1789) divides between "manner" and "style" beyond "simple" imitation, and *The Collector and His Circle* ends with a counting-up of artistic "styles," numbering not fewer than six, not counting subdivisions by emphasis and area of focus.[47] This apparently external descriptive enterprise is undercut, however, by some uses of the term "organ" that indicate Goethe's increasing commitment to idealist trends in philosophy at this time.

In the introduction to the mission statement of the classicist project, the introduction to its organ, the *Propyläen*, Goethe divides again between artistic "treatments" of the object to be constructed, naming the "spiritual" (which works through the internal conjunction [*Zusammenhang*] of the object), the "sensible" (which presents exciting, mild, and pleasant sense-impressions), and finally: "The mechanical, finally, would be that which through some bodily organ works on specific matters and thus confers being, actuality, on the work (*Arbeit*)."[48] Taxonomy mixes here with definition: the "organ," meant quite literally, prepares determinate materials for "work," giving the cipher through which the constructive activity can alone operate. This "organ" is close to the type of interventionary work admitted by Goethe in his reflections in *Simple Imitation*, where the "simple" imitation (which Goethe calls the "courtyard" of style) is mediated by "manner" (which binds the constructible-imaginary individual together in the manner of the artist's *Geist*). Finally, "style" performs a complete mediation of nature and spirit: "As simple imitation rests on calm existence and loving surroundings, manner grips an appearance with a lightly capacious mind, so *style* is based on the deepest fundaments of cognition, the essence of things, so far as it is allowed to us to know it this in visible and graspable forms."[49] This earlier writing—summarizing the experience in Italy in 1789—shows us the direction Goethe is taking: the mediation of the more abstract cognitive apparatus with the concreteness of intuition and feeling.⁵⁰ By presenting art in this holistic manner, Goethe's writing also trends toward the nature-philosophical point he would come to embrace: ultimately, reason's framework must come to be identical with that of nature. The metaphysical point—that reflective knowledge must be included in nature, or that knowledge must include its own justification—begins to emerge, parallel to a new use of the term "organ." Thus simple imitation of nature—vulgar mimesis—first gives way to mimetic ordering (or proliferating Herderian-Moritzian organs, encapsulated one within the other in Being) and then opens to the demand for a metaphysical attitude in artistic and scientific encounters with nature. The demand presented by Goethe's work around the time he befriended Schelling is for a mimesis of the subject-object relation itself, the concrete inclusion of reason in nature or nature in reason. And this was the stated task of the Romantic organ.

But if this demand existed, and if Goethe's work was only just coming into its oblique yet essential relationship to Idealism, a direct anticipation of his mature "doctrine" nevertheless occurred in 1798, precisely in *The Collector and His Circle*. In the debate between the philistine stranger and the philosopher—who is speaking in the name of "reason"—there is a clear tip of the hat to foundational, abstract thinking. As the caricatured *guest* speaks out heretically against classicism, the philosopher steps in, responding to admonitions about his metaphysicizing with more metaphysics. The *Gast* is satisfied with that artist who can pick out actual characteristics, simplify them, and produce a presentation of his concept. The philosopher objects, saying there is more to the story, and the collector states pregnantly: "Zum Versuche gehe ich mit."⁵¹ The philosopher describes what is lacking: "Through this operation there may have arisen in any case a canon, examplary, scientifically estimable; but satisfying to the mind (*das Gemüt*)."⁵² Here the collector breaks in again, strongly agreeing with this restriction: only a canon (a positive body of laws) can emerge from the abstract species-production that occurs through the representation of a concept. What is needed is "more" than this, and when the collector demurs, the philosopher picks up the thread: a full individual must be established, a circle of representation must be closed in the process, and this process must itself be that of reason, the only process satisfying the mind (*das Gemüt*). The *guest* loses his cool at this point, accusing the philosopher of speaking obscurely. The philosophers avers that none of this is philosophy, merely "various experiential matters," and the discussion continues:

> Guest: you call that experience which other cannot understand at all!
> I: There is a proper organ for every experience.
> Guest: Surely a particular one?
> I: Not a particular one, but it has to have a certain property.
> Guest: Which is?
> I: It must be able to produce.
> Guest: Produce what?
> I: Experience! There is no experience which is not produced, brought into being, created.[53]

The philosopher—who is not Goethe, but whose voice is now fully presented by the extradiegetical Goethe—anticipates Goethe's mature organological attitude. Experience itself is organological, a matter of organs and their plastic production, fully mediated by reason and only grasped in its essence through that process. Mimesis is self-reflexive and, in a sense we shall explore shortly, technological. The artist as the scientist must take account of philosophy, or at least of its subject, reason, the mediator of which cannot be excluded, but must itself produce and include by that production experience itself. And this production cannot occur abstractly, but instead by means of a concrete instrument or an organ.[54]

The Categories of Organs, or the Representation of Representation

Goethe had used the term "organ" in his essay "Metamorphosis of Plants" (1790) to denote the functional organic part, underlying both the type (unity) and development (becoming) of the plant.[55] The sense of the term, however, was altered by the essay's inclusion in the *Morphological Notebooks*,[56] which contain the beginnings of Goethe's broader organological thought, a conception of mimesis as the reproduction of representation itself. For Goethe, mimetic activity, scientific or poetic, is a doubling of representation's (already double) split between representing subject and represented object. The inclusion of the observer in the realm of observed—and of the represented in the realm of the representing—is recognizably Idealist in conception.

Goethe developed this line of argument further in his final biological efforts, especially his review of the "academy debate" between Cuvier and Geoffroy de St.-Hilaire and a late series of fragments engaged with Galen and Aristotle. This latter discourse puts the emphasis on the reproduction of the mimetic enterprise. It thus points to Goethe's organology as a social-scientific program for altering the world itself in collective reproduction of representation. Goethe's late biological classicism mixes with his long engagement with Idealism to produce a response to Romantic metaphysics, seeking to bind the latter's speculative enterprise to emergent disciplinary scientific institutions, mediating between comparative anatomy in Paris and Hegelian metaphysics in Berlin. Goethe thus insists that science should be what it in fact *is*: a determiner of a collective worldview inclusive of the citizens of that world. To participate in that determination is to inherit and form new organs of transcendental perception, to remake the world categorially in scientific concert.

With the name "morphology" came a cluster of theoretical writings, poetic and expository, which shifted the sense of "organ" in the *Metaphormophosis* retroactively, grafting the structural-developmental organ onto its philosophical counterpart. What emerged was a version of organology, and a response to Hegel's charge that Goethe's methods were without a self-reflexive instrument which could ensure their truth.[57]

Goethe's perspective on mimesis itself, on the scientific or aesthetic representation of objects, had gained in just that reflexivity by the time of the "Notebooks." He writes:

> Everything in the object which is in the subject +X
> Everything in the subject in the object +X lost . . .
> . . . .
> Concede power to the object
> Renounce . . . +
> Raise the subject with his + and don't recognize that +
> Everything which is in the subject is in the object, and still something more.
> Everything which is in the object is in the subject and still something more.
> We are lost or secure in a double way.
> If we concede the object its More and approach our subject More
> Let us insist on our subject.[58]

Epistemologically, the situation is one of mutual excess. The identity of subject and object is complemented by the difference between them, and this dialectical formulation characterizes representation as a re-production of that dialectical identity itself. Representation—on both sides of the subject-object divide—always represents that divide even as it produces the represented content. Mimesis is a "double infinite" built on the apparently simple mutual influence of subject and object. Self-reflexivity is built into representational knowledge, and Hegel's charge is answered. This is not only the proliferating organ-world of being and becoming on Herder's model (which would be open to the charge of external observation), but the intertwining relationships of both: "Hardly has he convinced himself of this mutual influence [of subject and object] than he becomes aware of a double infinite, in objects the manifold of being and becoming and relations crossing in a lively manner, in himself however the possiblity of an infinite training as he makes his receptivity and also his judgment skilled for new forms of absorbing and counteracting."[59] The subject's capacity to receive—just like its judgment—can experience on this basis infinite "training" or literally "outformation" (*Ausbildung*). On the one hand, receptivity is essential—Goethe is not an idealist in the sense that everything is meant to be *inside* representation.[60] And yet, even as a part of any possible object remains yet to be received, its inclusion in the crosshairs of dialectical self-representation shows that Goethe's progressive stance toward knowledge is developmentally Idealist: the circle of knowledge might find mediated and temporary completion in reason's organ. And that organ will have passed a Hegelian test, because it produces reflexive mimetic knowledge—the representation of representation itself. Indeed, the first passage above, found on the verso of a folio of the *Notebooks*, might be taken to be playing with Kant's "transcendental object = X," the constituted form of any object for us.[61] If this is the case, then Goethe's point is that that which we constitute is nevertheless an apparent excess, not in principle in some "beyond" but not necessarily in our organ already. The world is *able to be included*. The inclusion of *more world* is not contingent but appears so, and this parallax is essential. Our "tender empiricism"[62] insists—*pocht*—on the subject.

Goethe gives this representational apparatus the name organ, writing programmatically that "to grasp the truth a higher organ is needed than for the defense of error."[63] This organ for "higher" knowledge responds both to his own demand for an articulated method for naturalist observation,[64] and also to Hegel's critique. And yet "organ" is of different provenance from *organon*.

The classical commentary on the definitional problem of the "organ" is Aristotle's *De anima*. The philosopher uses "*organon*" in that work to specify that part (*morion*) which is possible where its object is actual.[65] The instrument works in a determinate field, and so must be suited to that field. Thus, for Aristotle, the senses are fields of possibility for their respective objects. The sense of touch is already suited to the touched—otherwise touching would not occur through this sense (that is, it would not be *this* sense).[66] With the help of an organ as the seat of possibility, the sense must deliver the formal element of perception to the understanding, without involving the material element. We see through the eye without having the seen in our eye; we represent wood without having corneal splinters. The organ separates the formal from the material but also unites both as a single object of perception. The "object" of the organ is subdivided into the representation and its being *qua* object. This is the most abstract determination of the process of knowing. The *possibly known* (in this passage in *De anima*, the sensible possibility) is transformed into the actuality of experience. The realm of the possible is delimited by the function of knowing, the *organon* itself.

Here Aristotle draws attention to a problem: where the sense-organs determine the specific laws of their fields of perception and thus delimit the areas of possible perception for themselves, this cannot possibly be the procedure for reason itself. For if thought itself had an *organon*, then it could only offer our knowledge those specificities suitable to the possibility of that tool. Knowledge would limit its field to the specifics of its instrument's function. We would be severely cognitively limited.[67]

Goethe was aware of the problem of the particularities of any organ, and thus tried to root knowledge in not "an" organ but organs as numerous as the objects they are meant to apprehend: "Every new object, viewed well, opens a new organ in us."[68] Aristotle himself had come to the conclusion that the understanding always *is* the object thought by that understanding.[69] The generality of thought demanded by metaphysics implies that there can be no particular and separating mediator—no actualizing means or instrument—between thought and its object. Yet thought does cover a field of possibility, one greater than any possible organ. That field of possibility is principally unlimited: cognition in general functions without mediator, and is thus noumenal, unmediated.

Goethe reflects both sides of this difficulty. He writes first of an organ, the eye, which he develops toward organological freedom on the basis of Aristotle's model. But he also talks of higher organs—categories, as we shall see—in dynamic cognitive development. Organology covers the spectrum of human cognition.

Perhaps the *Doctrine of Color*'s most famous passage tells us that the deeds of phenomena, not their essences, are to be observed and described, and continues: "Colors are deeds of the light, deeds and sufferings. In this sense we can expect from them information about light. Colors and light stand to each other, to be sure, in the most precise relation; but we must think both as belonging to the whole of nature: for nature it is nature as a whole [*sie ist es ganz*] that wants to reveal itself to the sense of the eye hereby."[70] The unity of the phenomenon must be established for scientific observation to have validity. The *derivation* of color will not do.[71] We need instead to place color in the whole (*das Ganze*) of nature, because only this inclusion can demonstrate the desired unity. Nature, however, reveals itself particularly through the sense belonging to the eye, and does so completely. An earlier redaction of this passage had phrased it: "Colors are all of nature revealed to the organ of the eye and to see right means to be right [*recht seyn*]."[72] This more compact version of the sentence shows that nature's holism can be divided by organs and nevertheless remain complete.[73] Indeed, the sentence is a sort of *précis* of the whole work: colors are nature itself, whole and revealed in a single organ. The possibility of error in perception parallels that of moral degeneration, and their simultaneous correction foregrounds the section on the "aesthetic-moral" effects of colors. Nature's totality is given under one aspect to the eye's *organ*. The term can be taken here as the physical eye (*appositive genetive*) *or* as a distinct capacity—on Aristotelian lines—for the realization of the eye's polarized field of possibility (*possessive genetive*). Goethe must at least include the latter sense in the former, for the organ's capacities develop just this Aristotelian model toward "freedom."

Goethe makes good on this promise of an organology of color in the section entitled "Totality and Harmony." The eye—as in Aristotle, suspended between two poles in a field (of vision)—takes the single impulse it receives from one end of that spectrum and complements it, running from one pole to the other, revealing the totality of color. This is most obviously seen, as Goethe claimed, in the after-image on a white surface, which produces the contrary color in the wheel. But this is only the first step: the organ, in fact, "sets itself into freedom by produces the contradiction of the singular thing force onto it and thereby a satisfying wholeness," because a "need for totality" is in-born to "our organ."[74] Goethe ascribes spontaneity to an organ both physical and metaphysical. The specificity of that organ (its field of possibility) is thus complemented by a function which, going immediately to totality, frees up perception for error and for truth. The organ is formed and forms toward freedom, a freedom which attaches also to the organs of knowledge.[75] Organology inserts itself between science's objects and philosophy's subjects, and extends its terminology into both. The organs of plants become the free organs of sense—and these pass into the *organs of knowing*.

All of nature flows formally in through the eye, which totalizes and harmonizes nature, freeing itself technically from mere receptivity. This eye, and the scientist possessing it, must have some means of making sense out of this colorful nature. In the *Hefte*, then, Goethe names that means "intuitive judgment" (*anschauende Urteilkraft*),[76] a term derived from Kant[77] but deviating from the letter of the Critical system.[78] *Intuition* for Kant is the representation of particulars without anything "general." It is therefore always of the senses, because the general can only be contributed to knowledge by the concept. On this basis, Kant polemicizes against any possible "intellectual intuition,"[79] that is, representation of noumena. But, as we have seen, Kant makes a famous regulative exception in the name of biology in *Critique of the Power of Judgment* §§76 und 77.[80] As I detailed above, the notion of intuitive understanding in that passage stands in stark contrast to the subsumptive judgments of the constitutive function of the understanding. Where the latter proceeds to construct the relation between concept and intuition, between part and whole analytically, the former conceives of the relation holistically or synthetically. For the constitutive understanding, wholes are aggregates; for the intuitive understanding, wholes are determinative moments in the development of a complex being characterized, finally, by its possession of organs.

Goethe calls his variant of this knowledge not "understanding," but "power of judgment" (*Urteilskraft*).[81] The understanding, for Kant, is the faculty of concepts, and Goethe replaces it terminologically with the faculty for unifying representations, of judgment in general, not limited to "discursive judgement."[82] The understanding is the unity of consciousness in the concept, while the power of judgment is that which unifies whatever representations—intuitions or concepts—are produced in that consciousness. The power of judgment is thus the practical element of the ideal, the activity of spirit itself.[83] Its forms determine the quality of knowledge and its content. And yet, for Kant, judgment's operation is second-order, working only on consciousness's representations, while the latter possess a possible (and always problematic) relation to an "outside."[84]

Because the unifying functions of judgments (the most general form of which Kant calls "categories") do not have an effect on any "exterior," for Goethe they can therefore become kinds of synthetic thought. This intellectual-historical conjuncture thus recommends that the doctrine of the categories in Kant be brought into conversation with the Aristotelian teaching on "organs." This is because the categories do the work of the organ: they divide the content of the represented from its form and determine knowledge as possible forms of whatever is represented. These forms are thus unities of the represented, and as forms they determine this content, the knowledge of which consists only in these forms.[85] For Kant they only determine this content formally, and the Critical question remains: whether we can have legitimate knowledge of the coincidence of these forms with any possible content. But for Goethe the content in the form is not external: a category becomes an organ in just that moment where the formal element of cognition itself becomes a synthesis of content and form. For Goethe, the judgment is the mediator that offers us the unities of nature we investigate in science and reproduce in art. We intuit nature not in the concept, but in the unities of the forms of judgment. The organs of these judgments are these forms themselves, which progressively transform themselves according to and in symbiosis with the "content" they treat. These forms—actually formations of organs—result from the process of science. Science is not merely the constitution of a world made image and law, but also the production of knowledge itself, the production of the meaning of the represented world and thus of that world itself. Science and culture are united socially in this judgment: we must, because we in fact do, form and formulate the world. Our constructive activity is a kind of creative imagination, and it literally creates experience, not in isolation from "objects" but in progressive mimetic reproduction of our very representation of them.

In progressive, investigatory reason *qua* organ, we continuously form new organs according to the syntheses demanded by the process of judgment. Goethe's famous "tender empiricism" is thus meant to remind the judging consciousness of its own formation (and mutual formulation) in being. The judgment unites intuitions and concepts into an order (*Ordnung*) called the "idea."[86] It then allows this order to be continuously transformed judgmentally according to intuitions that are parts of that judgment's unity. This mutuality allows in turn for a kind of intervention into nature. What is "tender" is the categorial transformation we undertake.[87]

We can now return to the "*Metamorphosis*," with its overtones overturned in its new, organological, context, and see its retroactively Idealist organ as a response to Hegel. Hegel's notion of the concept is the name for both the representational element of the cognitive process and the name for that process itself. The concept's double appearance in Hegel's taxonomy gives it the quality of the Romantic organ: it serves as the demarcator of the field of cognition even as it assigns rules to that field or provides the function in the process. It is thus the most general name for a "category"—it unites cognition in judgment. But it is also the object of that cognition, and so is also a category in the Aristotelian sense, ontological and formal-ideal simultaneously. The specifically Critical edge to this thought is its inclusion of the "material" of thought in the analysis of that thought, the treatment of content as formally implicated. The name for the unification of the form/content split, and the name for that split itself, is concept, and that concept is the "organon of truth."[88]

This dual role is precisely that played by the *organ* in the *Metaphormophosis*, read in the context of the theoretical apparatus of the *Morphological Notebooks*. Morphology alters the sense of metamorphosis. "Organ" comes to play the categorial role of Hegel's concept, neither only the functional part of an organic universe, nor merely that part reflected seamlessly into the ideal, a "force" underlying genius—as it had been for Herder and Moritz. There is a seam, precisely where the two organs meet. The subject-object split is reproduced, and thus becomes identity including difference.[89] Goethe's science becomes a mimesis of the order of representation itself. Goethe is strikingly close to Hegel in this conception, but his instruments of cognition are other than Hegel's. The intuitive power of judgment is an instrument that generates others, an "organ of organs" that is situated where "inner" and "outer" appear to (but do not) split. Form and instrument coincide.

If we accept that the insertion of this essay into the *Morphological Notebooks* alters the sense of its science—literally the manner of our knowledge of the phenomenon—then the organ comes to play the categorical role of Hegel's concept. The "organ" is not only the functional part of an organic universe. Nor, however, is it merely that part reflected seamlessly into the ideal, a "force" underlying genius. There is a seam, precisely where the two organs—recall the passage from *Faust*—meet. There is literally a subject-object split, and this split itself is mirrored as identity and difference in one. Goethe's science is a mimesis of the order of representation itself. If we look to the conclusion of Hegel's treatment of *Naturphilosophie* in the *Encyclopedia*, we see the true but limited parallel to Goethe's work, and thus resolve the ambivalence in Hegel's assessment's of Goethe:

> Because the inner side of nature [*das Innere der Natur*] is nothing other than the universal, so we are, when have thoughts, in this inside of nature with ourselves. If the truth in the subjective is the correspondence of the representation with the object, then the true means in the objective sense the correspondence of the object, the thing, with itself, that its reality is appropriate to its concept.[90]

Hegel's point is that, because of the reflexive yet ontological process of the concept in which we participate, there is a normative way of looking at real difference. Where the proliferation of organs obliterated that difference by including all organs in a single generic order, Hegel asserts it as real and conceptual simultaneously. Difference made contradiction is real, and this reality's contradictory nature allows for normative grasping of that reality. Goethe, it turned out, shared this viewpoint. Hegel's citational sense of Goethe was better than his stated opinion.

This real normativity in subject-object mimesis is the metaphysics of judgment, and without it, science—and history—lose their metaphysical quality. "Truth" is neither merely object-thought correspondence, nor the correspondence of elements of our cognition with each other according to a "canon": it is the conformity of the object to its reflection in reason, or, in his terms, to "itself." I have shown that Goethe is strikingly close to this conception, although his means for that transformation-based truth are other than Hegel's. And thus it is this difference in means—the line between two reactions to Romantic organology—that makes up the distance between the two thinkers.

The response to Hegel first answers to reason's demand for mediation of all known material, then goes a step further. For the organ, according to "der Philosoph" in *The Collector and His Circle*, must also produce experience.

Here Goethe's Idealism foreshadows his late turn to an ideal technology. The possibilities of conceptual intervention in and receptive inclusion of the world in development extend both to our speculative capacity and to our natural-scientific method. Scientific practice exceeds the bounds of contemporary disciplinary science; Idealism turns from representation to the reproduction of representation itself, and in so doing, aims to make intervention not only into nature but also into (nature's) history possible.

Reason's Hand: The Organ of Organs and Ideal Technology

To his arsenal of tools Goethe would add the hand that operates all tools. Drawing on the ancients at the very end of his life—precisely in the context of the academy debate—he would make technical development, reason's instruments, and even perception part of the same order of experience, belatedly developing a variant on Romantic organology.

In his last years, Goethe was both close enough to the mature Hegel and sure enough of his own quasi-philosophical approach to nature and history to level critiques back at Hegel, privately but with assurance that they would reach Hegel's ears.[91] Thus on 28 August 1827, he received the Hegelian jurist Eduard Gans in Weimar. The conversation focused on Berlin's university and Hegel's rise to dominance. Goethe was concerned to see that the philosophy current in Berlin did not exclude productive engagement with empiricists, repeating his critique of Hegel, asking if philosophy's categories did not need to change as history and science advanced.[92]

Gans responded, with specific reference to the Hegelian organ *Jahrbücher für wissenschaftliche Kritik* (to which Goethe had been invited to contribute), with the Hegelian dictum that philosophy is not an attempt to step out of time but to present its time in thought, to transform with that time. This, according to Gans's report, appeared to please Goethe. The conversation turned to specific examples in the periodical.[93]

Goethe's concern stemmed from his administrative sense for balance. Hegel's prowess notwithstanding, it was exceedingly rare to find one person with both synthetic and analytical genius. This was demonstrated, for Goethe, by the increasing tension between a synthesizer and an analyzer in Paris: Geoffroy de St.-Hilaire and Cuvier.[94] By referring to an example weighing ever more heavily on his mind, Goethe (perhaps unintentionally) brought his organological conversation with Hegel into the final register he would give it: the instrumental or technological side of its etymology. And here, for reasons I will explore below, his conversation left its Hegelian key and entered into more present and ancient dialogues, with Paris on the one hand and Aristotle and Galen on the other.

On 11 May 1828 Goethe had written to Soret to be reminded of the title of a work they had discussed: Augustin Pyramus de Candolle's *Organographie végétale*, which he then procured.[95] This work, expansive and dedicated to a functional but empiricist approach to the organs of plants, served Goethe as a sort of dictionary of French scientific usage at the time.[96] And Goethe himself was mentioned in the preface to the work: "Several German naturalists . . . among the moderns, the illustrious poet Goethe, have called attention to the symmetry of the *composition of plants*. . . . [O]rganography is the development of what holds in the symmetry of the partial organs."[97] Generally sympathetic to Candolle's attempt to balance functional comparisons of organs with type-comparisons across species and genera, Goethe thought that balance needed a philosophical means. He thus sought to connect Berlin and Paris through an organ. Even in Candolle's praise, he found cause for terminological concern: "*Composition*, a similarly unfortunate word, mechanically related to the previous mechanism . . . just as in art, when one speaks of nature, this expression is derogatory. The organs do not compose themselves as if previously finished, they develop themselves out of and into each other into a necessary being that grasps into the whole."[98] Organology thus perches inside evolutionary biology,[99] intervening in the debate between anatomical comparison (Cuvier) and physiological teleology (based, in Geoffroy's case, on a material monism).[100] Goethe's conceptualization of organs, however, is not merely an insistence on holism and organicism. Instead, it is based in a practical treatment of cognition's own power to represent and reproduce, both metaphysical and active. In a sense, even the organs of knowledge expand and contract, forming a necessary whole which—representing and reproducing itself as the order of being and of thought—intervenes in its own development.

Goethe thus enters a debate here on the brink of shifting dramatically. The role of the part's development in the formation of the animal, so centrally debated in the eighteenth century, would, in the next decades, come to a head in Britain with the advent of Darwin's theory of natural selection. As Robert Richards has argued, Goethe was closer than previously realized to Darwin's evolutionary concept, precisely because he allowed for external circumstances to influence formation. Here we see this played out at the level of the part and its function rather than at the level of organism.

The passage on *Naturphilosophie* in Hegel's *Encyclopedia*—which also included his critique of Goethe—rejects the possibility of species development.[101] Because essence is both concept and reality for Hegel, its development is not "mechanical," an external reordering of parts. The conceptual side of reality by its very nature demands that leaps in quality occur for essences to change. No extensive infinite is allowed, and gradualistic development is therefore rejected. A new species is another essence, conceptually and thus ontologically distinct from what went before. This applies all the way down to the smallest functional part, the organ.

If Goethe's response to Hegel was such that the constitutional overlap between thought and being made intervention (*das Eingreifen*) possible, this meant that he could allow for gradual progress in the organ-based observational process. The organ, which spreads out to "intervene" in the whole as it enters into being, is mirrored in a double-mimesis by the human organ, the evolution of which is *up to us*. The broadest sense of the word, as I pointed out in the introduction to this study, relates to its Indo-European (*uerg) and Greek (*ergon*) roots: the organ is metaphysical efficacity. Organology is *transcendental technology* because it conditions all technological possibility.[102]

Intervening in the debate between anatomical description and physiological teleology (based, in Geoffroy's case, on a material monism), Goethe drew attention, focusing on the French of the matter, to the problems of unity, type, and taxonomy. In this last effort to approach this area,[103] he unified the discourse around the classical problem of function in the term "organ." Pushing his Classical taxonomy into philosophical territory was Goethe's last scientific effort. Where morphology was one result, the "organ" offers us another view: to taxonomize with Idealism was to demarcate different tools for understanding and potentially altering any possible morphological object.

The problem is classical in two senses. First, it is indebted to the tradition of the ancients in biology, the "classics" (primarily Aristotle, Galen, and Hippocrates), whom Goethe now read one final time. But the problem is also related to his own "classicism," his aesthetic and scientific methodology, covering both external taxonomy and idealistic organological technique.[104] And this classical problem is therefore treated through the organological detour, adding in this way a new factor to Goethe's quasi-philosophy: a technology of sorts.

Five days before his death, Goethe wrote to Wilhelm von Humboldt: "Animals are educated by their organs, the ancients used to say, I add: humans too, they simply have the advantage of also educating their organs."[105] This dictum guides the central historical section of the "*Review of the Academy Debate*" and appears on a verso[106] transcribed from the *Maxims and Reflections*[107] and dated to September 1830. The presence of the cluster of *Maximen* on the paper on which the review was written, and its conceptually guiding role in the historical unfolding of the debate in the Academy, makes clear its centrality to Goethe's final theoretical efforts. The organ, in short, which distinguishes humanity is one in evolution (*Bildung*) of another sort: encompassing the "lower" soul and the physical organs, the higher operations of reason and judgment in their categories, it is a pedagogical or even technological organ, an organ for the formation of humanity and indeed human history.

Indeed, this line of thinking, brought to bear on the "volcanic eruption" of the Academy in 1830, is classicist in its source and its scope. The Aristotelian treatises to which Goethe returned were those on the *Animalia* (especially *De historia animalium*, *De partibus animalium*, and the *De anima*), treating of the physical, functional, and formal developments of animals in general. Aristotle's treatises therefore cover organ-formation in three distinct senses,[108] senses Goethe adopted and adapted to his doctrine.[109] For even as he read the hylomorphic accounts of organ-formation in the organic sphere, Goethe focused on the problem of the hand in this literature and reformulated the text he found, pushing his German into the realm of the technological.

Maxim 1192 (also on the verso of the *Rezension*) reads: "The ancients compare the hand to reason."[110] Galen, who here bases himself largely on Aristotle,[111] was the source of this speculative comparison. Goethe had read in the *De usu partium*, in a passage meditating on the anthropological lack of specific organs in favor of the generality or versatility of the hand, that:

> Indeed, Aristotle was right when he said that the hand is, as it were, an instrument for instruments [*organon ti pro organon*], and we might rightly say in imitation of him that reason is, as it were, an art for arts. For though the hand is no one particular instrument, it is the instrument for all instruments [*organon . . . pro panton estin organon*] because it is formed by Nature to receive them all, and similarly, although reason is no one of the arts in particular, it would be an art for the arts [*techne tis an sin pro technon*] because it is naturally disposed to take them all unto itself. Hence man, the only one of all the animals having an art for arts in his soul, should logically have an instrument for instruments in his body.[112]

The ancients, engaged in a debate about the human from the perspective of its parts, draw an analogy from the versatility of the hand (its general capacity to instrumentalize nature) to the generality of reason's grasp of all things. Goethe was not merely translating that passage in Galen, but proposing terms for a transfer of the whole discourse of the ancients on the hand as technology ("die Alten sagten . . . "). Aristotle, on whom Galen is clearly drawing, had written in the *De anima*: "It follows that the soul is analogous to the hand; for as the hand is a tool of tools [*organon estin organon*], so the mind is the form of forms [*eidos eidon*] and sense the form of sensible things [*aisthesis eidos aestheto*]."[113] The local debate from Aristotle on is with Anaxagoras, who had claimed that the hand had made the human intelligent. For Aristotle as for Galen, the reverse is true: the hand is a suitable instrument for the animal possessing reason. Note how Galen plays in the last quoted sentence on the beginning of Aristotle's politics: *the human is that animal having reason*. But here, reason turns out to be the "*techne* of all *techne*," the origin of capacity, and this is where the technological analogy of the hand is rooted. If Goethe adapted an organology from the Romantics, it would be here, in the analysis and exploration of this analogy between two conditions of possiblity, spiritual and material, and their determination of the human.

Goethe finished his cluster of *Maximen* with a translation of the climactic sentence from this passage from Galen: "Reason is the art of arts, the hand the technique [*Technik*] of all handicraft."[114] Goethe has, at the very least, reduced the complex syntax in Galen to the pith of complementarity. But he has also done more than this, taking the key repeated phrase, *techne pro technon* (for reason), and rendering it in the antiquated sense of "art" in its root in *artificialis/künstlich*. Reason (*logos*) is constructive, even technological, an artifice which builds not arbitrarily (by *composition*) but necessarily, progressively including more nature in its constructive apparatus and literally giving it new form, creating new categories within it. That would be a technological metaphysics, mixture of making and necessity. The hand then becomes "technology," too, and the translation of organ into *Technik*. "Technik" means the *knowledge* of handicraft before the twentieth century, not the artifacts of that craft.[115] But Goethe's translation here anticipates the combination of both by equating apparatus and knowledge of production in the "hand" and makes finally clear that Goethe goes beyond Hegel's framework. Beyond the instrument for truth is the instrument for making truth. Organ is the instrument of instruments, a reflexive technology spanning the spectrum of human being, which turns out to be human making. The mirroring of the organ from reason's capacity to intervene in nature into the technology of the versatile hand is not merely analogical. The generality and productivity of each is isomorphic with that of the other—they interact in what the Romantics would have called "higher realism." If Goethe has an "anthropology," it is *technologia transcendentalis*, or organology. Reason's organ is not a specific logic, not merely an organon. Or rather, it is precisely that *organon for metaphysics*[116] that Kant had once called for. And yet it is more than that. Redoubling mimesis and creating a dialectical and constructive relationship between thought and being, it also trends toward the practical. And we should not think of "practical" in a vulgar empiricist sense here, but instead as the ideal interaction in reality which Goethe's organ is meant to make possible. For Goethe, the possibility of the "externality" of agents of change in the evolutionary process is maintained permanently as a possibility, precisely because the organ of reason is prepared to learn something principally new and then *alter* that newness in its reproduction of representation. Mimesis—as the reproduction of representation itself—holds the possibility of a shift in the represented in reserve. The cascade of organs—intellectual, sensible, instrument—subtends any developmental schema and points to the possibility of its refashioning.

"Tender empiricism" brushes the world with new organs, new categories, painstakingly reproducing its split between subject and object, and changing the form—*eidos*—(and thus content) of that split through reproduction. The hand, the sense, and Reason itself thus describe an organological human with dialectical ability, one who can progressively include and alter the world in development. This organology thus reproduces the order of representation and becomes productive within it—*physis* and *techne* are underpinned not by Galen's *logos* but by Aristotle's self-knowing *nous*. Goethe's "translation" of this discourse pushes self-knowledge (in Hegelese: *das Selbstwissen des Geistes*) into production: the encapsulated organs of form confront the genuinely gradual developments of a processual world. Taxonomy treats the proliferation of forms in development, but is complemented and harmonized by an Idealism founded in Reason's categorial organs. Organology thus offers the vision of a *technological freedom*. Nature is delicately forced to mean something, and our human organs match the forward march of technological and disciplinary isolation with a thus meaningful resistance.

Novel Organs: Coda on the Tools of Bildung

At the beginning of the eighth and final book of Goethe's *Wilhelm Meister's Years of Apprenticeship*, Wilhelm has achieved his goal, or so it seems. His theatrical education has drawn to a close, and he has been made aware that his actions and destiny have been under the control, thoughout, of the Tower Society. He has been presented with a written account of his own life and a letter detailing various platitudinous suggestions for the correct manner of living. He has, for the first time consciously, a son in Felix. His *Bildung* is both complete and, as he feels, just beginning. The narrator tells us: "Wilhelm saw nature as though through a new organ."[117] It is the only occurrence of "organ" in this sense in the novel, and as we know, Goethe would later write of "new organs" and their development in his scientific methodology. Here, *Bildung* is described as resulting in an organ that will be further developed or educated: it enters, new, at the moment of self-realization through the medium of the mysterious institution of the Tower Society. The deadlock in which Wilhelm has stood is, for the moment, resolved. The world is actually changed: the organ remains, in 1796, metaphorical. Goethe would slowly adopt this term in its Romantic sense, altering the very doctrine he assimilated, over the course of the last two decades of his life.

The novel's organ (1796) falls into the same period as that of *The Collector and His Circle* (1798). Goethe was searching for new literary means by which to combine idealism and realism,[118] classicism and classification. I have been arguing that it is the organ that bridges these gaps, although Goethe makes this explicit only late in his life. In *Wilhelm Meister*, it adumbrates his cautious reaction to Romantic organology. And although the term "organ" might be taken as a metaphor (*wie durch* actually makes the *seeing* the metaphor), perhaps the most literal sense is correct: Wilhelm has gathered the educational experience to have orientation in the world. Although he will now continue to learn, the "organ" he has gained is precisely the instrument of that learning. Perhaps: but the organ might also be more conflictual, less pacific. The organ might be the eyes of the Tower Society itself, the ability to see one's life mediated not by destiny but by institutions.[119] A third meaning is also possible: the biological organ itself, the "new organ" that Goethe would later speak of and which he would open (see the chapter epigraph) for his generation. It was the combination of these three meanings that allowed Goethe's considered reaction to the Romantic doctrine in the 1820s. Wilhelm's new organ is the result of experience and education, the ability to orient himself within his larger destiny—this goes some way to explaining the narrative that completes the *Years of Apprenticeship*, and the utterly undramatic *Journeyman Years*. Wilhelm does not control his destiny, but he has the ability to intervene in it. Economic adulthood comes with a demystification of "life" as an unknown and violent trajectory. An organ that allows for this kind of renunciation of the dramatic—a renunciation literally of the theater—would have to be institutional, would have to allow the protagonist the economic abilities he scorns in his bourgeois friend Werner in book 1. It is, in this sense, the organ that allows for passage from apprenticeship to journeying. As with the chemical doctrines invoked in *Elective Affinities*, Goethe's economic model is both real and slightly outmoded. Integration of individual and institution, the novel suggests, still operates on the model of the artisan. There can be no division of labor among the subject's capacities, only the fundamental discontinuity between the organizing powers of the Tower Society and the drive of the subject. To be on both sides of this division— to be tragic hero first and the chorus who tells the story second, and to have *both* capacities facing the still uncertain world—is Wilhelm's passage into maturity. This is the literary telling of the end of both epic and drama,[120] and the passage into a new kind of narrative, one that must represent the straddling of the divide between the individual judgment and the stubborn totality of experience *as one term*, as the hero.[121] Organs in the novel are literally organological, because they both unite and divide the hero even as they make that united and divided hero into the representational content of a literature that remains methodologically committed to more than representation. The literal meaning of "organ" is thus not biological. The novel organ—the new organ—is the function that allows for the narrative bridging of continuity and discontinuity, of life and career, of ideal and real. That is, after all, precisely the organ Schlegel would call for in the *Dialogue*. The narrative that covers this gap is the one Schlegel would also choose to place at the center of Romantic notions of prose, *Wilhelm Meister*:

> The born drive of this completely organized and organizing work to form itelf into a whole, expresses itself in the greater as in the smaller masses. No pause is accidental and meaningless; . . . everything is means and end [*Mittel und Zweck*] . . .[122]

Schlegel invokes Kant's definition of the organ, the simultaneous cause and effect of the whole in which it is a part. It is all too easy to think of this determination as "organicist," and Romantic literary criticism as following the impulse of biological rhetorics—although Schiller accused Goethe of quite the opposite, of introducing a machinic finish in the Tower Society. The reconstruction of organology disallows either model, in fact, because it shows that the problem addressed here is functional. True, all is both means and end, and a single part can be taken as both means and end itself. At *every scale*, the novel is machines or organs all the way down. But this is not to take the terms of an uncertain biology and cast them over formal concerns in prose-production. In Schlegel's terms, it is instead to take the precision of the organ—the tool for metaphysics, the moment where speculation becomes real—and to find it in the very construction of a novel that also takes depiction of that organ as its task. Perhaps the last book of *Wilhelm Meister* is organological, the first experiment in the Romantic style with the capacities of the term. If that is so, then it is Goethe's redoubling of terms—the synthesis of subject and object, as of object and subject—that allows for the excess to emerge as the experience of the *Bildungsroman*. We are not swallowed up in destiny; we do not overcome nature's flow. We *encounter* the real, and we might have organs to guide this encounter, to constitute the real differently. That difference covers the distance from individual to institution. That experience of the novel is hardly the stuff of drama, but it forms in its determination of its epoch the fundamental continuity of experience between the life in the novel (Wilhelm) and the novel in life (his reader: Schlegel). The organological novel constitutes this complex subject as its audience, creating the subject it constructs internally, and thus becomes an instrument in the flow of cultural history.

NINE

Instead of an Epilogue: Communist Organs, or Technology and Organology

> The whole history of a "thing," an organ, a practice can in this way be an extended chain of signs of new interpretations and correction, the causes of which need not cohere among themselves, rather circumstantially simply accidentally follow on and separate from one another. The "development" of a thing, a practice, an organ is thus nothing less than its *progressus* towards a goal, still less a logical or the shortest *progressus*, achieved with the smallest expenditure of force,—but instead the sequence of more or less penetrating, more or less independent processes of overcoming playing out on it.
>
> —NIETZSCHE, *On the Genealogy of Morality*

In Samuel Butler's Victorian satiric utopian novel *Erewhon* (1872), the protagonist translates an Erewhonian treatise on machines into English for the reader. The three chapters covering this topic—the "Book of the Machines"—consist mostly of this translation, with the occasional note that part of the treatise was "too obscure" for the translator. The treatise attacks the existence of machines in Erewhon's society, arguing that they have evolutionary properties like animals and humans, and that while they currently stand at the stage of having only "stomachs," they will likely soon develop reproductive systems and supplant humans as the highest form of life. The treatise has only one dissenter, who argues that humans simply *are* machines, that their limbs lie around them, disjointed from their bodies: the human is the "machinate mammal."[1] The Luddite treatise wins, and due to a certain washing-device's invention date, all machines dating up to 271 years before the change are eliminated.

The "Book of the Machines" is a rewriting of a short treatise that Butler wrote during his four years (1860–64) of self-imposed exile as a sheep farmer in New Zealand. That treatise, "Darwin amongst the Machines," contains the same central argument for the possibility of machine evolution. In *Erewhon*, this is presented as a digression from the treatise. The protagonist is smoking his tobacco pipe, and an Erewhonian interlocutor points to a small nub on the bottom of the bowl, saying that it is a "rudimentary organ." Darwin is here literally among the machines: "rudimentary" is the designation of those organs that no longer bear their function, according to the *Origin of Species*. The protrusion on the pipe's bowl, the Erewhonian (correctly) speculates, must have originally served to keep the pipe's heat from whatever surface it might be placed on. By time of writing, however, it is well on its way to mere ornamentality.

The two treatises share a concept of organs that extends beyond the organic body. They share the sense that the machine-human interface is not a hard-and-fast categorial boundary. Gilles Deleuze and Félix Guattari put it this way: "There is a Butlerian manner of carrying each of the arguments to the point where it can no longer be opposed to the other. . . . *He shatters the vitalist argument by calling in question the specific or personal unity of the organism, and the mechanist argument even more decisively, by calling in question the structural unity of the machine.*"[2] The two arguments effectively cancel the distinction between machines and animals, making the conceptual basis—their respective unities—irrelevant in favor of their functions. Butler serves as a key textual source for Deleuze and Guattari's notion of a "body without organs," a nonunified element of resistance to the functional and ultimately normative grooves established by organs.[3] Evading traditional metaphysical categories, Deleuze and Guattari argue beyond the framework of whole and part, suggesting that the source of resistant energy is not in the universal or the generalities of being, but in the "body" that, bearing the very organs its works, reduces them continually to the nontotality in which they function. Strange among their designations of this "body without organs" is the notion of capital, which is the resistant body of the capitalist: "It produces surplus value, just as the body without organs reproduces itself, puts forth shoots, and branches out to the farthest corners of the universe. It makes the machine responsible for producing a relative surplus value, while embodying itself in the machine as fixed capital."[4] The argument that cancels the antinomy between the organic and the mechanic results immediately in the problem of work in the machine-driven environment. Conceptually, machines and "organic" beings are no different from one another. What then becomes of the life of labor, of what Marx calls "living labor"? What is the source of value when machines might be taken to be "doing work"? By linking the experience of machines to the problem of financing the factory, *Anti-Oedipus* lays out the crucial background on which Marx had constructed what I will here call an economic metaphysics—a kind of late version of organology. If machines and humans are interchangeable—if their parts might all be said to be "at work"—then the very source of capital is called into question. Darwin, organs, and value: these would be the terms of Marx's inheritance of the Romantic doctrine.

Human Nature Techne: Marx and Darwin

Friedrich Engels spoke the following words at Karl Marx's funeral:

> Just as Darwin discovered the law of development of organic nature, so Marx discovered the law of development of human history: the simple fact, hitherto concealed by an overgrowth of ideology, that mankind must first of all eat, drink, have shelter and clothing, before it can pursue politics, science, art, religion, etc.; that therefore the production of the immediate material means, and consequently the degree of economic development attained by a given people or during a given epoch, form the foundation upon which the state institutions, the legal conceptions, art, and even the ideas on religion, of the people concerned have been evolved, and in the light of which they must, therefore, be explained, instead of vice versa, as had hitherto been the case.[5]

This is only one of many such analogies, self-comparisons drawn between the writing duo and the arch-naturalist as soon as his *Origin of Species* was published in 1859. It would seem that Marx and Engels understood themselves as scientists in a transposed sense, ridding not biology but a still-nascent social science of theology. The law that Darwin had discovered (natural selection) had been conceived in explicit opposition to another type of thought that veritably required God: thought based primarily upon the unity of the species. If the idea of the species preceded the exploration of its development, then no amount of equivocation about comparative structures (Cuvier) or morphology (Goethe; Geoffroy de St.-Hilaire) could help the scientist. The biological real would have to be construed as history itself for there to *be* biology. It appeared that Nature would have to take on the burden of the argument. And just so, according to the apparent intent of Engels's words at Marx's funeral: even the species-determining characteristic of abstract thought—even, finally, the ability to be alienated—could not justify, in the end, sticking to the ultimately abstract language of Feuerbachian materialism, with its stated collective species-being, its sensually rooted higher capacities, its static forces. Even this "human" had been conceived primarily as species-idea—the brute reality of its needs, and the social history of its attempts to deal with those needs, had remained undiscovered until Marx. *Naturalism* seems to be the watchword of this analogy between giants.

Not so—or rather, if what Marx and Darwin share is a "naturalism," it is one that looses nature from its ontological moorings and shifts the naturalist debate from problems of laws of being to problems of contingency arising from mundane purpose. That is: so far from a set of immutable laws, this shared picture of nature—and human nature—presents us with the history of nature as continual repurposing, as the struggle of partial entities to become units, as the depersonalized uses of parts for purposes of survival and propagation. In other words, Marx and Darwin were writing about technology, or a technologized nature, the parts of which were organs.[6]

Marx, of course, was not so naïve as to confuse nature with technology (nor was Darwin)—yet the new relationship between human and nature he perceived having taken partial and monstrous shape around him in the Industrial Revolution called for the two concepts to be brought closer to each other. Here is Marx's version of the Darwin analogy:

> Before his time, spinning machines, although very imperfect ones, had already been used, and Italy was probably the country of their first appearance. A critical history of technology would show how little any of the inventions of the 18th century are the work of a single individual. Hitherto there is no such book. *Darwin has interested us in the history of Nature's Technology, i.e., in the formation of the organs of plants and animals, which organs serve as instruments of production for sustaining life. Does not the history of the productive organs of man, of organs that are the material basis of all social organisation, deserve equal attention?* And would not such a history be easier to compile, since, as Vico says, human history differs from natural history in this, that we have made the former, but

not the latter? Technology discloses man's mode of dealing with Nature, the process of production by which, he sustains his life, and thereby also lays bare the mode of formation of his social relations, and of the mental conceptions that flow from them. Every history of religion, even, that fails to take account of this material basis, is uncritical. It is, in reality, much easier to discover by analysis the earthly core of the misty creations of religion, than, conversely, it is, to develop from the actual relations of life the corresponding celestialised forms of those relations. The latter method is the only materialistic, and therefore the only scientific one. The weak points in the abstract materialism of natural science, a materialism that excludes history and its process, are at once evident from the abstract and ideological conceptions of its spokesmen, whenever they venture beyond the bounds of their own speciality.[7]

What is "historical materialism," and why is Darwin at least apparently brought under its umbrella? What does it mean for Marx not to exclude but include "the historical process" in "scientific method"? And why would doing so necessarily involve a history of "technology," natural or otherwise?

In a letter to Ferdinand Lassalle (1825–1864) of 16 January 1861, Marx again confirmed the importance of Darwin's antiteleological stance, but complained that "one has to put up with the crude English method of development."[8] By casting Darwin as part of a general trend in British letters, Marx recalls the terms of his attacks on classical political economy in the persons of Adam Smith (1723–1790) and David Ricardo (1772–1823). Marx rarely disagrees with these figures insofar as they *describe* the state of the politics of economy. Rather, he thinks that they introduce descriptively accurate terms that fail to comprehend the generative laws at work. They operate very much the way Ptolemaic astronomers did, while Marx proposes his method as that of Copernicus. Thus, for Marx, method and politics are immediately linked. The production of a representation of the totality of the economy in its present historical form is the tool for any political intervention—*Capital* itself. And so with Darwin: Marx's judgment cuts both ways. The description is accurate, but more than description is needed.

Marx did not lack for sympathetic passages in *The Origin of Species*. Darwin had argued for natural selection on the example of artificial selection, requiring his reader to substract the personification of nature from the analogy once complete: "So again it is difficult to avoid personifying the word Nature; but I mean by nature, only the aggregate action and product of many natural laws, and by laws the sequence of events as ascertained by us. With a little familiarity such superficial objections will be forgotten."[9] The fit between means and ends emerges qualitatively on the basis of an infinite chain of events restricted only by the "aggregate" nature. The metaphor, however, is stubborn:

> Man can act only on external and visible characters: Nature, *if I may be allowed to personify the natural preservation or survival of the fittest*, cares nothing for appearances, except in so far as they are useful to any being. She can act on every internal organ, on every shade of constitutional difference, on the whole machinery of life. Man selects only for his own good; Nature only for that of the being which she tends.[10]

The reason Darwin remains with his metaphor, which can never fully disappear, is because the concept of "use" remains essential. The fit between means and end may have no guiding principle,[11] but it remains a relation between means and ends. This is what Marx means by the "natural technology": our knowledge of nature's ability to regulate these relations.

Marx cites Darwin only twice in *Capital*, the second time on the definition of organs:

> And as long as the same part has to perform diversified work, we can perhaps see why it should remain variable, that is, why natural selection should have preserved or rejected each little deviation of form less carefully than when the part has to serve for one special purpose alone. In the same way that a knife which has to cut all sorts of things may be of almost any shape; whilst a tool for some particular object had better be of some particular shape. Natural selection, it should never be forgotten, can act on each part of each being, solely through and for its advantage.[12]

Thus both participate in the metaphorical narrowing between *physis* and *techne* in the nineteenth century. Now it is not merely the artificial analogy of selection, but also the variable functions of tools, that provides the leap. Darwin has considerations here explicitly taken from those of Geoffroy de St.-Hilaire on the number in proportion to the variability of the use of organs: the more the same organ is repeated in the individual, the more likely it is to be performing different kinds of work. Darwin is writing biology as organology, and Marx is extending that use into the human world. Tools, like organs, are the concrete artifacts of economic development, that is, history. They concretize but leave residual potential in the curve of that history.

Darwin's metaphor is powerful for the description of nature at large, but Marx was dissatisfied with the remainder of personification he found there. In the chapter entitled "Law of Variation," Darwin took issue with various "idealist" laws of type and arrangement of organs, including among them the "law of compensation" of "the elder Geoffroy and Goethe," in which the "economy" of nature is imagined as a credit-and-debit sheet, as a zero-sum of forces in interaction forming animal bodies over time. Darwin does not reject this law, merely casting it in terms of natural selection, with its tendency to economize, again for "use." The problem of "use," however—as it was for the political economists—is that its very concept is split. On the one hand, "use" means the end that is proposed in fashioning a means. On the other hand, use is the way the means is consumed—*used*. Darwin's law of variation insists that true organs develop as the result of naturally selective use-values. These *values*, as I would call them by analogy to political-economic discourse, are quite different from the minor shifts Darwin must also acknowledge come from *using* organs. The following passage points up the difficulty Marx had in reading the *Origin*: "On the whole, I think we may conclude that habit, use, and disuse, have, in some cases, played a considerable part in the modification of the constitution, and of the structure of various organs; but that the effects of use and disuse have often been largely combined with, and sometimes overmastered by, the natural selection of innate differences."[13] This is the "crude method of development": in showing that species must have developed, must have become arranged as series of organs over vast stretches of time, Darwin develops a single conceptual tool—that of "natural selection"—with great explanatory power. But the concept itself is constituted of two importantly differing senses of "use," and this blurs the concept's descriptive value. When Darwin confronts this problem, he has to admit that there is probably some distinction between "habit, use and disuse" and "natural selection" as mechanisms of speciation or organ-development. Darwin defers this problem to further experimental and paleontological research. Marx does not take issue with the science at stake, but with the method of its expression, which is first and foremost conceptual and descriptive. Nor can it be otherwise: Marx does not want to change the representational content of Darwin's biology, but thinks a different method is demanded for the history of the human. Ultimately, that method would be required for science, too—but there is little reason to take issue with the self-consciously incomplete work of a Darwin. The focus, for Marx, had to be on nature's *techne* as it was expressed in the human.

Marx—who read deeply in both physics and biology—does not operate, I want to argue, primarily between the "force" of the physicists and the metabolisms of proto-naturalists.[14] He was hardly insensitive to the ongoing struggle between biology and physics for primacy in the natural sciences. Yet his thought, which comprehends these developments, is located rather between the material and immaterial expressions of what he here calls "technology."[15] To write the history of human organs—a project Marx hinted at several times over the course of his career—required confronting the very problem of philosophical method with the descriptions of the scientists. It required the development of an organ to respond to the crisis in manufacturing that reduced humans to organs of factory works, literally organic elements of the contingent and increasingly autonomous development of machines.

For this reversal to occur, the human must already have a strange relationship to nature, and Marx never wavers in this description. The human is a universal extension of the problem of nature. Nature, to be sure, is immanent: no type, not even a *species-being* of a Feuerbachian stripe, has a knowable source. Philosophical searches for *origins* are excluded—this is why the characterization of Marx as a "materialist" needs further explanation, and always has. Marx's anthropology is naturalistic, but his naturalism is intentionally hemmed in by anthropological method. The human is the point on which the differentiation "organic/inorganic" hinges:

> Nature is the inorganic body of the human, namely nature so far as it is not itself the human body. That the human lives from nature means: nature is his body, with which he must remain in constant interaction [*Prozeß*] in order not to die. That the physical and spiritual life of the human must cohere with nature, means nothing other than that nature coheres with itself, for the human is a part of nature.[16]

The human is a part of nature: Marx is committed fully to this description. And yet "nature" is also the extension of man's body beyond its organic boundary. It is both the object he cognizes and consumes and that which lies beyond his capacities to do so. Nature is both alien and proprietary. But this very determination only comes from the "universality" of the human *as animal*:

> The universality of human appears practical even in the universality that makes all of nature to its inorganic body, as much insofar as it is 1. an immediate means of life, just as to the extent that it is 2. the material, the object, and the tool of its own life-activity.[17]

Nature is, both theoretically and practically, the immediate means of life and simultaneously the material, object and tool of what the human does (*Lebenstätigkeit*, "life's activities"—given the context, perhaps: "life's employment"). The human lives entirely *in* nature, yet makes nature the object and instrument of his own activity, both theoretically and practically. What seems a contradiction—and what has given rise to so much debate—is really a methodological restriction. The human is the difference between the organic and the inorganic, the difference between body and nature—or, as we shall see, organ and machine—*as a part of nature*. The distinction between the organic and the inorganic is merely of human origin, a necessity imposed by the universality of human activity. There is no difference between human nature and nature, or rather, this difference is merely of human imaginary origin. Pointing to this truth, however, has little effect. The difference has deeper roots than the notion of its negation. But this was Marx's point to begin with: humans as nature merely are this universal difference, and their knowledge of the relations between means and ends—their *Technologie*—is simply an extension of this fact. Their organs determine but also allow their activity, which is ultimately work. To develop a "spiritual organ" to intervene in the technical processes of social construction and domination was Marx's goal from 1844 on, culminating in the method of *Capital*. That organ was Marx's mature method itself, and it stands in the tradition of Romantic organology.

This inheritance clarifies the Marxian project—methodological shifts underlie the otherwise strikingly consistent content of Marx's writings. But it is more than simply Marx's mysterious and shifting allegiances—to Feuerbach, to Hegel, *not* to Hegel, and so on—and indeed, more than his apparent shifts on questions of labor, machines, or humans. It is the movement of these ideological positions with respect to a shift in method that matters. This shift has to do with the relation between concepts and their expression—that is, with writing, with depiction. When Marx writes about organs, he is at his most Romantic, not because he is "passionate" about the development of machines (for this is where "organ" pops up), but because the term relates his views to his method.

There has been, of course, endless debate on Marx's development.[18] I tend to see this problem as one of method above all.[19] In the early works, Marx is concerned, as his Young Hegelian peers were, to state the program for a good politics based on various intentional distortions of Hegel's system.[20] *Capital*, on the other hand, is committed to dialectical method as the substantial connection of discourse and reality. It does not seek to represent the future, but to penetrate the laws of the present deeply enough to provide them with a counterweight. If the anticapitalist Owl of Minerva flies only at dusk, it nevertheless thinks the next day might be constituted differently because of its flight. *Capital* is thus also an organological response to Hegel, as we are about to see. Like Goethe, Marx proposes that the dialectical method might be used to change the rules—of the natural world as well as the social.

Marx's relation to the problem of force—the problem of the expression of the conflict of order, inherited from biological debate and philosophy after the Herder-Kant polemic—reflects not only his knowledge and commitment to contemporary science, but also his attempts to include the rigor of philosophical (Romantic) method into his philosophy of economy. His commitments to the politics of social domination, to the cause of the worker and the problem of constitutive capitalist inequalities, did not change from the 1840s through the 1870s. Instead, what changed was his method for addressing these problems. In short: he moved from a projection and description of the ailments of society to a genetic presentation of those problems as relative necessities. This genetic model confronted the problem of force directly. Rather than hypostasizing forces or merely asking after their local causes, however, Marx increasingly immersed the question about force into a philosophical style meant to allow force the redirection of those forces. In other words, by subordinating representation to a moment within philosophical and social communication, Marx combined the panoptic method of the dialectic with a syncretic presentation of societal ills. This method of presentation allows for genetic analysis and understanding of historical formations of (economic) society, while also presenting its necessity with a view toward *replacing* that necessity. *Capital*, then, would be an ideological weight meant to displace the simultaneously cognitive and material vector of capitalist society. A literalized organ would be the lever on which this method turned.

This methodological intervention had, of course, massive consequences: it is not merely "really existing socialism" that has convinced us for so long to take Marx so seriously. I want to suggest here that another reason we have taken—and must continue to take—Marx seriously is that he uses organology to address a material condition in which the Romantics did not yet live. The Romantics inhabited a liminal space between the problem human/nature and the problem human/technology. To be sure, both problems existed when Romanticism came on the scene. But this is precisely the point: the condition of nonknowledge of the real laws of nature, and the concomitant crumbling of confidence in metaphysics, coincided in Romanticism with the question of radical construction, the question of the human capacity to invent and implement a new order of things. For Marx, the latter concern is identical to the former, because there is no question that humans can institute a new order. That order is in ubiquitous evidence, even for the young Marx. And the central question around which a post-organology would have to circle became, therefore, the mediation of nature (work) by the necessary struggle between worker and machine. The precarious disintegration of the old natural order, combined with the scintillating emergence of the possibility of a new social order, formed the backbone upon which Romantic metaphysics grew. The historical crisis of the cognitive and social problems associated with these developments had given way to a first crystallization when Marx began to write. Whatever the historical status of what became known as the Industrial Revolution, it must be clear that Marx addresses himself to a condition in which the mechanical arts have taken on a life of their own, have become ordering instances in what had appeared to the Romantics (in an equally material yet highly precarious condition) an eminently orderable world.

Senses of Organology: Human and Communist Organs

Max Horkheimer, writing at the outbreak of the Second World War and in the passage from the Institute of Social Research to the formation we now call the Frankfurt school, recognized a hidden organological agenda of the early Marx's project. The mission statement of the Frankfurt school, "*Traditional and Critical Theory*" (1937), divides famously between a traditional theory caught in an attempt to describe everything the skeptical subject can come to know, and a critical theory which realizes itself as the legitimate product of the class-relation as it emerges from the mode of production. This second, historicized theory treats its very objects and its means of perception not as static, but as the result of social development (of work). Horkheimer writes:

> The facts which our senses bring to us are societally preformed in two ways: through the historical character of the observed object and the historical character of the perceiving organ. Both are not only natural, but formed through human activity; the individual still experiences himself in perception as perceiving and passive. The contradiction of passivity and activity which appears in the theory of knowledge as the dualism between sensibility and understanding, does not hold for society in the same measure as for the individual. Where the individual experiences itself as passive and dependent, society, which is nevertheless composed of individuals, is an active subject, even if it is also unconscious and thus inauthentic.[21]

The appearance of passivity—the sense of an external limitation not produced by the self—is the result of social work (as it had been for Fichte). The passage is a précis of the larger argument. Even with his sophistication, Kant has reified a felt contradiction by retaining a view from the subject (Horkheimer goes on to read the *epigenesis* metaphor with respect to the emergence of ideology from the infrastructure). To the subject, to be sure, the *organs* are passive, static, receptive. From the social standpoint of critique, however, they are also active, because they are the result of a history. This means not that they are themselves dynamic in reception, but that this very receptivity is active in the sense of *form-giving*. These are the ambivalent organs of an ideology, and the task is to critique their offerings, to make them match up with the content of the class struggle.

This notion of ideological organs is ultimately derived from Marx, as we will see momentarily. But it was a powerful strain of thought in the circles of the Frankfurt school. Its formulation goes back at least to Walter Benjamin's famous essay entitled "The Work of Art in the Age of Its Technological Reproducibility": "Then film came along and exploded this prison-world with the dynamite of tenths of seconds, such that we now blithely undertake adventurous journeys between their widely-strewn ruins. Space extends itself in the close-up, movement in slow-motion."[22] The categorial banality of the everyday is redoubled and exploded by the technological organ of the aesthetic, film. The tricks of the medium's presentation cut, in Benjamin's terms, surgically into the complex of our perception, altering time and movement fundamentally. If we look forward to Horkheimer's essay of the next year, we can see an organological strand in this ideologically critical media theory. That is not incidental, of course: as far back as Herder it had always been clear that media need *organs* to sustain any investigation of them.

Horkheimer, writing with Theodor Adorno, would later give another striking formulation to the thought: "The achievement that the Kantian schematism had still expected from subjects, namely to relate the sensuous manifold anticipatorily to fundamental concepts, is taken over from the subject by industry. It runs the schematism as its first service to the client. There is supposed to be a secret mechanism in the soul that already makes preparations of the immediate data so that they fit into the system of Pure Reason. Today the secret unraveled."[23] The schematism is the empirical subsumption of intuition under concept—the application of the epigenetic quality of pure reason. Transcendental schemata must be generated, for Kant, in order to allow for this mediation between the *a priori* and the *a posteriori* to occur. These schemata allow for the formation of empirical correlates (the form of a plate, say, or a dog, in general). Schematism allows the *Critique of Pure Reason* to pass from possible experience to actual experience.

Film, for Horkheimer and Adorno in 1945, changes all that. Or rather, it changes the source of schematic generation. Film takes over the task of making the categories applicable to empirical intuitions, literally because it replaces the mechanism for intuition-delivery, displacing it from the senses in relation to their world to the senses in relation to a mechanically generated and completely controlled simulation of a world. The film destroys the mythical secret of the soul's schematism (the "ground of sensibility," into which we cannot see, lacking *suitable tools*). Shifting the construction and constructability of the object of the senses, film fulfills the aesthetic task of creating a "world within a world." But it also shifts the burden of aesthetic content—the generation of ideology—to Hollywood, to the culture industry. Ideology is thus buried deeper than previously possible: it is delivered directly to the disenchanted soul, which understands its immediacy as myth. Filmic Enlightenment returns of itself to its opposite, and Benjamin's optimism is undermined. The Frankfurt school inherited a strand of organology, but found no hope in it for a salutary politics.[24]

Returning to *Traditional and Critical Theory*, however, we find a striking deorganicizing version of the claim that also ultimately goes back to Marx:

> The human physiological sense apparatus has itself already been largely working in the driection of physical experiments for a long time. The manner in which pieces are severed and gathered in perceiving consideration, how singular things go unnoticed, others are brought forward, is just as much the result of the modern mode of production as the perception of a man from some clan of primitive hunters and fishers is the result of his conditions of existence and also admittedly of the object. In this matter we could reverse the proposition that tools are extensions of human organs so that organs are also extensions of instruments.[25]

Critical theory grasps, at its very starting point, an essential element of the organological program. On the basis of the Marxist theory of modes of production (and this not accidentally, as we shall see in a moment), tools and organs become interchangeable—Ernst Kapp's notion that tools are extensions of organs is reversed, universally the functional sense of both. *Work* is here reduced to its most abstract concept (in keeping with Marx's elaboration in the *Grundrisse*): *techne* as the condition of a world with humans in it means that organs are the result of work, and that their object-determining work is in turn a source for the production of a better ideology. This moment of optimism in 1937 was drowned out by war, and especially the technologies of mass murder, by the time Horkheimer and Adorno reformulated the doctrine in 1945. Still, the version of organology Horkheimer had drawn on for his early definition of critical theory was Marxist down to the very text.

330   *After Organology*

The passage indeed picks up two parts of Marx's organology: on the one hand, the idea of a history of the human senses, and on the other, the technical interchangeability of "organ" and "tool." Marx gave voice to the first in his *Economic-Philosophical Manuscripts* (1844), and the second in volume 1, chapter 13, of *Capital* (1867).

The *Manuscripts* are a curious mix of Feuerbach and Hegel. We can read Feuerbach in Marx's attachment to the human, and Hegel in such formulations as "communism is the standpoint of the negation of the negation."[26] Feuerbach also speaks of organs, even active ones—the sense is the "organ of the *absolute*."[27] But Marx's text, as we shall see, is shot through with a dynamism in the *organ* that corresponds to the dynamism he would demand of materialism in response to Feuerbach. At least in 1844, that dynamism is framed in Hegelian terms.[28] And the term "organ," which runs throughout the section on "private property and communism," is of other (Romantic) provenance.

The abstract character of industrial work has, for Marx, stripped humans not only of their possessions through immiseration but also of the specifically human property of their very perceptions. To reclaim the commonality of perception is an essential part of the project of the establishment of a non-abstract communism. The latter requires the formation of literally communistic organs, both in the social sense (institutions as organs) and in the physiological sense (common organs of metaphysical and ethical perception conducive to socialist politics).[29] The first statement of the post-organological program (from which Horkheimer seems to be drawing) reads:

> For not only the 5 senses, but also the so-called spiritual senses, the practical senses (will, love, etc.), in a word *human* sense, the humanity of the senses first becomes that through the existence of its object, through *humanized* nature. The *formation* [*Bildung*] of the 5 senses is the work of the whole of world history to date. The *sense* caught under raw practical need also has a *limited* sense . . . thus the objectification [*Vergegenständlichung*] of the human species [*Wesen*], with respect to theory as well as practice, belongs to making the *senses* of the human *human* just as to creating a corresponding *human sense* for the whole wealth of the human and natural being [*Wesen*].[30]

Note that this statement contains the essential characteristics of Romantic organology. The five senses are the precipitate of the history of humanity;

Instead of an Epilogue 331

they are essentially related to the practical senses and the spiritual "senses"; and they are used ultimately to make the world human and those senses themselves proper to humanity. "Human sense" runs the two senses of sense through each other: it is both the field of potentiality of humanity's self-recognition (as the bearer of historical and infinite fields of potentiality), and the rhetorically milder understanding for other humans. The work of history must be matched by a new kind of work if this sense is to be developed—smelling and hearing, as well as the higher senses, must be made instruments of the human. But then, once they are *humanized*, their instrumentality will be made organ.[31]

This project is necessarily communistic, since the human is defined in Feuerbachian terms as the generality of its essence. This generality is alienated, of course, for the young Marx, alienated through the process of abstract work and the privatization of property. The specific cocktail of perception and legality that ownership of objects[32] represents hampers the free development of the organs of human perception, creation, and social being. Thus:

> The abolition of private property is thus the complete *emancipation* of all the human senses and qualities; but it is that emancipation precisely because theses senses and qualities have become *human*, both subjectively and objectively. The eye has become the *human* eye, just as its *object* has become a societal, *human* object stemming from humans and for humans. The *senses* have thus immediately become *theoreticians* in their practice. The relate themselves to the *thing* for the thing's sake, but the thing itself is an *objective humanized* relating to itself and to the human and vice versa. The need or enjoyment have therefore lost their *egoistic* nature and nature its mere *usefulness*, because use has become *human* use. So also have the senses and the enjoyment of other humans become my *own* assimilation [*meine eigne Aneignung*]. Besides these immediate organs there thus form *societal* organs, in the *form* of society, so e.g. activity immediately in society with others etc. has become an organ of my *expression of life* [*Lebensäußerung*] and a way that *human* life assimilates.[33]

Private property is not merely a legal problem—it is an epistemological problem. In the perception of a possessed object, something goes missing not only from the object but also from the field of possibility represented by the literal organ, the eye. The senses have become immediate theorists, because perception itself is not innocent of the production processes that stand behind its objects. The development of the senses—itself a product of social historical work in common—is thus perverted by the privately possessed, we might almost say, by *private perception*. The atomistic ego is the result of this process, and its relations are those of mere instrumentality. But for Marx, there is no going back. The senses should not return to their unadulterated state. They should be built and educated further, in the direction of "humanity." This means taking a further step, one anticipated by the Romantics and despaired of in the Frankfurt school: creating social (and ultimately socialist) organs in the forms of society. We might think of Goethe's imagined scientific community in this context. Marx thinks of such communal (and communist) organs as the bearers of superstructural organization. In a sense not given by his work of the following year, *The German Ideology* (1845), such organs would be ideological in a positive sense.[34] They would have been the noninstrumental organs of the content of the state.

What Horkheimer had identified—the historicity and produced nature of the organs of perception—was thus derived from an early hope of Marx himself. And Horkheimer had also recognized, perhaps unintentionally, another part of Marx's organology. When the former spoke of the interchangeability of organs and experimental tools (as extensions of one another), he was calling upon the nonutopian discourse of *Capital*.

Working the Organ

Marx's magnum opus makes the premise of political economy that human work is a horizon beyond which methodology should not seek to see. The central chapter of *Capital* on the process of production and the process of valuation presents the means, and then the relations of production as concrete descriptions of work in general, and then work in capitalism. The production process requires three factors: work, the object of work, and its instruments.[35] Because work is the dynamic essence of the species "human," it is the condition and definition of the production process. The kind of work Marx has in mind is not restricted by any historical determination other than the existence of humans. This work has duration, is "purpose-oriented," paradigmatically producing uses. And it is only through work that the human emerges as the human, defines itself historically in the process of evolution (as we shall see presently with Engels). The human stands in a "metabolic" relation with nature[36]—not merely "part" of nature, but potentially all of nature. The ability to work nature generally is the *differentia specifica* of the human.

This means that the human must be defined as *having organs*, specifically "instruments" that exceed the inherited organs of the body. This is because work requires precisely what all functions require: an executor, a tool. Tools do not differ essentially from organs, because they are identified as historical signifiers of function both inside bodies and out. Marx thus extends his notion of nature as the "inorganic body" of the human to make tools organs and vice versa: "The first thing of which the labourer possesses himself is not the subject of labour but its instrument. Thus *the natural becomes one of the organs of his activity, one that he annexes to his own bodily organs*, extending his natural form, in spite of the Bible. As the earth is his original larder, so too it is his original tool house."[37] Both tools and organs are fossils of production, of the goal-oriented process of human work and the adaptive process of natural selection. Indeed, as we shall see shortly with machines, the *means of production* are always a reflection of materialized human attention.[38] The coexistence of tools and organs is essential to the world in which humans live (that is, have worked). By working through the means of organs and then instruments, the human chooses and alters an object which thereby becomes a "means," literally an *inorganic organ*, either a means of sustenance or a tool, an organ that is not given to but created by the organ-having human. *Homo laborans* cannot make an ultimate distinction between tools and organ—at best, the sense of which "belongs" and which doesn't is provisional.[39] It is thus not quite "having organs" that defines the human, but the *work* that organs allow, that ultimately allows for the extension and alteration of organs, and for their indistinction from tools. This is what I take Marx to mean when he criticizes Feuerbach for having a static materialism: the generality of human existence, its capacity to cognize a broad range of different contents, is ultimately rooted not in the specific structure of its sense-apparatus, but in the very plasticity that is built into the very function-bearing apparatuses (organs) that it possesses. The human is organological because the human works, and Marx's method is organology as political-economic philosophical method.

That method capitalizes on one of organology's most important promises: to eliminate the primacy of the distinction between the organic and the mechanical. For Marx, this allows for an alternate history that makes Darwinian evolution and human history continuous yet separated by the generalized capacities of the human animal. Marx quotes Benjamin Franklin to the effect that the human is the "tool-making animal."[40] The instruments of work are the evidence of human history as analogized to the evolutionary process, but they differ because it is not their result—the survival of the animal—but their very existence that determines the periods of human history. Marx writes:

> The use and fabrication of instruments of labour, although existing in the germ among certain species of animals, is specifically characteristic of the human labour process, and Franklin therefore defines man as a tool-making animal. Relics of means of work [*Arbeitsmittel*] have for the judgment of extinct economic forms of society the same importance as do the structure of relics has for the knowledge of extinct animal species. It is not the articles made, but *how they are made, and by what instruments*, that enables us to distinguish different economic epochs. Instruments of labour not only supply a standard of the degree of development to which human labour has attained, but they are also indicators of the social conditions under which that labour is carried on.[41]

The analogy cuts both ways. On the one hand, Darwin's attention to adaptive mechanisms, which allowed him to turn away from notions of species pregiven (by God or nature), also here allows for an analytical entry point into the construal of human history at large, not according to ideals but according to the functions that delimited the production process. On the other hand, however, Marx is, as we have seen, not happy with the "crude method" that allowed for two concepts of use to be conflated and then distributed willy-nilly throughout the theory of natural selection. Darwin's method would, by analogy, tend to the tools that allowed for certain classes to flourish (that is, the organs that allowed certain species to survive). Yet it would have no capacity to explain the marginal influence of habit and disuse. Whether an organ has an adaptive function or not fades into a kind of methodological tautology, and the designation of "rudimentary" and "nascent" organs is left conceptually underdetermined. To be fair, Darwin does not *want* to tell a story about why which organs became adaptive and what caused others to recede into uselessness. But in the analogy, this cannot be obviated: *human* history can't be told without reference to the narrative through-line that is already contained in the duration of attention materialized in work by means of organs and tools. The existence of the means of production tells us a great deal, but we also need to understand the *relations of production* as they exist materially. Marx draws the metaphor to the point of attenuation: "Among the instruments of labour, those of a mechanical nature, which, taken as a whole, we may call the bone and muscles of production, offer much more decided characteristics of a given epoch of production, than those which, like pipes, tubs, baskets, jars, &c, serve only to hold the materials for labour, which latter class, we may in a general way, call the vascular system of production."[42] There are, of course, no nonmaterial artifacts. But there are traces of the relations within the extended production process in the artifactual record. Our understanding of human evolution as the elaboration of the organ-based, tool-creating work that defines that very human includes not only the production process but also the valuation process. And indeed, the entirety of economic history (in Marx's terms, the only topic at all) can be understood as the relation of inorganic to organic parts, of different types of functions bearing different meanings and different possibilities. The potential for the body to be altered in the course of time underscores the fine anatomical and analytical approach that *Capital* takes, insisting at every moment on the revolutionary potential, the utter plasticity, of the body its takes as its subject. The current artifactual record, Marx is arguing, makes the body the subject of a dominance relation—the expropriation of surplus-value—that is mystifying to the casual observer, intuitive to those living in its thrall, and utterly changeable.

The method that would change it is the one of *Capital* itself, and it is in this sense that the book is an organ. The method is a familiar one: the commodity, for example, houses the contradiction of use- and exchange-value, and it is this very realized contradiction that generates the circulative process and allows for the capitalist mode of distribution. Underlying commodity exchange, however, must be the possibility of profit. If the production process is *in general* the organ-based, tool-creating labor of humans, it is from this very process that the capitalist must draw profit. In other words: the capitalist must find a way to do more than trick consumers into paying high prices. He must find a soft point in the system of production that allows him to profit. Marx's proposal is as well-known as it is elegant: labor-power is that use-value that generates exchange-value (through the production of useful commodities). The capitalist regards labor-power as merely one more thing bought and placed into the production process, as though the factor "labor" had become merely an instrument of production. This illusion is essential to the fiction that the product belongs to the owner, not to the labor. Labor must be a "thing" among others for the capitalist to be able to demand return on his investment,[43] by making a spectral organic body out of the abstract labor he has reduced to a machine: "By turning his money into commodities that serve as the material elements of a new product, and as factors in the labour process, by incorporating living labour with their dead substance, the capitalist at the same time converts value, i.e., past, materialised, and dead labour into capital, into value big with value, a live monster that begins to 'labor,' as if it had love in its body [*als hätt' es Lieb' im Leibe*]."[44] Capital *appears* to work—the capitalist claims his profit as return on his "work." Capital *is*, however, the appearance of life—or of humanity, as working—in monstrous form. The words "als hätt' es Lieb' im Leibe" come from the scene in Goethe's *Faust* in Auerbach's cellar, from a song that the tavern-goers sing among themselves about the poisoning of a rat. Marx will use this phrase in both the *Machine-Fragment* and in the chapter of *Capital* on machines: it stands in for him for the process of the mechanized presentation of work taking on its own life.[45] The "monster" that the automatic factory becomes is appearance of work in a form not created by nor suited to human life. The species human, as organ-bearing worker, confronts the projection and perversion of his very bodily constitution, the functional and productive body, *not* in the machines he makes to aid in production, but in the alternate life these machines take on, in the shift from workers using machines to machines using workers. The latter state of affairs, which I argue is the materialization of the relations of production as the means of production, leads Marx to suggest that, in the semi-metaphorical muscular and vascular systems of the economic period called capitalism, humans are the organs of a larger whole called the machine.[46]

Before turning to Marx's characterization of the economic-evolutionary epoch known as capitalism—which, crucially for my exposition of the organology of his method, comes to its apex in the automatic factory—I want to note that Marx's analogy to Darwin leaves a question unanswered, a philosophical gap between animal evolution and economic history. The question is: how did the human emerge from an evolutionary process and *replace that process with another* "technological" evolutionary process?

Marx and Engels divided their labor. Marx was to write the history and economy of humans, and Engels was to answer philosophical and scientific questions independently. The "dialectical materialism" he devised was the subject of intense debate during the Second International, and largely fell to the wayside of Marxist theory after the publication of the early works in the 1920s and 1930s. The use of Marx's economic philosophy to speculate about nature seemed an absurd project, one that could produce little more than an externalized Hegelian structure, a "dialectics of nature" (as Engels intended to title his magnum opus, left incomplete) that reads very much like the poorer attempts of some Young Hegelians to construct the history of nature on the reduced Hegelian schema, thesis—antithesis—synthesis.[47]

Engels's short fragment, *"The Part Played by Labor in the Becoming-Human of the Ape,"* is meant to present the orthodox opinion on precisely this gap. How did the human emerge from the animal kingdom? The descent of man, Engels proposes, cannot be a matter of his given organs or their arrangement, for the human is not only too similar to the primate for this, but is in fact less capable than his animal brethren in this respect. Instead, the hand—the organ shared with the monkey—must have "become free" in the course of human evolution: "Thus the hand is not only the organ of labor, *it is also its product.*"[48] The line could be taken from *Capital*: it applies the logic of work, the categorial flexibility of its factors, to the human body itself. The hand is not merely an organ; it is also the product of work. In other words, the hand is not merely the means of production—nor is it the means of means, as it was for Aristotle, Galen, and Goethe—it is also the tool constructed *by* work. This position of the hand, as both condition and product of work, is generalized by Engels to two other essential human organs: the brain and language. The brain and the hand develop in parallel as organs: the record of their mutual advance is the archaeology of tools (as it still is today). There is no hope, however, of deciding which of these has causal primacy, that is, the still topical question of whether the hand's articulation drove the complexification of the brain or vice versa. The organ, in each case, is both condition and product: it is not the origin, which cannot be found in a static arrangement of parts, even functional ones. The origin is the dynamic yet material process called work.

Nor can the origin of human difference be his social tendency, shared by primates as well. Perhaps the weakest part of Engels's argument is the following: "In short, the becoming human got to the point where they *had something to say* to each other. This need created an organ for itself."[49] Yet the point remains: speech is not the distinctive marker of humans, not the organ of reason or the *differentia specifica* of *homo loquens* (recall Herder), but instead a function created for and adapted to work on nature. Hand, brain, larynx: these are not the basis but the concrete historical expression of human work. And even on the philophical side of the division of labor between Marx and Engels, there is no answer to the question of human origins. Or, rather, there is no metaphysical answer to that question. If there is an anthropology here, it is a functional one, based in the insight of organology. There are only functions and qualities, parts and wholes, and the history of their interactions. Engels shares the loss of method, in the story of the symbiosis of organs and tools, that the *Naturphilosophen* had before him. The projection of organs as tools, the indifference of their definitions and the undermining of the metaphorics of mechanism and organism, was not enough for the Romantics, as it was not enough for Marx. There had to be some way of intervening in the dynamic states of affairs the philosophy also represented as its content, and the method of intervention needed to overlap with this represented content. The point of overlap was, as it had been for the Romantics, the organ.

Intelligences of the Machine: Humans as Organs

The question of machinery in the factory was taken, from its emergence in the early nineteenth century, as a variant on the question of the division of labor.[50] The distribution of individual tasks to different workers had both simplified labor and increased its output; it seemed that transferring even more finely granulated parts of tasks onto machines would not only free workers from the exertion of performing tasks themselves (thus making work itself easier) but also shorten the workday, creating leisure time. David Ricardo, who had argued these points in the first edition of his *Principles of*

Economy and Taxation (1817), later revised both his book and his opinion. The third edition of 1821 held that machines simplified work but also made its intensity in one activity greater, wearing on the bodies of workers. The simplification lowered wages, since anyone could perform the tasks with minimal training. Finally, the automatization process simply eliminated some jobs. Ricardo concluded that this problem would have to be taken care of by the larger market—workers would have to be absorbed into other industries.[51]

Marx responded to this style of argumentation with the method he had developed in the first chapters of *Capital*, namely, from the perspective of the abstract yet real unity of the production and valuation processes. It goes without saying that he disputed the first version of Ricardo's machinery chapter:

> It is an undoubted fact that machinery, as such, is not responsible for "setting free" [*Freisetzung*] the workman from the means of subsistence. It cheapens and increases production in that branch which it seizes on, and at first makes no change in the mass of the means of subsistence produced in other branches. Hence, after its introduction, the society possesses as much, if not more, of the necessaries of life than before, for the labourers thrown out of work; and that quite apart from the enormous share of the annual produce wasted by the non-workers. And this is the punchline of economic apologetics! . . . Thus he saves himself from all further puzzling of the brain, and what is more, implicitly declares his opponent to be stupid enough to contend against, not the capitalistic employment of machinery, but machinery itself.[52]

Do not destroy the machines, Marx says, but the manner or method of the machines, and you destroy the evolutionary epoch of the economy called capitalism. Neither destruction of all machines (à la *Erewhon*) nor the mere sense that they are being used for capitalistic purposes is enough. The point is to root out the very relations of organs and their deployment that make up capitalism. The goal is to shift the relation between the musculature and the vascular system, to alter the body of human history.

For this reason Marx presented machinofacture as the *materialization of the relations of production as the means of production*. Talk of "monsters" and the "undead" in the factory is not merely metaphorical: it addresses simultaneously the worker's body and discipline in the factory *and* the particular constellation of dominance that guides the construction and use of that factory. In fact, these elements cannot be separated, especially in the automatic factory. Once machines use humans as functional parts of their own production process, the relations of production have crystallized into the very means of production. The social dominance of the capitalist is embodied in the instruments of labor, which are no longer merely implements. The reversal of mechanical and organic elements, and the potential dehumanizing evolution of the machine, had to be met with more than a metaphorical description or a polemical prescription. It had to be met with organology.

The rise of machinofacture has two primary consequences. First, it reverses the relationship between human and machine, exacerbating the already abstract nature of divided labor. Second, and because of this, it makes abstract *who* is laboring—indeed, Marx devotes more than half of his writing about machinofacture in *Capital* to problems of labor by women and children. What might seem contingent—capitalists try to exploit all possible forms of labor, and are resisted slowly by parliamentary regulation (which Marx meticulously documents)—is not: who works is a matter of social categories, and the machine allows, for example, many tasks to be carried out by children that could not have been before. Although essential to the analysis of the present, child and female labor are treated under the umbrella of the larger shift in which workers are used by machines, in which workers become the organs of those machines, and in which factories take on emergent properties beyond the control of their supposed masters. It takes place under the aegis of the *depersonalization* or deindividuation of work, and this is the difference, for Marx, between the division of labor and machinofacture:

> The performance capability of the tool is emancipated from the restraints that are inseparable from human labour power. Thereby the technical foundation on which is based the division of labour in manufacture, is swept away. Hence, in the place of the hierarchy of specialised workmen that characterises manufacture, there steps, in the automatic factory, a tendency to equalise and reduce to one and the same level every kind of work that has to be done by the minders of the machines [*die Gehilfen der Maschinerie*]; in the place of the artificially produced differentiations of the detail workmen, step the natural differences of age and sex.[53]

The reversal of machine and human in the automatic factory does more than the division of labor. The latter makes work abstract; the former *abstracts the worker himself*, who is no longer necessarily "himself." In other words, it threatens to dislocate work from the very notion of the worker, the person who works.

Marx adopts the notion that the relation between worker and instrument is reversed in the automatic factory from the British economist Andrew Ure, whom Marx calls "the Pindar of the factory system."[54] Ure supplies Marx with the notion that factories might be organ-filled, that machines might have organs at all:[55] "It is in a cotton mill, however, that the perfection of automatic industry is to be seen; it is there that the elemental powers have been made to animate millions of complex organs, infusing into forms of wood, iron, and brass an intelligent agency."[56] Ure completed the chiasm, however, in favor of the disciplinary use of humans as the organs of that very machine. The factory, he writes, "involves the idea of a vast automaton, composed of various mechanical and intellectual organs, acting in uninterrupted concert for the production of a common object, all of them being subordinated to a self-regulated moving force."[57] The very ideologue of machinofacture had proclaimed the irrelevance of the organic/mechanical distinction for modern work. It was an instance of political economy from Britain that Marx could not deny. Nor was it enough to polemicize against it. Marx drew its consequences philosophically in *Capital*, suggesting that its theorization was one, if partial, instrument for its dismantling.

For Marx, the advent of the machine-driven factory represents the industrial mode of production finally achieving its normative form. Marx is treating such developments as the cotton gin and the spinning jenny, replacements for hand-driven machines that increase productivity through an increase and differentiation of the abilities of fixed capital. The machine is divided into three parts: the force-impulse mechanism, the transmission mechanism, and the mechanism of tools. According to Marx, the third of these is the key, and a shift in its constitution has brought the factory into its own—it has technized the site of work. Older developments had, of course, changed the source of energy, for example through the use of animal or water power to lend force to the machine for operating. The complexity of transmission mechanisms, which guide and transfer the force to the functional parts of the machine, has also increased. But neither the origin of force nor the design of the transfer have changed the fundamental character of the machine, and thus of work. This change depends on the last, on the character of the tools of the machine—that change brings about the Industrial Revolution.[58]

The workshop had undergone a similar transition, from the site of isolated workstations to a functional unity in greater production because of the division of labor. But even the resulting factory still relied on juxtaposed elements, not in its use of human labor, but in its use of machines. These machines were replacements for human hand-operations: they worked on individual areas of application, each functioning on the above model. The factory realizes its own concept, however, only when the machines are united such that laborers are only guiding their activity. When this happens, the individual machines become functional parts of a greater factory-machine—just as the factory was a greater unity of the workshop.[59] And just as individual workers became functional parts of the factory, the tool-mechanisms of individual machines become, in the greater factory-machine, organs: "Just like many tools form the organs of a machine of labor, many machines of labor now simply form uniform organs of the same mechanism of motion."[60] Tools become organs in the completion of the concept of the Industrial Revolution. The complexity of the factory now takes on the fullness of its concept as a higher unity: it becomes organic. If there was indecision in the eighteenth century about where organs could properly be located—from Leibniz to Reil one had spoken of machine-organs—Marx points to the reemergence of organological unity on the other side of supposed machinic reduction. What might have seemed a technicization[61] and machinification in the simple terms of mechanism takes on, in its character as and determination of work, the characteristics of the humanity Marx had sought twenty years earlier. The tool was once human, but in machinofacture it becomes a functional part of a unity that is greater than human: "The tool is . . . not suppressed by the machine. From the dwarf-tool of the human organism there extends in reach and number the tool of a mechanism created by the human. Capital makes the worker labor, rather than with a hand-tool, with a machine that guides its own tools."[62] Organ becomes the necessary resurgent designator of this apparent return, on the other side of abstraction and degradation, to a higher unity in the workplace. If organics returns, it is not as a model for biology, but for a technological process that is neither human nor obviously under the control of the human will.

The resurgence, however, has a dark side. The machine-factory intensifies labor even as it simplifies it, forcing women and children into the factory in greater masses (Marx notes that, due to their relative lack of education, they are effectively sold as slaves by men in this context). The reemergent organic unity of the machine is monstrous:

> A structured system of machines, to which motion is communicated by the transmitting mechanism from a central automaton, is the most developed form of production by machinery. Here we have, in the place of the isolated machine, a mechanical monstrosity [*mechanisches Ungeheuer*] whose body fills whole factory-buildings, and whose demon power, at first veiled under the slow and measured motions of his giant limbs, at length breaks out into the fast and furious maelstrom of his countless working organs.[63]

Organics is not organology: Marx's vision threatens here to deliver the humanity of humans to the machines' replacement-organs. It seems that the structure and material of work will cannibalize the worker. The robust organology of the *Manuscripts* seems to have been replaced with a terrifying anticipatory version of Horkheimer's claim that organs and tools have become extensions of each other. The relations of production, once materialized as the means of production (as fixed capital, here), ramify themselves. The relations gain a material basis that delimits the possibilities of development of future relations—a fact Marx is ambivalent about. The point, however, is that the *Ungeheuer* is not merely the machine, like the famous monster that swallows lines of workers in Fritz Lang's *Metropolis*. It is the *specific combination* of relations and means, of form and instrument, that has spun out of control. Any thought based on notions of individual-social interface will fail to grasp this problem in its material urgency. When machines externalize systems of production, schematizing cherished metaphysical patterns and engraving those patterns on the social, new thought is called for. For that, the organological strain of *Capital* was needed.

In a move that will be adopted throughout the Marxist tradition in reaction to refinements of the capitalist system that threaten the ability to resist, Marx points to the cooperative character of the machine-factory:

> The implements of labour, in the form of machinery, necessitate the substitution of natural forces for human force, and the conscious application of science, instead of rule of thumb. In manufacture, the organisation of the social labour process is purely subjective; it is a combination of detail labourers; in its machinery system, modern industry has a production-organism that is purely objective, in which the labourer becomes a mere appendage to an already existing material condition of production. In simple co-operation, and even in that founded on division of labour, the suppression of the isolated, by the collective, workman still appears to be more or less accidental. Machinery, with a few exceptions to be mentioned later, operates only by means of associated labour, or labour in common. Hence the co-operative character of the labour process is, in the latter case, a technical necessity dictated by the nature of the means of labour itself.[64]

In one sense, then, the hope for the humanization of work and of the organs of that work from the *Manuscripts* is suggested in its dual subjective and objective character. Cooperation is no longer a contingent element of factory-life—it is a technical necessity. Technicization has brought with it not only immiseration and partial irrelevance of the specificity of human work, but also the necessity of community, the necessity of planning to force the organs of production to work in concert. The *telos* of the organic factory should emerge, then, from the internal planning of what would eventually receive the name *soviet*—this "collective" is premised on the notion that the means of production must be appropriated because they are the materialization of the relations of production. Cooperation is *ipso facto* part of factory life. If this is so, machines present the *rules for mediating* between the musculature and the vascular system of the animal body that is, in its current expression, capitalism. In other words, what is needed is a science that administrates the evolution of economic history. And the source of such a science remains obscure, as it did for Marx. What follows is a suggestion of how organology contributes to the construction of that desired science.

There is a tendency to interpret Marx's writings on machines as reflecting an incomplete element of his analysis. Because Marx devoted a very important fragment of the 1857 *Grundrisse* to machines, and then wrote extensively in notes on the problem in the 1860s, they appear as diagonal to his relatively complete *Capital* as does finance capital, as Engels pointed out in 1894.[65] If, in one mood, Marx seems optimistic about the forced cooperation of the factory—as Lenin would later be optimistic about the forced "cooperation" and increased organization of monopoly capital—one strain of interpretation will argue that machines are one such source of optimism. They are, after all, precipitations of collective cognitive and social effort, the visibilization of science. Thus: "With machinery—and the atelier based in its use—the dominance of past labour over living labor gains not only social truth, expressed in the relation of capitalist and worker, but instead, so to speak, *technological* truth."[66] *Technologie* is both the knowledge and practice of production. The term comes from the cameralist tradition, and Marx means it as knowledge of practice and the ability to regulate that practice. The domination of worker by capitalist—the unequal distribution of profit, and the bodily exploitation in which it has its origins—is physically present in the machine. This is not mystical: the machine represents and applies this particular form of domination. Its sociotechnical reality is the expression of this domination.

To be sure, this allows for what would later be called "critical theory." The following passage, which I quote at length, could well be taken from Horkheimer's writing:

> The application of *natural agents* [*English in original*—LW]—in a way its assimilation into the body of capital—coincides with the development of *science* as an independent factor of the production process. Just as the production process becomes the *application of science*, so vice versa does science become a factor, so to speak a function of the production process. Every discovery becomes the basis of a newer invention or new, improved methods of production. Only with the capitalist mode of production are the natural sciences delivered up to the immediate production process, while on the other hand the development of production delivers the means to the theoretical subjection of nature. Science gains the name of being the means of wealth-production: means of enrichment. Only in this mode of production can there arise practical problems that can only be solved scientifically. Now for the first time experience and observations—and the *necessities* [*Eng.*] of the production process itself—on a hierarchy that scientific application allows and makes necessary. *Exploitation of science*, of the theoretical progress of humanity. Capital does not create science but it exploits it, assimilates it to the production process. Thus at the same time *separation of science*—as *science applied* to production—from *immediate labour*, where in the earlier stages of production,

limited amount and experience is connected with labour itself, does not develop as an independent power separate from it, and thus on the whole also transcends the traditionally practiced collection of recipes extending itself on a small scale (experientally based acquisition of the *mysteries of each handicraft* [*Eng.*]. Hand and head not separated.[67]

The production process has become capitalistic, and the very relation of capital is altered by its revolution of its instruments. Again: science here does not contribute to the general good of human society, but is made into an instrument of profit, *joining intellectual and physical labor* outside the body of the individual worker. Thus an organ is created for capitalism that exceeds the scale of the worker's body, that takes its leave of the individual on which the traditional notion of work is based. The machine is neither merely science applied nor the retrospective innocence of Adam Smith's labor-divided workshop: it is biopower as defined by Foucault. It operates on general phenomena (here: the distribution of values) at the level of their generality. The purpose of the technologies of power Foucault dubs "biopolitical" is "not to modify any given phenomenon as such, or to modify a given individual insofar as he is an individual, but, essentially, to intervene at the level at which these general phenomena are determined, to intervene at the level of their generality."[68] The machine-factory is in the first instance the instrument of bodily discipline, its regime tooled to modify workers' bodies to make its production more efficient. The method of *Capital* reveals it to be something far more sinister: the crystallization of a science developed in the period of craftsmanship, but under the guidance of the capitalist's interest. It is the precipitate of a social relation that conditions the direction and valuation of work, and so the beginning of biopower. By taking on the determinate characteristics of social organization and materializing them, it emerges as a new determinate instance in the arrangement of production and valuation. It introduces finance capital (which Marx sometimes calls "fictitious capital") as all too real, and a determinate force in market processes.

The mechanical complex which develops changes not the source of work (which has always marshalled forces of nature and animals, and so on), and not the transmission mechanism (the plow, for example, which turns the mule's strength into furrows), but instead the *Werkzeugmaschine*. It changes work at the point where the human controls the application of the tool to the material—and reverses the application. The human becomes the organ of the machine; the machine is the collapse of both organics and mechanics *and* physical and intellectual labor. This is why it takes on unexpected forms: our knowledge is limited by those categories. Mechanics is originally applied to further a social relation already in existence, but it has the effect of dominating that very relation and bringing it into its own essence, as Marx repeatedly says. Machinery is the relations of production as the means of production, and as such presents the supercession of organic or mechanical bodies, and of merely intellectual or merely physical labor—thus the partial optimism of the notion of "cooperation." But thus also the devastating pessimism about the world of capitalist organs; its body is malformed, in permanent attack on its lesser members. It is not ill so much as devolved, and the full question of its plasticity is now posed, according to Marx's metaphor, in its bones.

It is not clear that there can be an "answer" to this problem—isolating it theoretically is the bulk of what *Capital*'s self-imposed limits allow. But it remains the methodological prism through which one must approach the ever-expanding simultaneous problems of individual oppression and emergent qualitative direction at the level of social totalities. The fundamental recognition is that these are *of a piece*, and that their relative whole is *itself* an "organ" in the specialized sense I have explored in these pages. The method proposed by Marx, it is no longer too much to say, is a kind of economic metaphysics, a metaphysics in which the concrete relations and means of social production are contained in material functions that operate according to incomplete and often utterly destructive dialectical shifts. The question's answer must, in this setting, seem tired, since we have retreated from it for so long, and seen our only attempts at it go so deeply awry. But it remains stubborn: only social organs, ultimately communist organs, can serve—the kind that can allow moral experience to be colored, in Novalis's sense, for the aesthetic to adumbrate metaphysical shifts, as in Hölderlin, ultimately for the sense, which Marx shifts to technology from Schelling's art, that we live in and must urgently administer functional organs, as syntheses of freedom and necessity, and their emergent properties. This action is at a higher level than that of the individual, in fact, is constrained exclusively by supra-individual factors. But then, Romantic organs were never individuals, never subjects. Political Romanticism does not act by occasion but promises, in the concrete yet norm-producing location that is the organ, to become more than individual. Romantic organology calls for a collective organ. It is hard to think that we have built it; it is the more pressing to return to its original articulation critically.

Organ's Techne

Organology, I have been arguing, is the employment of the complex etymology of the root *organ*—to establish a productive relationship between intuitions of the world—law-based representational contents—and philosophical method. Its prehistory in the eighteenth century, coeval with the crisis in metaphysics attendant upon especially Newton's rearrangements of the scientific world, provided the ground on which the Romantics would build a metaphysics for "modernity," understood as the absence of continuity between the world and the human. This transition was also marked by the upheaval now called the Industrial Revolution, and the Romantics were well aware of the liminal space they occupied in this respect. Everywhere the "world" was of uncertain meaning, even as its laws were progressively discovered and determined. This world, however, was simultaneously populated by man-made yet unpredictable beings: technology was on the scene. As we have seen, even the professedly resistant Goethe had to take this into account, to come to terms with the metaphysics the Romantics had inserted into the heart of this historical transformation. Where representation and method, intuition and ability, coincided—this was the point the Romantics designated with their term "organ." Organology was the attempt to address the pressure point of input and output, of perception and action, even of self and society. Ultimately, as I have argued, Goethe chose to contribute to this philosophical construction, laying weight on its collective and scientific side, suggesting a cultural history of engaged science. If Romantic organology was modern metaphysics, Goethe seemed to point the way toward applied organology. Where Novalis had reduced representation to a factor in a transcendental technology, Goethe implicitly accepted this move even as he deepened strategies of representation. Organs were neither the organic bodies they belonged to nor the machines their cognitive counterparts conceived and created—they were the interface between such contradictory elements. They were subjects and objects, self and world, ego and society. They were problems as much as principles.

If we survey the various homes the term "organ" found after Romanticism—in sociology with Émile Durkheim;[69] or in philosophies of technology from Ernst Kapp[70] to Gilles Simondon and beyond; from Georg Simmel[71] to Martin Heidegger[72] to Humberto Maturana and Francisco Varela[73]—we will find that they circle around transcendental thought in the Critical tradition and its encounter with material formed by human activity. The term has remained central to philosophical efforts to come to terms with the ghostly body of metaphysics in the post-natural and post-sovereign world. The characteristic technological viewpoint of the Romantics, their conviction that new forms of cognition and action could be combined to transform the natural and historical world(s), was the specific inheritance of Marx, a layer of discourse that has remained in a kind of power, since it stands for a view of the world that is paradigmatically caught in the loop of becoming-transcendental, of knowing, being, and the disruptive role making plays in their traditional relations.

This tradition has crystallized in the politics of psychoanalytic Marxism in the present. Jacques Lacan construed the Freudian libido as an organ—both as part of the organism and as an instrument,[74] and goes on to speak of a "false organ"[75] (the "lamella") at the center of the drive cycle.[76] Organ once again articulates the material basis of transcendence, the point where the empty center of the drive disconnects from the apparently meaningful instrument of the libido, the represented desire. Slavoj Žižek has situated himself against Deleuze and on the side of Lacan's organology,[77] showing that ongoing debates rely on organological philosophical premises, as the crux of the political construal of the metaphysics arising from the technological and social-historical situation globally. Organology cuts across the divides between transcendental and immanent that dog interpreters of social processes, and this becomes *even more true* as those processes become more complex. I hope this study has contributed one more tool for this kind of analysis.

I have been arguing that the antinomy organic/mechanical did not ultimately drive the Romantic impulse. Contrary to appearances, the attempt to combine epistemological method with intuitive or even scientific content led, in the generation of Romanticism, to organology, to the attempt to construct the moment in which cognition and action are both rooted. In short—as is especially clear with Schelling and Novalis—organology is characterized by a different antinomy, that of freedom and necessity. Hölderlin's dialectical organs were used to reify this antinomy and make it the substance of a new metaphysics. Yet necessity is in some sense still attached to nature—even with Schelling, for whom nature becomes free necessity and art necessary freedom. The provenance of the felt contradiction is the opposition between cognition and action, both cast in relation to "being"—the point comes from Kant (metaphysics of nature, metaphysics of morals) as well as from Fichte (empirical I, transcendental I). Organology abandons this origin—nature and self become interchangeable elements in the work of technological metaphysics. Perhaps this is clearest if we remember the young Schelling's notion that a higher intuition is the general tool (*organon*) for a world populated by artificial objects (indeed, *art* objects as the organs of that intuition). Romantic terraforming, however, retained a sense of futurity, of hope: collective attempts at guiding the world populated by art would operate in, but be unhampered by, the final difference between freedom and necessity. Organology should make a world permanently alterable, because within and in a certain sense beyond the inherited contradictions of old metaphysics.

Goethe had cautioned against dreaming too lightly in this direction—rather, he thought, employ the insights of organology in the institution of science. Organology would have to be patient as its institutions were built, as it encountered the freedom/necessity divide in the political and social sphere of scientific endeavor.

Marx's problem represents another, equally important, shift in this philosophical trajectory. His philosophy of technology, in particular, reacts to the shift of the problem of contingency and necessity onto a set of objects *specifically constructed to administrate this problem*. Where humans had appeared to make machines to relieve the intensity of labor and provide rational direction in the administration of collective life, their machines now took on emergent properties and forms. The machine world cannibalized the human, became, as Marx would put it, *organic*. But the machine was not "actually" organic, just as the worker was not actually mechanical—instead, Marx brought his writing to bear on the appearance of unpredictable or contingent necessity in the very realm made to extend and complete human freedom. The new constraint was materialized freedom itself. This problem, with its basis in the shift from manu- to machinofacture, required an approach at the level of its material complexity. Marx, of course, would always insist on Hegel as the origin of this approach. Hegel had to be "turned right side up"—or, as Feuerbach had it, the dialectic had to be rooted in the sensuous materiality of human being. But Marx was dissatisfied with what he saw as a projection of dialectical process into the representational content of a static concept of the human. Dialectics had to remain a part not of the representation of human history, but of the method of political economy. In other words, political economy had to establish the means to represent the necessity of necessity's appearance in the heart of freedom. To describe that shift would, however, be to project the dialectic as if on a screen—to see history, and perhaps even nature, as embodying ideal shifts, complex narrative twists. Marx needed a weapon against this kind of thinking. He needed a conception of dialectics that would allow the ideal (written) production of this shift in necessity with a view to changing it. To reproduce its conditions in the abstract; to represent with a view to altering; to find the pressure point where the real condition met the emergence of its representation; to reduce the material condition to its possibility, even if—indeed because—that possibility were precisely the determination of freedom as necessity—Marx perhaps did not know it, but he was practicing organology.

If we recall the terms Blumenberg set forth for a history of technology, we can see here that his characterization of Marx misses the mark. The text of *Capital* reveals, in light of Romantic organology, much more than the identification of conditioning circumstances for technological development. This study promised to examine how a *technological imagination* (rather than *will*) would step out of history, as Blumenberg puts it:

> The history of technology will have to be, above all, also the history of the emergence of technology *out of history itself*. Whether and how technological arises from a determinate and new understanding of reality and of the position of the human within that reality will have to be the theme of an intellectual history of technology, which will not merely collect and register self-construals of technological activity and authorship, but which instead will allow the motivations of a lifestyle aimed towards technology and underpinned by that technology to become graspable.[78]

Technology is a transcendental condition, because it is the remainder and the material sign of our activity in the world. In Romantic organology, we can read the reflexive moment of techno-imagination, one that does not merely play with new images, but literalizes and transcendentalizes the conditions of its own thought. In so doing, the Romantics offer a kind of response to the conditions of disciplinarity and technicization. Technological will might be taken, with Marx, to be the hallmark of human activity and the ground of philosophy alike. Alternatively, the specificity of technological will in Blumenberg's sense might appear earlier than the technological imagination. Even if that is true, however, a second technological will is the result of the connection of metaphysics and politics in Romantic organology. Whatever the status of the origins of the modern technological moment the will to make the abstract concrete—and the means to impose technologies of the spirit onto the apparently given world—will have been the locus of a modern metaphysical imagination, which finds a metaphysical motor in a signal contribution of the Jena Romantics. That is, I take it, an unexpected Romanticism, and the possibility of historical dialogue with it for presentist purposes is for all that the richer, and perhaps the more pressing.

ACKNOWLEDGMENTS

This book has had many benefactors who have made it better than I could have alone; I am grateful to all of them. They include institutions, like the Comparative Literature and Literary Theory Program at the University of Pennsylvania; the US Fulbright Commission, which sent me to the Humboldt University in Berlin in 2011–12; and the German Department at New York University, which has generously supported me in innumerable ways. They also include many people, starting with my *Doktormutter* Catriona MacLeod, to whom my debt can surely never be repaid, and other advisers, Simon Richter, Warren Breckman, Paul Guyer, Gunnar Hindrichs. The intellectual rigor and sheer enthusiasm of my colleagues at NYU—Ulrich Baer, Andrea Dortmann, Christiane Frey, Alys George, Eckart Goebel, Avital Ronell, and Christopher Wood—are, I hope, reflected in some way in what follows, something that could be said of Jacques Lezra and Richard Sieburth, too. To Thomas Lay, Stefanos Geroulanos, and Todd Meyers, I am thankful for unflagging attention to detail and unfailing support. For material aid and intellectual engagement, thanks to Sean Larson. Jocelyn Holland, Adrian Daub, Gabriel Trop, Stefan Börnchen, John Smith, and Horst Lange all read drafts and made generous comments. Some mixture of personal and intellectual gratitude is due Will Krieger, Nick Theis, Peter Kuras, Caroline Weist, Dan DiMassa, Ian Cornelius, Matt Handelman and Jeffrey Kirkwood. To Jeffrey and Matt I owe a return on their patience and capacious guidance. To my family—Abigail Reid, Bruce Weatherby, Nora Weatherby—a similar debt of patience, but this one again surely unrepayable. This book is dedicated to them. And to Maya Vinokour, thanks for all of the above, for the life from which this work precipitates.

NOTES

INTRODUCTION

1. René Descartes, *The World and Other Writings*, trans. Stephen Gaukroger (Cambridge: Cambridge University Press, 2004), 99. Unless otherwise noted, translations are mine.
2. René Descartes, *A Discourse on Method*, trans. Ian MacLean (Oxford: Oxford University Press, 2006), 6.
3. See Descartes, *The World*, xxiii, and xxxvi, on the convention of using the "hypothetical method" to get around the Inquisition.
4. Ibid., 99.
5. Cited in Thomas Fuchs, *The Mechanization of the Heart: Harvey and Descartes*, trans. Marjorie Grene (Rochester, NY: University of Rochester, 2001), 1.
6. Descartes, *The World*, 99.
7. Ibid., 48; cf. Aristotle, *De anima*, 2.1.413a 8–9.
8. Descartes, *Discourse*, 46.
9. Ibid., 46–47.
10. Descarteś, *The World*, 169.
11. Ibid., 168. Cf. in the later "Description of the Human Body," 170, "organs or springs [*organes ou ressorts*]."
12. Descartes, *Discourse*, 46–47.
13. See Fuchs, *The Mechanization of the Heart*, 19ff.
14. This physiology shifted to an enclosed body with various systems of internal circulation slowly, crystallizing in the modern consensus in the eighteenth century. See Albrecht Koschorke, "Physiological Self-Regulation: The Eighteenth-Century Modernization of the Human Body," *MLN* 123, no. 3 (2008): 469–84. Cf. Koschorke, *Körperströme und Schriftverkehr: Mediologe des 18. Jahrhunderts* (Munich: Fink, 2003), 54–66.

355

15. See R. J. Hankinson, "Philosophy of Nature," in *The Cambridge Companion to Galen*, ed. Hankinson, 210–42 (Cambridge: Cambridge University Press, 2008), 230.

16. Ibid., esp. 225 and 227: "Galen's teleology is particularly uncompromising; where Aristotle will often describe a structure (or product) as a 'residue,' useless in itself, but a necessary by-product of something which is teleologically explicable, Galen does so far more sparingly, preferring to discern actual functions in apparently purposeless parts." Galen uses the Greek *organon* for parts, but in newer translations, this is construed as "instrument" because each part carries out an instrumental function (more on this distinction in the context of Aristotle below). See, for example, Margaret Tallmadge May's translation of *On the Usefulness of the Parts of the Body* (Ithaca: Cornell University Press, 1968), esp. 67–68.

17. Galen, *On Antecedent Causes*, ed. R. J. Hankinson (Cambridge: Cambridge University Press, 1998), 92–93ff.

18. See Hankinson, "Philosophy of Nature," 226.

19. This holds throughout the eighteenth century in German. Cf. Johann Jakob Bodmer and Johann Jakob Breitinger, *Schriften zur Literatur*, ed. Volker Meid (Stuttgart: Phillip Reclam 1980), 31,102. Johann Joachim Winckelmann, *Winckelmanns Werke in einem Band* (Berlin and Weimar: Aufbau, 1969), 144, after Robert Jütte, *Geschichte der Sinne: Von der Antike bis zum Cyberspace* (Munich: Beck, 2000), 164.

20. Hans Vaihinger, *The Philosophy of 'As If': A System of the Theoretical, Practical and Religious Fictions of Mankind*, trans. C. K. Ogden (London: Routledge, 1968), 1.

21. Ibid., 6; translation modified.

22. See Samuel Taylor-Alexander, *On Face Transplantation: Life and Ethics in Experimental Biomedicine* (New York: Palgrave Macmillan, 2014); and Samuel Taylor-Alexander, "How the Face Became an Organ," http://somatosphere.net/2014/08/how-the-face-became-an-organ.html.

23. The term was used in the 1790s to mean the study of parts of plants (*Planzen-Organologie*), and once, in a mystical sense, to indicate the usability of all parts and regions of the world for hermetic purposes. See Karl Gottlieb Feuereisen, *Pflanzen-Organologie oder Etwas aus dem Pflanzenreiche: Insonderheit die sonderbaren Würkunge des Nahrungssaftes in den Gewächsen* (Hannover: Pockwitz, 1780). This work refers to organs as "vessels" (a common equivalence before the 1790s) and allows the function of the organs to be not only subordinate to the internal telos of the plant's life but also the physico-theological guidance of God. God is far more central in Karl von Eckartshausen's *Probaseologie oder praktischer Theil der Zahlenlehre der Natur: Ein Schlüssel zu den Hieroglyphen der Natur* (Leipzig: Gräff, 1795), who treats "organology" as the

"workshop of forms" in nature and a key to divine presence (see 203–27). Novalis may have borrowed elements from this conception, but neither of these works anticipates the Romantic program in its extent and philosophical ambition.

24. See his seminal essay "Machine and Organism," in *Knowledge of Life*, by Georges Canguilhem, trans. Stefanos Geroulanos and Daniela Ginsburg (New York: Fordham University Press, 2008), 75ff.

25. Bernard Stiegler, *For a New Critique of Political Economy*, trans. Daniel Ross (Malden: Polity, 2010), 34; "Allgemeine Organologie und positive Pharmakologie (*Theorie und "praxis"*)," in *Die technologische Bedingung*, ed. Erich Hörl, 110–47 (Frankfurt a.M.: Suhrkamp, 2011).

26. Simondon, *Du mode d'existence des objets techniques* (1959; Paris: Flammarion Aubier, 1989), 65.

27. Compare the excellent study by Jan-Peter Pudelek, *Der Begriff der Technikästhetik und ihr Ursprung in der Poetik des 18. Jahrhunderts* (Würzburg: Königshausen und Neumann, 2000), which argues for the technicization of aesthetics and a departure from the fixed imagination of an *ars* over the course of long eighteenth century. See also John Tresh's excellent account of "mechanical Romanticism," *The Romantic Machine: Utopian Science and Technology after Napoleon* (Chicago: University of Chicago Press, 2012).

28. Friedrich Schlegel, *Dialogue on Poesie*, in Schlegel, *Kritische Friedrich-Schlegel-Ausgabe*, ed. Ernst Behler, Jean Jacques Anstett, and Hans Eichner (Munich: Schöningh, 1958–), sec. 1, vol. 2 (1967), 314. Hereafter cited as *KFSA*; subsequent citations are in the form 1.2.314 and refer to section, volume, and page number.

29. On different types of "system"-notions, see Paul Franks, *All or Nothing: Systematicity, Transcendental Arguments and German Idealism* (Cambridge: Harvard University Press, 2005), esp. 84ff., where "holistic monism" is differentiated from the "derivation monism" of Karl Leonhard Reinhold—Kant's first popularizer—and those in dialogue with him.

30. Because this study revolves around this term, I will always and only translate *das Organ* as "organ," modifying translations where necessary—as it often will be—to retain this historical-semantic precision.

31. *KFSA* 1.1.264; emphasis in original; cf. KFSA 1.18.143.

32. Ibid., 1.2.256.

33. Friedrich Schiller, *Sämtliche Werke*, vol. 5, *Erzählungen, theoretische Schriften*, ed. Wolfgang Riedel (Munich: Hanser, 2004), 254.

34. In fact, neither the Latin edition nor the complete German translation handed in is extant. See Peter-André Alt, *Schiller: Eine Biographie. Band 1. 1759–1791* (Munich: Beck, 2000), 156-57, and what follows for the specific intervention in physiology, especially the critique of Haller.

35. Schiller, *Sämtliche Werke*, 5:658; emphasis in original.
36. Ibid.
37. See Simon Richter, introduction to *The Literature of Weimar Classicism*, ed. Richter (Rochester, NY: Camden House, 2005), 3–45.
38. Schiller, *Sämtliche Werke*, 5:611.
39. Leonard Wessell makes a similar point: "Schiller's relativization of objective or Greek art and his concomitant reevaluation of modern poetry acted as a fresh impetus to Schlegel's thought by presenting him with the antipodes of an aesthetic antithesis whose dialectical resolution resulted in romanticism" (Wessell, "Schiller and the Genesis of German Romanticism," *Studies in Romanticism* 10, no. 3 [1971]: 191). The title is a sort of genre of article: cf. Arthur O. Lovejoy, "Schiller and the Genesis of Romanticism," *MLN* 35, no. 1 (1920): 1–10.
40. Ernst Behler emphasizes that the third term is merely the synthesis of the other two: "Schlegel's literary theory must be seen as an attempt to unite these two antagonistic aesthetics, to find a synthesis of the antique and the modern, the Classical and the Romantic" (Behler, "The Origins of the Romantic Literary Theory," *Colloquia germanica* 2 [1968]: 117). Note how this means that Romanticism is both a historical factor and the vehicle of that history's overcoming. In other words, Romanticism is both term and function of its own historically conceived synthesis.
41. See Jörg Henning Wolf, *Der Begriff "Organ" in der Medizin* (Munich: Fritsch, 1971), 9.
42. Johann Heinrich Campe, *Wörterbuch zur Erklärung und Verdeutschung der unserer Sprache aufgedrungenen fremden Ausdrücke: Ein Ergänzungsband zu Adelung's und Campe's Wörterbüchern* (Braunschweig: Schulbuchhandlung, 1813), 449–50 (qtd. in Wolf, *Der Begriff*, 14).
43. Johann Heinrich Campe, *Wörterbuch zur Erklärung und Verdeutschung der unserer Sprache aufgedrungenen fremden Ausdrücke: Ein Ergänzungsband zu Adelung's und Campe's Wörterbüchern* (Braunschweig: Schulbuchhandlung, 1801), 498.
44. See Ernst Cassirer, *The Philosophy of the Enlightenment*, trans. Peter Gay (Princeton, NJ: Princeton University Press, 1979), 3–37.
45. What Jacobi calls "the being in all existence [*das Sein in allem Dasein*]" (Jacobi, *Werke*, ed. Klaus Hammacher und Wolfgang Jaeschke (Stuttgart: frommann-holzboog, 1998), 4.1:87).
46. I thus intentionally exclude some uses of the term that deserve their own studies—for example, the use of the term to discuss the voice in theatric and especially operatic settings. Cf. John Tresch and Emily I. Dolan, "Toward a New Organology: Instruments of Music and Science," *Osiris* 28 (2013): 278–98. I also generally leave out the sense given to periodicals and

newspapers as the "organs" of certain governing bodies, etc. The full range of these, including the development of the political term, can be found in Otto Brunner et al., eds., *Geschichtliche Grundbegriffe: Historisches Lexikon zur politisch-sozialen Sprache in Deutschland*, vol., 4, *Mi-Pre* (Stuttgart: Klett-Cotta, 1997), 519–621; see also Ethel Matala de Mazza, *Der verfaßte Körper: Zum Projekt einer organischen Gemeinschaft in der Politischen Romantik* (Freiburg: Rombach, 1999), esp. 131ff.; cf. Reinhard Saller, *Schöne Ökonomie: Die poetische Reflexiion der Ökonomie in frühromantischer Literatur* (Würzburg: Königshausen und Neumann, 2007), 68ff. See also the excellent treatment of fictional corporate bodies and the problem of possession by Stefan Andriopoulos in *Possessed: Hypnotic Crimes, Corporate Fiction, and the Invention of Cinema*, trans. Peter Jansen and Stefan Andriopoulos (Chicago: University of Chicago Press, 2008), esp. 42ff. I also do not treat the discourse on "new organs" that arose in conjunction with Franz Mesmer's doctrines, the so-called "animal magnetism." On Jean Paul's and E.T.A. Hoffmann's engagements with the same, see Gerhard Neumann, "Romantische Aufklärung: Zu E.T.A. Hoffmanns Wissenschaftspoetik," in *Aufklärung als Form. Beiträge zu einem historischen und aktuellen Problem*, ed. Helmut Schmiedt and Helmut J. Schneider (Würzburg: Königshausen und Neumann, 1997), 106–48.

47. See Frederick Beiser, *German Idealism: The Struggle against Subjectivism, 1781–1801* (Cambridge: Harvard University Press, 2002), esp. 349–465; cf. Beiser, *The Romantic Imperative: The Concept of Early German Romanticism* (Cambridge: Harvard University Press, 2003), 131–53.

48. Wolf, *Der Begriff "Organ,"* 38–44.

49. Wolf maintains that Aristotle never gives a definition of *organon*, although he often defines *meros/morion*, "part" (ibid., 17–18). But Aristotle uses the same conception of the term across a number of metaphorical applications, and thus provides us with something like a working or base definition. The *De partibus animalium* is, of course, *peri zoion morion*.

50. The usage is extended to biological phenomena, where it is not metaphorical (rather, it is systematic) but not *exclusively characteristic of this field*. As Justin E. H. Smith puts it: "an organon or organ is for Aristotle "an entity whose essential nature is to be telic, an entity whose being is to be for the sake of that activity of which it is the organ." The whole animal, what we today would call an 'organism,' can be seen as the 'global organ,' as distinct from the 'local organs' that contribute to the telos of the global organ. In animals, the telos of the global organ is living, and this telos includes the lesser telos of, for example, the eye, which is seeing" (Smith, *Divine Machines: Leibniz and the Sciences of Life* [Princeton, NJ: Princeton University Press, 2011], 107).

51. Aristotle, *Generation of Animals*, trans. A. L. Peck (1942; Cambridge: Harvard University Press, 2000); hereafter cited as GA 734b22–23.

52. Wolf collects Aristotle's uses of this analogy (*Der Begriff*, 16ff.).

53. Aristotle, *On the Soul/Parva Naturalia/On Breath*, trans. W. S. Hett (Cambridge: Harvard University Press, 1935). Present translation from J. A. Smith, in Aristotle, *Basic Works*, ed. Richard Peter McKeon (New York: Random House, 2001); hereafter cited as *De anima* 424a.

54. *De anima* 424a–b; translation modified: *aistheterion*, which is translated as "organ of sense," is correct according to present usage, but insufficient for terminological-historical purposes.

55. A thorough compendium of works that use some version of this terminology to describe various parts of the arts curriculum in the Early Modern period can be found in Gorgio Tonelli, *Kant's "Critique of Pure Reason" within the Tradition of Modern Logic: A Commentary on its History*, ed. David H. Chandler (New York: Olms 1994), 133–58.

56. Aristotle, *Problems*, 30.5.955b; Aristotle, *Problems II, Books XXII–XXXVIII*, trans. W. S. Hett (Cambridge: Harvard University Press, 1965), 170–71.

57. *Problems*, 30.5.955b, pp. 172–73.

58. *De anima* 429a/19.

59. Ibid., 429a; translation modified from "organ" to "tool." As A. P. Bos has argued, "organ" is a "totally unacceptable" translation-choice for Aristotle, whose view is a "cybernetic instrumentalism" that conceives of "bodily tools" but in no way prefigures nineteenth-century organicism. See A. P. Bos, *The Soul and Its Instrumental Body: A Reinterpretation of Aristotle's Philosophy of Living Nature* (Leiden: Brill, 2003), 69–88. Aristotle is conceivably also responding to a Platonic view here, as expressed in the *Timaeus* at 45a–b, where we hear of *organa* "of forethought for the soul."

60. Analogically—and determinatively for the tradition—the *body* was the *soul*'s *organon*. So *De anima* 415b/18: "All natural bodies are organs of the soul" [*panta gar ta phusika somata tes psuches organa*]. Wolf mentions this topos in *Der Begriff*, 20. Again, the analogy is allowed because it rests on the conceptual background of the *techne-physis* comparison. And the definition is retained: the form (*morphe*) must be actualized in matter (*hule*) by some means.

61. *De anima* 432a.

62. Cf. Reinhard Löw, *Philosophie des Lebendigen: Der Begriff des Organischen bei Kant, sein Grund und seine Aktualität* (Frankfurt a.M.: Suhrkamp, 1980), 50–52. Löw suggests that Aristotle's method is based in the organ as the unit of bodily-intellectual experience, and that therefore the identity of the known and knowledge is organ-based. The suggestion of an instrument of knowledge in this sense is tied to the possession of organs.

63. Friedrich Wilhelm Joseph von Schelling, *Historisch-kritische Ausgabe* (Stuttgart: frommann-holzboog, 1976–), vol. 9, ed. Hans Michael Baumgart-

ner, p. 328. Hereafter Schelling is cited as Schelling, *HKA*; subsequent citations are in the form 9.328 and refer to volume and page number.

64. Novalis, *Novalis Schriften: Die Werke Friedrich von Hardenbergs*, ed. Paul Kluckhohn and Richard Samuel, 3rd ed. (1960; Stuttgart: Kohlhammer, 1977–); hereafter cited as Novalis, *HKA* 3.385; *AB* 642.

65. Novalis, *HKA* 3.466; *AB* 1075.

66. Paul de Man, *Aesthetic Ideology*, ed. Andrzej Warminski (Minneapolis: University of Minnesota Press, 1996), 179; cf. Werner Hamacher, "Position Exposed: Friedrich Schlegel's Transposition of Fichte's Absolute Proposition," in *Premises: Essays on Philosophy and Literature from Kant to Celan* by Hamacher (Stanford: Stanford University Press, 1999), 222–61.

67. Or at least, it is not organicist. The understanding of biological history newly articulated by Tobias Cheung—which consists in a massive shift of the notion of functional agents in plastic individuals acting between internal and external milieux—is much closer to the organological program I spell out here. See Tobias Cheung, "Organismen: Agenten zwischen Innen- und Außenwelten, 1780–1860," transcript (Bielefeld, 2014), e.g. 14.

68. I also hope that my efforts here will make dialogues possible with such excellent studies as those by Robert Mitchell, *Experimental Life: Vitalism in Romantic Science and Literature* (Baltimore: Johns Hopkins University Press, 2013); and Denise Gigante, *Life: Organic Form and Romanticism* (New Haven: Yale University Press, 2009), with whom I could not agree more on the following score: The Romantic "project was not hermetic or escapist but anchored in a strong conviction that aesthetic power can have real-world transformative capacity" (48).

69. *Commentarii societatis regiae scientiarum Gottingensis*, vol. 2 (Göttingen, 1753), 116.

70. Albrecht von Haller, *Primeae lineae physiologiae in usum praelectionum academicarum* (Göttingen: Acad. Bibl., 1751), 295.

71. Ibid., 247.

72. On this group, see Laura Otis, *Müller's Lab: The Story of Jakob Henle, Theodor Schwann, Emil du Bois-Reymond, Hermann von Helmholtz, Rudolf Virchow, Robert Remak, Ernst Haeckl, and Their Brilliant, Tortured Adviser* (Oxford: Oxford University Press, 2007).

73. Theodor Schwann, *Mikroskopische Untersuchungen über die Ueberein-stimmung in der Struktur und dem Wachstthum der Thiere und Pflanzen* (Berlin: Sander, 1839), iv; cf. xv: "a common principle of development for the elementary parts of all organisms." Cf. Otis, *Müller's Lab*, 62.

74. Schwann, *Mikroskopische Untersuchungen*, 196.

75. Schwann saw that cells made up widely varying macroscopic parts of bodies, including nerves (see ibid., 177, where organs build themselves out of "seemingly indifferent" cellular structures).

76. *Pace* the view that Schwann, in particular, was a mechanist rejecting all forms of organicism. See Lenoir, *The Strategy of Life: Teleology and Mechanics in Nineteenth-Century German Biology* (Chicago: University of Chicago Press 1989), 112–56.

77. Gigante, *Life*, 36–40.

78. Thus the miniaturization of the investigation is not a linear narrative of the development of biology as a science, but more likely one part of an unfolding controversy over form, environment, and mechanism. Lynn Nyhart argues that method was attached to form in biology, and that this raised the question: "What was the appropriate practical level at which one should seek the causes of form—in the cells, the tissues, the organs, the organ systems, the organism as a whole?" (Lynn Nyhart, *Biology Takes Form: Animal Morphology and the German Universities, 1800–1900* [Chicago: University of Chicago Press, 1995], 3). Nevertheless, the organ seems to have lost its conceptual lustre in the drive to find smaller units of causality.

79. Its development of Leibnizian language about individuals may have contributed to the cell theory itself, however, documenting how Romantic thought contributes strains of scientific and literary modernism that are continually erased in the very constitution of the latter. See Canguilhem, Knowledge of Life, 151–52.

80. Thomas Kuhn, *The Essential Tension: Selected Studies in Scientific Tradition and Change* (Chicago: University of Chicago Press, 1977), 66–105.

81. Ernst Cassirer, *The Problem of Knowledge: Philosophy, Science and History since Hegel* (1950; New Haven: Yale University Press, 1978), 17–19.

82. Hans Blumenberg, *Paradigmen zu einer Metaphorologie* (Frankfurt a.M.: Suhrkamp, 1998), e.g., 13.

83. See esp. "Nachahmung der Natur: Zur Vorgeschichte der Idee des "schöpferischen Menschen," in Hans Blumenberg, *Wirklichkeiten in denen wir leben: Aufsätze und eine Rede* (Stuttgart: Reclam, 1981), 55–103. The thesis is to be found in many of Blumenberg's works, including his magnum opus, *The Legitimacy of the Modern Age*, trans. Robert Wallace (1983; Cambridge: MIT Press, 1999).

84. Blumenberg locates this emergence in the *Literaturstreit* with Breitinger and Bodmer. Bodmer saw Milton as producing a "world" through a "metaphysische Handlung." See Blumenberg, "Nachahmung der Natur," 91.

85. Blumenberg's major treatment of this topic is *Lebenszeit und Weltzeit* (Frankfurt a.M.: Suhrkamp, 1986).

86. Hans Blumenberg, *Shipwreck with Spectator: Paradigm of a Metaphor for Existence*, trans. Stephen Rendall (Cambridge: MIT Press, 1997), 83.

87. Hans Blumenberg, *Zu den Sachen und zurück*, ed. Manfred Sommer (Frankfurt a.M.: Suhrkamp, 2002), 19–43.
88. Ibid., 19.
89. Ibid.
90. Ibid., 20.
91. Ibid.
92. Ibid., 42–43.
93. Blumenberg, *Paradigmen*, 193.
94. Hans Blumenberg, *Die Lesbarkeit der Welt* (Frankfurt a.M.: Suhrkamp, 1986), 256.
95. Hans Blumenberg, *Arbeit am Mythos* (Frankfurt a.M.: Suhrkamp, 1979), 619–20.
96. Intentionally, indeed cheerily—see Blumenberg's response to Dieter Henrich's charge that he is "pre-Critical" at the Poetics and Hermeneutics Colloquium of June 1962, in *Nachahmung und Illusion: Kolloquium Gießen Juni 1963, Vorlagen und Verhandlungen*, ed. H. R. Jauß (Munich: Eidos, 1964), 226.
97. This is not to say that Blumenberg never writes about Romanticism. Perhaps his most compelling reading of Romanticism can be found in Blumenberg, *Die Lesbarkeit der Welt*, 233–81. This reading does not, however, consider Romantic philosophical method in the larger history Blumenberg describes.
98. See his posthumous *Geistesgeschichte der Technik: Mit einem Radiovortrag auf CD* (Frankfurt a.M.: Suhrkamp, 2009).
99. Hans Blumenberg, "Lebenswelt und Technisierung unter Aspekten der Phänomenologie," in *Wirklichkeiten*, 7–55.
100. See Husserl, *Krisis*.
101. See Martin Heidegger, "The Question Concerning Technology," in *Martin Heidegger: Basic Writings from "Being and Time" (1927) to "The Task of Thinking" (1964)*, ed. David Farrell Krell (New York: Harper and Row, 1977), 283–319.
102. Cf. F. Scott Scribner's *Matters of Spirit: J. G. Fichte and the Technological Imagination* (University Park: Pennsylvania State University Press, 2010).
103. Michel Foucault, *The Foucault Reader*, ed. Paul Rabinow (New York: Pantheon, 1984), 176–77.
104. Michel Foucault, *The Order of Things: An Archaeology of the Human Sciences* (1971; New York: Vintage, 1994), 217–50, 263–80.
105. Cf. Shirley Roe, "The Life Sciences," in *The Cambridge History of Science*, ed. David C. Lindberg, Mary Jo Nye, Roy Porter, and Theodore M. Porter, vol. 4, *Eighteenth-Century Science*, 397–417 (Cambridge: Cambridge University Press, 2003).
106. Foucault, *Order*, 347.

107. Ibid., 242.

108. Ibid., 243.

109. As is clear from the section on "The System of the Principles of Pure Understanding," B188/A148–B198/A158, especially at B197/A158: "The conditions of the *possibility of experience* in general are at the same time conditions of the *possibility of the objects of experience*, and on this account have objective validity in a synthetic judgment *a priori*" (*Critique of Pure Reason*, trans. Paul Guyer [Cambridge: Cambridge University Press, 1998], 283).

110. Foucault, *Order*, 245.

111. Ibid., 247. Fichte and Hegel are the first attempts to unite these fields, and Husserl follows.

112. Azade Seyhan, *Representation and Its Discontents: The Critical Legacy of German Romanticism* (Berkeley: University of California Press, 1992); cf. Martha Helfer, *The Retreat of Representation: The Concept of Darstellung in German Critical Discourse* (Albany: SUNY Press, 1996).

113. Ibid., 2, 8.

114. As Jocelyn Holland has recently written with respect to Novalis, "Hardenberg's appeal for man to be an instrument of the self can be interpreted as the next logical step: it completes the transition from a definition of man based on representation to one based on function" (*German Romanticism and Science: The Procreative Politics of Goethe, Novalis and Ritter* [New York: Routledge, 2009], 88).

115. Foucault, *Order*, 264.

1. METAPHYSICAL ORGANS AND THE EMERGENCE OF LIFE:
FROM LEIBNIZ TO BLUMENBACH

1. Kant, *CPR* A viii.

2. Cf. Christian Wolff, *Erste Philosophie oder Ontologie*, ed. Dirk Effertz (Hamburg: Meiner, 2005), § 1, 18.

3. Gottfried Wilhelm Leibniz. *Sämtliche Schriften und Briefe; sec. 4, vol. 6, 1695–1697*, ed. Friedrich Breiderbeck et al. (Berlin: de Gruyter, 2008), p. 789. Subsequent citations are in the form 4.6.789 and refer to section, volume, and page number.

4. Ibid.

5. Ibid., 4.6.790.

6. See John Neubauer, *Symbolismus und Symbolische Logik: Die Idee der* ars combinatoria *in der Entwicklung der modernen Dichtung* (Munich: Fink, 1978).

7. "Leibniz used the words organisme (only in singular and in French), organismus (in Latin) and organisation (in French) to characterize the divine

mechanism of organic bodies" (Tobias Cheung, "From the Organism of a Body to the Body of an Organism: Occurrence and Meaning of the Word "Organism" from the Seventeenth to the Nineteenth Centuries," *British Journal for the History of Science* 39, no. 3 [2006]: 325).

8. The microscope is particularly important here, as it contributes to the equivalency "machine = organ" at a microstructural level for Leibniz. "If Aristotle's was a biological metaphysics, Leibniz's was thoroughly *micro*biological" (Justin E. H. Smith, *Divine Machines:* Leibniz and the Sciences of Life [Princeton, NJ: Princeton University Press, 2011], 97). By deepening the machinic or "organic" structure of bodies to their (infinitely) small parts, Leibniz sought to argue against the notion that bodies were like human-made machines—they were instead divine.

9. Leibniz's early work is marked by commitment to the so-called "complete concept" doctrine, which emphasizes that substances and accidents are like subjects and predicates, and holds that each individual being has a "complete concept" (including its past and future) which could be known—and is known by God. The later metaphysics shifts focus to the force-dynamics within substances, and ends with the articulation of the monadology. "Organ" holds throughout. See Donald Rutherford, "Metaphysics: The Late Period," in *The Cambridge Companion to Leibniz*, ed. Nicholas Jolley (Cambridge: Cambridge University Press, 1995), 128.

10. See Steven Nadler, ed., *Causation in Early Modern Philosophy: Cartesianism, Occasionalism and Pre-established Harmony* (University Park: Pennsylvania State University Press, 1993), in which Eileen O'Neill's historical overview "Influxus physicus" (27–57) is especially helpful; cf. Rainer Specht, *Commercium mentis et corporis: Über Kausalvorstellungen im Cartesianismus* (Stuttgart–Bad Cannstatt: frommann-holzboog, 1966).

11. See John W. Yolton, "The Three Hypotheses," *Locke and French Materialism* (Oxford: Clarendon, 1991), 10–38.

12. The "*vis viva*" debate surrounded the Cartesian principle of the conservation of motion. Leibniz was able to show that this was false and held that "force" was conserved. The thought-experiment in §17 of the *Discourse* is meant ultimately to show that motion, conserved, would simply peter out. See Roberto Torretti, *The Philosophy of Physics* (Cambridge: Cambridge University Press, 1999), 33–36.

13. Leibniz, *Philosophical Texts*, trans. R. S. Woolhouse and Richard Franks (Oxford: Oxford University Press, 1998), 216.

14. Gottfried Wilhelm Leibniz, *Theodicee, das ist, Versuch von der Güte Gottes, Freyheit des Menschen, und vom Ursprung des Bösen*, ed. Johann Christoph Gottsched (Hannover and Leipzig: Verlag der försterischen Erben, 1763), 192–93.

15. But cf. Frederick Beiser. *Diotima's Children: German Aesthetic Rationalism from Leibniz to Lessing* (Oxford: Oxford University Press, 2009), 53–54, which argues that Wolff establishes a direct connection between *techne* and cognition.

16. See §8ff. of the *Discourse on Metaphysics*, in Leibniz, *Philosophical Essays*, trans. Roger Ariew and Daniel Garber (Indianapolis: Hackett, 1989), 35–69.

17. Or the essence of an individual, its matter-embodied form, which was equated with "occult quality" after Descartes; see Lewis White Beck, *Early German Philosophy: Kant and His Predecessors* (Cambridge: Belknap Press of Harvard University Press, 1969), 213.

18. Leibniz, *Philosophical Essays*, 139.

19. Ibid.

20. "An organic body is contrasted with a mere machine to the extent that there is literally no lower limit to its mechanical composition" (Smith, *Divine Machines*, 98).

21. One of the great ironies of the larger biological debate is that the preformationists used the term "evolution" to describe their nongenerative viewpoint, halting its use for the epigenetic line that led to Darwin. See Richards, *Romantic Conception*, 211–12.

22. Leibniz, *New Essays*, 55.

23. Ibid., 141.

24. Leibniz, *Philosophical Texts*, 210, translation modified.

25. Leibniz, *New Essays*, 142.

26. Ibid.

27. See Smith, *Divine Machines*, 106.

28. Where Descartes had held that what we conceive clearly and distinctly *is* true, Leibniz holds that by an increasing process of making representations distinct we arrive at the truth. See Robert Mcrae, "The Theory of Knowledge," in *The Cambridge Companion to Leibniz*, ed. Nicholas Jolley (Cambridge: Cambridge University Press, 1995), 177–78.

29. Relying on the *scientia intuitiva* of Spinoza's *Ethics* book 2, proposition 42, scholium. See Leibniz, "Meditations on Knowledge, Truth, and Ideas," in *Philosophical Essays*, 23–28.

David Wellbery, *Lessing's Laocoon: Semiotics and Aesthetics in the Age of Reason* (Cambridge: Cambridge University Press, 1984), 232.

30. Ibid.

31. Cf. ibid., 6–7.

32. Leibniz, *Philosophical Essays*, 41.

33. For the textually complicated history of this claim, see Giorgio Tonelli, "Leibniz on Innate Ideas and the Early Reactions to the Publication of the Nouveaux Essais (1765)," *Journal of the History of Philosophy* 12, no. 4 (October 1974): 437–54, esp. 441.

34. G. W. Leibniz, *New Essays on Human Understanding*, trans. Peter Remnant and Jonathan Bennett (New York: Cambridge University Press, 1981), 53.

35. Ibid., 54–57. Even the preestablished harmony is meant to be clarified at the level of exposition by these perceptions (56). Because they make for the "confused" matter of perception, their clarification or *distinctivization* puts us on the road to intuitive adequate knowledge—properly God's, for us an ideal.

36. Ibid., 72.

37. Ibid., 74.

38. Ibid., 75.

39. Ibid., 84.

40. Ibid., 80.

41. To point out how precisely analogous the preformationism in metaphysics is, we can call to mind the statement, from *De natura ipse*, that "things have been given a certain ability, a form or force (such as we usually call a 'nature'), from which the series of phenomena follows in accordance with the dictates of the original command" (Leibniz, *Philosophical Texts*, 213). Spoken here against occasionalism with its discrete interventions by God, the doctrine amounts to the produced "tendency" or disposition at the level of force and substance. Again, however, it is not metaphorical in the metaphysical context. A "nature" is both its own property and a gift (from God).

42. "However, it was only from around 1830 that the individual 'organism' became a recurrent technical term in various research fields" (Cheung, "Organism of a Body or Body of an Organism," 335).

43. Johann Christoph Adelung, *Verusch eines vollständig grammatisch-kritischen Wörterbuches der hochdeutschen Mundart mit beständiger Vergleichung der übrigen Mundarten, besonder aber der Oberdeutschen. Dritter Theil, von L–Scha-* (Breitkopf: Leipzig, 1777), 613–14.

44. Johann Samuel Traugott Gehler, *Physikalisches Wörterbuch oder Versuch: Erklärung der vornehmsten Begriffe und Kunstwörter der Naturlehre mit kurzen Nachrichten von der Geschichte der Erfindungen und Beschreibungen der Werkzeuge begleitet in alphabetischer Ordnung. Dritter Theil von Liq bis Sed* (Schwickert: Leipzig, 1790), 388.

45. Aristotle, *De generatione* 730b.

46. See Wolf, *Der Begriff*, 19ff.; cf. Justin E. Smith, ed., *The Problem of Animal Generation in Early Modern Philosophy* (Cambridge: Cambridge University Press, 2006), 1–21.

47. Which buttressed the fall of scholastic metaphysics and simultaneously established a world "within" the phenomena as a new area of study. For a general history, see Catherine Wilson, *The Invisible World: Early Modern Philosophy and the Invention of the Microscope* (Princeton, NJ:

Princeton University Press, 1995). For an examination of the implicit metaphysics which emerged with the advent of this instrument, see Hartmut Boehme, "The Metaphysics of Phenomena: Telescope and Microsope in the Works of Goethe, Leeuwenhoek and Hooke," in *Collection, Laboratory, Theater: Scenes of Knowledge in the 17th Century* (Berlin: de Gruyter, 2005), 355–94; cf. Christiane Frey, "The Art of Observing the Small: On the Borders of the *subvisibilia* (from Hooke to Brockes)," *Monatshefte* 105, no. 3 (2013): 376–88

48. The imperfection of the instrument allowed preformation a longer life span than it might otherwise have had. Von Haller's arguments against Wolff's version of epigenesis came increasingly to rely on the "transparency" or invisibility of preformed organs, an argument that failed as the instrument gained in precision. See Shirley Roe, *Matter, Life, and Generation: Eighteenth-Century Embryology and the Haller-Wolff Debate* (Cambridge: Cambridge University Press, 1981), 83–84.

49. See Peter Hans Reill, *Vitalizing Nature in the Enlightenment* (Berkeley: University of California Press, 2005), 56–71.

50. Johann Friedrich Blumenbach, *Über den Bildungstrieb* (Göttingen: Dieterich 1791), 32–34; emphasis in original.

51. Ibid., 32–33.

52. Ibid., 32.

53. See James Larson, "Vital Forces: Regulative Principles or Constitutive Agents? A Strategy in German Physiology, 1786–1802," *Isis* 70, no. 2 (1979), 241.

54. See the classical accounts in François Duchesneau, *La physiologie des lumières: Empirisme, modèles et théories* (Boston: Nijhoff, 1982); and Karl Rothschuh, *Geschichte der Physiologie* (1953; Berlin: Springer, 2012).

55. Caspar Friedrich Wolff, *Theorie von der Generation* (Berlin: Birnstiel, 1764), 131–32, 210. My thanks to Sarah Eldridge of the German department at the University of Tennessee, Knoxville, for pointing out these passages to me.

56. Blumenbach, *Bildungstrieb*, 13–14. Blumenbach's predecessor in defending epigenesis, Caspar Friedrich Wolff, describes the division in this way: "Whether organic bodies of nature are evolved from an invisible state to a visible one, or are truly produced" (Wolff to Haller, 6 October 1766, in Roe, *Matter*, 166).

57. Newton, *Opticks*, 376ff.

58. Albrecht von Haller, *Umriss der Geschäfte des körperlichen Lebens* (Berlin: Haude und Spener, 1770), 224.

59. Albrecht von Haller, *Primae lineae physiologiae* (Göttingen: van Rossum, 1758), 158; my emphases.

60. I treat Haller at more length in "Das Innere der Natur und ihr Organ: Von Albrecht von Haller zu Goethe," *Goethe-Yearbook* 21 (2014): 191–217.

61. See Haller, *Sur la formation du coeur dan le poulet* (Lausanne: Bousquet, 1758), 173, qtd. in Roe, *Matter*, 68.

62. Or a soul; cf. Andrew Cunningham, *The Anatomist Anatomis'd: An Experimental Discipline in Enlightened Europe* (Burlington, VT: Ashgate, 2010), 383.

63. Cf. Christopher Young and Thomas Gloning, *A History of the German Language through Texts* (London: Routledge, 2004), 248–49, suggest that "organ" would still have appeared as a *Lehnwort* in 1790 in Goethe's *Versuch, die Metamorphose der Pflanzen zu erklären*.

64. Immanuel Kant, *Kant's Gesammelte Schriften* (Berlin: Reimer, 1900–); hereafter cited as Kant, AA V, 373–74.

65. Immanuel Kant, *Critique of the Power of Judgment*, trans. Paul Guyer and Eric Matthews (Cambridge: Cambridge University Press, 2000), 245; translation modified. See Roe, *Matter*, 158–73.

66. Helmut Müller-Sievers, *Self-Generation: Philosophy, Biology and Literature around 1800* (Stanford, CA: Stanford University Press, 1997), 5. Cf. John Zammito, *Kant, Herder, and the Birth of Anthropology* (Chicago: University of Chicago Press, 2002), 472. Roe, *Matter*, 87, situates the controversy at the level of explanation of phenomena, not differences in observation or technique.

67. There is much debate on Kant's role in the founding of modern biology. Lenoir, *Strategy of Life*, sees Kant at the beginning of a functionalist biology that seeks as much as possible to approach life by way of its mechanical function. Robert Richards (*The Romantic Conception of Life: Science and Philosopohy in the Age of Goethe* [Chicago: University of Chicago Press, 2002]) takes direct issue with Lenoir and pleads against Kant's case as contributing to biology whatsoever, since he makes it a "non-constitutive" science.

68. Blumenbach, *Bildungstrieb*, 79–80.

69. Löw, *Philosophie*, 175–80.

70. Kant, AA XIII, 400.

2. THE EPIGENESIS OF REASON: FORCE AND ORGAN IN KANT AND HERDER

1. On the "eclectic" school and its pietistic opposition to Rationalism, see Lewis White Beck, *Early German Philosophy: Kant and His Predecessors* (Cambridge: Belknap Press of Harvard University Press, 1969); on the Lange-Wolff controversy, see Jonathan Israel, *Radical Enlightenment* (Oxford: Oxford University Press, 2001), 541–63.

2. Kant, *CPR* B167.

3. Ibid., B1; translation modified.

4. Ibid., B861/A833.

5. Jennifer Mensch has recently argued that this metaphor played an essential role in the development of the Critical system, reversing the tradition of arguments about the role of organics *within* that system. See Mensch, *Kant's Organicism: Epigenesis and the Development of Critical Thought* (Chicago: University of Chicago Press, 2013), 12.

6. Kant, *CPR* B862/A834.

7. See ibid., B xvii–xxxviii, where Kant shifts the metaphor onto the architecture of the *Critique* as a book.

8. Ibid., B200/A161, the "table of principles."

9. Ibid., B823/A795.

10. Ibid., B824/A796.

11. Ibid.

12. Ibid., B25/A11; ibid., B77–8/A53.

13. See Ernst Cassirer, *Kant's Life and Thought*, trans. James Haden (New Haven: Yale University Press, 1981), 43, 55.

14. Immanuel Kant, "M. Immanuel Kants Nachricht von der Einrichtung seiner Vorlesungen in dem Winterhalbenjahre, von 1765–1766," in Immanuel Kant, *Vorkritische Schriften bis 1768/2 (Werkausgabe Band I/II)*. vol. 2, ed. Wilhelm Weischedel, 905–17 (Frankfurt a.M.: Suhrkamp, 1960), 912–13.

15. Giorgio Tonelli, Kant's "Critique of Pure Reason" within the Tradition of Modern Logic: A Commentary on Its History, ed. David Howard Chandler (Zürich: G. Olms, 1994), see 37ff.

16. Ibid., 38.

17. Synthesis was one of Kant's perennial concerns; see Paul Guyer, *Kant and the Claims of Knowledge* (Cambridge: Cambridge University Press, 1987), 1–25.

18. Johann Heinrich Lambert, *Neues Organon, oder Gedanken über die Erforschung und Bezeichnung des Wahren und dessen Unterscheidung vom Irrtum und Schein* (Wendler: Leipzig, 1764), "Vorrede," (pages unnumbered).

19. Lambert based his moderate Rationalism on insights gathered from Euclid, namely the inclusion of intuitive materials in construction. See Beck, *Early German Philosophy*, 404.

20. This was an early concern of Kant's; see *Vorkritische Schriften* 1:478–80, 479–81.

21. Kant, AA X, 52.

22. For a general account of the particular categories in the pre-Critical period, see Heinz Heimsoeth, "Zur Herkunft und Entwicklung von Kants Kategorientafel," *Kant-Studien Ergänzungsheft* 100 (1970): 109–32.

23. He uses this term in the same letter; AA X 64–66. Proceeding from "simple concepts," Lambert claims that the possibility of their connection must be included within them. Thus, the genetic investigation of these

simples and the manners of their connection will be the *organon* itself, determining what is true (alethiology), separating truths and falsehoods in perception (phenomenology), determining word use and the rules for symbolic cognition (semiotics), and (chronologically first, and closest to Kant's concerns), establishing the general rules of thought (dianoiology).

24. Defined in the *Notes*: "Transcendental logic deals with cognitions of the understanding with respect to their content, but without determination with regard to the way in which objects are given." (This is from the Duisburg Nachlass, note 4675, 20 May 1775 [10:182].) Immanuel Kant, *Notes and Fragments*, trans. Paul Guyer et al. (Cambridge: Cambridge University Press, 2005), 163.

25. Kant, *CPR* B303/A247.

26. See Michael Friedman, *Kant and the Exact Sciences* (Cambridge: Harvard University Press, 1992), 1–55.

27. See Guyer, *Claims*, 18ff.

28. See Kant, *CPR* B74/A50.

29. Giorgio Tonelli, "Die Umwälzung von 1769 bei Kant," *Kant-Studien* 54 (1963): 369.

30. Ibid., 371–72. As late as *Inaugural Dissertation*, Kant still considered pure intellection as a potential organon; see AA II, 397–98.

31. Kant, *CPR* B81/A56.

32. Some of what follows has appeared in Leif Weatherby, "The Romantic Circumstance: Novalis between Kittler and Luhmann," *SubStance* 43 (2014): 46–66.

33. See, for this development, Dieter Henrich, *Between Kant and Hegel: Lectures on German Idealism* (Cambridge: Harvard University Press, 2008).

34. The most famous response is that of Salomon Maimon, to whom Kant responded (through Marcus Herz) with a summary of the "B" argument. See Kant AA: XI, 48–55.

35. See Paul Guyer, "The Deduction of the Categories: The Metaphysical and Transcendental Deductions," in *The Cambridge Companion to Kant's "Critique of Pure Reason,"* ed. Guyer (Cambridge: Cambridge University Press, 2010).

36. Kant, *CPR* B132.

37. Ibid., B117/A84.

38. I borrow these terms from the so-called Pittsburgh Hegelians, and especially John McDowell, for example, *Mind and World* (1994; Cambridge: Harvard University Press, 1996).

39. Kant, *CPR* B167.

40. See Robert John Richards, The Romantic Conception of Life: Science and Philosophy in the Age of Goethe (Chicago: University of Chicago Press, 2002), 207–38, 307–13.

41. See the "New System of Nature," in Gottfried Wilhelm Leibniz, *Philosophical Essays*, ed. Roger Ariew and Daniel Garber (Indianapolis: Hackett, 1989), 138–45.

42. What makes the generation of the categories unique is that although they are generated (both as rules for synthesis and as discursive concepts) only under empirical conditions, their content is determined independently of these empirical conditions and, indeed, is an a priori condition for the generation of any representation of empirical objects at all. See Béatrice Longuenesse, *Kant and the Human Standpoint* (Cambridge: Cambridge University Press, 2005), 29. I think that what is "developed" epigenetically is the judgment's result: experience.

43. Cf. Paul Guyer, "The Deduction of the Categories: The Metaphysical and Transcendental Deductions," in *The Cambridge Companion to Kant's "Critique of Pure Reason,"* ed. Guyer, 118–51, 140.

44. See ibid., 121, 123.

45. Kant, *CPR* B138–39; my emphasis.

46. See Kant *CPR* B141; in Longuenesse's formulation: "synthesis (of intuition) for analysis (into concepts) for synthesis (of these concepts in judgment)" (Longuenesse, *Kant and the Human Standpoint*, 22).

47. The question of priority—whether analytic or synthetic unity must come first—does not change the double and generative role judgment is accorded in this process. See Kant, *CPR* B133; cf. Longuenesse, *Kant and the Human Standpoint*, 37.

48. Kant, *CPR* A845/B873; cf. Mensch, *Kant's Organicism*, 123–24.

49. See Fichte, *Versuch einer neuen Darstellung der Wissenschaftlehre* (1797/98), ed. Peter Baumanns (Hamburg: Meiner, 1984); and the *Wissenschaftslehre nova methodo 1798/99*, ed. Erich Fuchs (Hamburg: Meiner, 1994). The view I am paraphrasing here is the highly influential reading given by Dieter Henrich, "Fichte's Original Insight," trans. David Lachterman, in *Contemporary German Philosophy* 1 (1982): 15–53.

50. Henrich makes this case for Hölderlin in his magisterial *Der Grund im Bewußtsein: Untersuchungen zu Hölderlins Denken* (Stuttgart: Klett-Cotta, 1992), arguing that Hölderlin exploits the structure he finds in Fichte to articulate self-knowledge as a springboard for knowledge of the world.

51. This opposing view—that the recursive structure of self-consciousness plunges us into a nonfoundation without orientation—is made virulent by Manfred Frank on the example especially of Novalis in his *Philosophical Foundations of Early German Romanticism*, trans. Elizabeth Millàn-Zaibert (Albany: SUNY Press, 2004), 151–77.

52. See Frederick Beiser, *German Idealism:* The Struggle against Subjectivism, 1781–1801 (Cambridge: Harvard University Press, 2002), 435–65, 465–597.

53. Cf. Guyer, "Transcendental Deduction," 121, 123.
54. See ibid., 149.
55. In this section, I am building on the excellent work done by Mensch, *Kant's Organicism*, drawing on the other side of the metaphor—the "organon"—to show the form and constitution of the mind's organs as a consequence of the epigenesis metaphor.
56. It was Jacobi who first drew attention to this problematic, stating famously that he could not enter the Critical system with the thing-in-itself, but could not remain in it without it. See "Über den transzendentalen Idealismus," appended to *David Hume über den Glauben oder Idealismus und Realismus* (Breslau: Loewe, 1787), 209–30.
57. Kant, *CPR* B12/A8; translation modified, my emphasis.
58. Ibid., B95/A70.
59. Ibid., B99–100/A74.
60. Ibid., B184/A144–45.
61. Ibid., B185/A145. Guyer translates "sum total of time" (Kant, *Critique*, 276).
62. Time is emphasized here because the modalities express the relationship to the "inner sense," which is characterized by time as transcendental form (where "outer sense" includes space as well). In the *Anthropology*, Kant points out that "inner sense" is not, as the outer senses are, characterized by multiple *organs*, but that instead "the soul is the organ of the inner sense" (Kant, AA VII, 161).
63. Kant, *CPR* B 266/A219.
64. Ibid., B357/A300.
65. Ibid., B359/A302.
66. Ibid., B385–86/A329.
67. Ibid., B379/A322ff.
68. The three absolute wholes are those of propositions in respect to the *self* (rational psychology), *objects* (cosmology), and *all things* (the phrase is taken from Wolff's *Deutsche Metaphysik*, but here implies God—rational theology). See ibid., B391/A334ff.
69. I think it likely that Herder's use influenced Kant's, *pace* Jörn Henning Wolf, *Der Begriff* Organ in der Medizin (Munich: Fritzsch, 1971), 38–43, who gives Kant pride of place.
70. *Encyclopédie, ou dictionnaire raisonné des sciences, des arts et des métiers, etc.*, ed. Denis Diderot and Jean le Rond D'Alembert, University of Chicago, ARTFL Encyclopédie Project (Spring 2011 edition), ed. Robert Morrissey, http://encyclopedie.uchicago.edu/.
71. Ibid.
72. See Peter Hanns Reill, *Vitalizing* Nature in the Enlightenment (Berkeley: University of California Press, 2005), 33–71.

73. Ibid., 186–99.

74. *Locus classicus*: Friedrich Meinecke, *Historism: The Rise of a New Historical Outlook* (London: Routledge and Kegan Paul, 1972).

75. Herder's earliest engagements with the issue of force come from his time in Königsberg under Kant's tutelage. Kant's lectures of that period—indeed, those extant in Herder's hand—are filled with considerations of force, especially as that concept plays a role in mind-body-relation problems. See Herder's earliest-known philosophical fragment, the *Versuch über das Sein*, which places the intensive feeling of force at the center of speculation. This consideration remained determinative throughout Herder's career, as has been convincingly argued in a careful reading of that fragment by Marion Heinz, *Sensualistischer Idealismus: Untersuchungen zur Erkenntnistheorie des jungen Herder (1763–1778)* (Hamburg: Meiner, 1994).

76. Johann Gottfried Herder, *Werke in zehn Bänden*, vol. 6, *Ideen zu einer Philosophie der Geschichte der Menschheit*, ed. Martin Bollacher (Frankfurt a.M.: Klassiker 1989), 169; hereafter cited as Herder, *Ideen*, followed by page number.

77. Cf. Reill, *Vitalizing*, 1–17, 199–237.

78. Herder speaks of "the analogy of nature"; see, e.g., *Ideen*, 176: "Let's leave metaphysics aside; we want to observe analogies of nature." Since humans stand in the center of this analogy, our view of being and our methods for investigating it coincide in the *organ* of thought. See John Zammito, "Herder, Kant, Spinoza und die Ursprünge des deutschen Idealismus," in *Herder und die Philosophie des deutschen Idealismus*, ed. Marion Heinz (Amsterdam: Rodopi, 1997), 107–45.

79. Jean-Baptiste-René Robinet, *De la nature (IV)* (Amsterdam: van Harrevelt, 1761), 78.

80. For Leibniz's general impact in France, see W. H. Barger, *Leibniz in France, from Arnauld to Voltaire: A Study in French Reactions to Leibnizianism, 1670–1760* (Oxford: Clarendon, 1955).

81. See, for example, Arthur Lovejoy, *The Great Chain of Being: A Study of the History of an Idea* (1936; Cambridge: Harvard University Press, 1964), 277–83.

82. Robinet, *De la nature*, 78.

83. See H. B. Nisbet, "Herder, Goethe, and the Natural 'Type,'" in *Publications of the English Goethe Society vol. XXXVII: Papers Read before the Society 1966–67*, ed. Elizabeth M. Wilkinson et al. (Leeds: Maney and Son, 1967), 81–119; it is not Robinet's cosmology but his standout use of "organ" that interests me here.

84. Herder receives only one mention (280) in Lovejoy's chapter entitled "The Temporalizing of the Great Chain of Being," in *The Great Chain*, 242–88.

85. Herder, *Ideen*, 87–88.
86. Ibid., 88.
87. Ibid., 88–89.
88. Ibid., 167–70, applied in similar terms to language (sound and word are like organ and force) at 182.
89. Ibid., 171.
90. For Herder's use of Haller's irritability/sensibility distinction, see Simon Richter, "Medizinischer und ästhetischer Diskurs im 18. Jahrhundert: Herder und Haller über Reiz," *Lessing-Yearbook* 25 (1993): 83–97.
91. Herder, *Ideen*, 174.
92. Ibid., 175.
93. Johann Gottfried Herder, *Vom Erkennen und Empfinden der menschlichen Seele*, in Herder, *Werke in zehn Bänden*, vol. 4, *Schriften zu Philosophie, Literatur, Kunst und Altertum*, ed. Jürgen Brummack and Martin Bollacher (Frankfurt a.M.: Klassiker 1994), 363. This assertation shows us that the human analogy is based on the biblical version of the *phusis-techne* analogy, God's making the human in his image. See ibid., 361. Cf. Herder, *Ideen*, 38.
94. It appears that Leibniz had an edition without the metaphorical *tanquam* preceding the phrase. See Alexandre Koyre and I. Bernard Cohen, "The Case of the Missing Tanquam: Leibniz, Newton & Clarke," *Isis* 52, no. 4 (1961): 555–66. See also Hans Blumenberg, The Legitimacy of the Modern Age, trans. Robert M. Wallace (Cambridge: MIT Press, 1983), 80–82.
95. Herder, *Ideen*, 183.
96. Ibid.
97. See ibid., 667.
98. Johann Gottfried Herder, *Werke in zehn Bänden*, vol. 8, *Schriften zu Literature und Philosophie, 1792–1800*, ed. Dietrich Irmscher (Frankfurt a.M.: Klassiker, 1998), 321. The phrase occurs in Hamann's own *Metakritik* (1784): "language, the only, first, and last organon and criterion of reason, with no credentials, but tradition and usage" (Johann Georg Hamann, *Writings on Philosophy and Language*, ed. Kenneth Haynes [Cambridge: Cambridge University Press, 2007], 208). Hamann cites Herder's *Essay* at 121, and his citation is from the fall of 1772. Herder first uses the concept of language as the *organ of reason*, which Hamann cites and then includes in his objections to Kant. These objections, which give us the important notion that the *a priori* and the *a posteriori* meet in language, are then used again by Herder in his *own Metakritik* (1799), now uniting his own phrase with his friend's linguistic objection to Kant. *Tradition* is given a different sense in Herder's *Ideen* (1785)—not merely what is handed down, but the organic unities of nations. The phrase was picked up by Wilhelm von Humboldt and used in his language-theoretical writings. See Siegfried J. Schmidt, *Sprache und Denken als*

sprachphilosophisches Problem: Von Locke bis Wittgenstein (The Hague: Nijhoff, 1969). Use of the metaphor continues up to Noam Chomsky, whose early conception of a language-acquisition device in evolution was often characterized as a "language organ."

99. Herder, *Abhandlung über den Ursprung der Sprache*, in Johann Gottfried Herder, *Werke in zehn Bänden*, vol. 1, *Frühe Schriften 1764–1772*, ed. Ulrich Gaier (Frankfurt a.M.: Klassiker, 1985), 733.

100. See Charles Taylor, "The Importance of Herder," in Charles Taylor, *Philosophical Arguments* (Cambridge: Harvard University Press, 1995), 79–100; cf. Michael Forster's *After Herder: Philosophy of Language in the German Tradition* (Oxford: Oxford University Press, 2010). Taylor argues that language establishes an index of the feeling of truth, while Forster points to the tradition after Herder that instead emphasized its plasticity.

101. Herder, *Ideen*, 181.
102. Kant, AA XIII 400.
103. Ibid., AA VIII, 61.
104. Ibid., AA VIII, 61–62.
105. Ibid., AA VIII, 57.

3. THE ORGAN OF THE SOUL: VITALIST METAPHYSICS AND THE LITERALIZATION OF THE ORGAN

1. Alexander von Humboldt, "Life-Force, or the Genius of Rhodes," trans. Leif Weatherby, *Yearbook of Comparative Literature* 58 (2012): 168.

2. Ibid., 167.

3. Alexander von Humboldt, *Ansichten der Natur* (Darmstadt: Wissenschaftliche Buchgesellschaft, 1987), 323ff.

4. This is considered the *ad quem* of the discourse by Carsten Zelle, "Sinnlichkeit und Therapie: Zur Gleichursprünglichkeit von Ästhetik und Anthropologie um 1750," in *"Vernünftige Ärtze": Hallesche Psychomediziner und die Anfänge der Anthropologie in der deutschsprachigen Frühaufklärung*, ed. Carsten Zelle (Tübingen: Niemeyer, 2002), 6–24.

5. Ernst Platner, *Anthropologie für Aerzte und Weltweise* (Leipzig, 1772), XVIff., also cited by Zelle, "Sinnlichkeit und Therapie," 6–7.

6. See Carl Friedrich Kielmeyer, *Ueber die Verhältnisse der organischen Kräfte unter einander in der Reihe der verschiedenen Organisationen, die Geseze und Folgen dieser Verhältniße*, with an introduction by Kai Torsten Tanz (Basilisken-Presse: Marburg an der Lahn 1993); Johann Dietrich Brandis, *Versuch über die Lebenskraft* (Hannover: Verlag der Hahn'schen Buchhandlung, 1795); Johann Christian Reil, *Von der Lebenskraft* (Leipzig: Johann Ambrosius Barth, 1910); Alexander von Humboldt, *Aphorismen aus der Physiologie der Pflanzen*, trans. Gotthelf

Fischer (Weimar: Voß und Compagnie 1794), a translation of Alexander von Humboldt, *Florae Fribergensis specimen plantas cryptogamicas praesertim subterraneas exhibens /Accedunt aphorismi ex doctrina physiologiae chemicae plantarum. Cum tabulis aeneis* (Berline: Rottmann, 1793); Johann Wilhelm Ritter, *Beweis, dass ein beständiger Galvanismus den Lebenprocess in dem Thierreich begleite. Neben neuen Versuchen und Bemerkungen über den Galvanismus* (Weimar: Verlag des Industrie-Comptoirs, 1798); Samuel Thomas Soemmerring, *Über das Organ der Seele* (Königsberg: Friedrich Nicolovius, 1796).

7. Reil, *Lebenskraft*, 21.

8. Ibid., 4.

9. Humboldt, *Aphorismen*, 135.

10. Reil himself continues to speak of the "machine of nature," which has organs. See, e.g., *Lebenskraft*, 21.

11. Kant's proposal is to divide the medieval "higher faculties" (theology, law, and medicine) according to the object of their social role in producing human happiness. The theological faculty aims at general happiness; the juridical at bourgeois welfare; and the medicinal at the well-being of the body. I characterize the philosophical faculty as "public" (in keeping with Kant's famous definition from 1784) because it is the furthest from being instrumentalized by the state, and thus forms the autonomous center of the university. See Kant, AA VII, 20–23. The parallel with the *Enlightenment* essay is clear, as has been noted by Paul Guyer (*Kant* [New York: Routledge, 2006], 291–93).

12. Kant, AA VII, 22–23.

13. Ibid., AA VII, 36–37.

14. We might include *Von einem neuerdings erhobenen vornehmen Ton in der Philosophie* (AA VIII, 387–423), and *Über eine Entdeckung, nach der alle neue Kritik der Vernunft durch eine ältere entbehrlich gemacht werden soll* (AA VIII, 185–253).

15. See John H. Zammito, *Kant, Herder, and the Birth of Anthropology* (Chicago: University of Chicago Press, 2002), 3.

16. The period between the metaphysical discourse which focused on the soul and the mid-nineteenth-century positivism which discovered localization of brain functions is characterized by a striking openness, and it is in this indeterminate atmosphere that the *organ* emerged as the term for the instrumentalization of the body by the soul. See Michael Hagner, *Homo cerebralis: Der Wandel vom Seelenorgan zum Gehirn* (Frankfurt a.M.: Suhrkamp, 2008); and Jan Verplaetse, *Localizing the Moral Sense: Neuroscience and the Search for the Cerebral Seat of Morality, 1800–1930* (New York: Springer, 2009); cf. Caroline Welsh, *Hirnhöhlenpoetiken: Theorien zur Warhnehmung in Wissenschaft, Ästhetik und Literatur um 1800* (Freiburg: Rombach, 2003).

17. Kant already uses Platner's later term *Nervengeist* to describe the transmission of Cartesian *ideas materiales* in the *Dreams of a Spirit-Seer* (AA II, 345). Kant also talks about organs of the brain (339), distinguishing them from sense-organs (in an early use of "organ" for the senses in German) and naming the central organ a *Sensorium der Seele*.

18. His 1772 *Anthropologie* had used the term "organ"—as we have seen— and even *Seelenorgan*, but the renewed effort devotes some fifty paragraphs to the problem of the *Seelenorgan*, compared with only a handful in the earlier edition. See Ernst Platner, *Neue Anthropologie für Ärzte und Weltweise mit besonderer Rücksicht auf Physiology, Pathologie, Moralphilosophie und Aesthetik* (Leipzig: Crusius, 1790), 58–91. Note that Reil, too, had made an uncommonly direct claim that the brain itself was the organ of the soul. See Reil, *Lebenskraft*, 48: "That the organ of the soul (the brain) and not the nerves are the peculiar tool of representations, is indeed irrefutable." The preceding pages have an account of nerve's action in relay to the brain and the muscles, which are the reactors that allow for sensation and voluntary motion after intital excitation.

19. The 1772 edition contains a discussion of the sense-organs, and even a discussion of the *Sitz der Seele*, but does not emphasize the term "organ of the soul." The organs work in both versions by mediating their formal impressions to the *Nervengeist*, which communicates directly to the brain. The brain-marrow (1772) or the nerve-spirit (1790) is the *sensorium*, the basis of all perception and indeed the soul's point of communication with the body and the world.

20. Platner, *Neue Anthropologie* §175, p. 58.

21. Ibid., §186, p. 61.

22. Ibid., §200ff. defends *influxus physicus* (or rather assumes it, referencing his *Aphorisms*), and attacks Kant's Critical system explicitly for complicating this seemingly straightforward point.

23. Cf. Reil's attempt to mediate: *Über die eigentümlichen Verrichtungen des Seelenorgan*, in Johann Christian Reil, *Gesammelte kleine physiologische Schriften* (Vienna: Doll, 1811), 1–159.

24. Platner, *Neue Anthropologie* §177, p. 59.

25. See Hagner, *Homo cerebralis*; and also, on the separation of the two problems, Werner Euler, "Die Suche nach dem Seelenorgan: Kants philosophische Analyse einer anatomischen Entdeckung Soemmerrings," *Kant-Studien* 93 (2002): 453–80.

26. The extensive imagination of these cavities and their relation to acoustics is explored in Welsh, *Hirnhöhlenpoetiken*; see 55–70 for Soemmerring.

27. Soemmerring was aware of this discursive transgression, as is clear from Soemmerring, *Seelenorgan*, 38; cf. Samuel Thomas Soemmerring, *Ueber*

das Organ der Seele (1796), ed. Manfred Wenzel (Basel: Schwabe, 1999). I cite the first edition throughout.

28. Soemmerring, *Seelenorgan*, 43.

29. Ibid., 44.

30. See, in Gunter Mann et al., eds., *Samuel Thomas Soemmerring und die Gelehrten der Goethezeit* (New York: Fischer, 1985): Peter McLaughlin, "Soemmerring und Kant: Über das Organ der Seele und den Streit der Facultäten," 191–203, esp. 192–94; and, on Soemmerring and Heinse, Manfred Dick, "Der Dichter und der Naturforscher: Samuel Thomas Soemmerring und Wilhelm Heinse," 203–29.

31. See Manfred Wenzel, "Johann Wolfgang von Goethe und Samuel Thomas Soemmerring: Morphologie und Farbenlehre," in *Gelehrten der Goethezeit*, ed. Mann et al., 11–35, *Seelenorgan* discussed briefly at 28.

32. See McLaughlin, "Soemmerring und Kant," 194ff. For Kant's more general contacts with the scientists in Göttingen, see Timothy Lenoir, "Göttingen School," Studies in the History of Biology 5 (1981): 11–205.

33. Cf. McLaughlin, "Soemmerring und Kant," 198–99.

34. See Karl Ameriks, *Kant's Theory of Mind: An Analysis of the Paralogisms of Pure Reason* (1982; Oxford: Oxford University Press, 2000), esp. 27ff.

35. Soemmerring, *Seelenorgan*, 82.

36. Ibid. Kant's afterword is printed in the text of the *Seelenorgan*; see also the letters between the two in Kant, AA XII.

PART II: ROMANTIC ORGANOLOGY: TOWARD A TECHNOLOGICAL METAPHYSICS OF JUDGMENT

1. The speed of philosophical development after Kant is breathtaking, but the immense amount of work done in the last fifty years helps to summarize some of its broader stakes. The central work done by Dieter Henrich and his follows—called "constellation work"—has produced the basic narrative of the path "between Kant and Hegel" (see Henrich, *Between Kant and Hegel: Lectures on German Idealism* [Cambridge: Harvard University Press, 2008]). Manfred Frank has argued passionately and rigorously for the inclusion of Novalis (and, to a lesser extent, Schlegel) in Henrich's pantheon (which already included Hölderlin centrally) (see Frank, *The Philosophical Foundations of Early German Romanticism*, trans. Elizabeth Millàn-Zaibert [Albany: SUNY Press, 2004]). Meanwhile, Frederick Beiser's work has put Frank's interpretive framework into question, at least for Romanticism after about 1797 (see Beiser, *German Idealism: The Struggle against Subjectivism* (Cambridge: Harvard University Press, 2002]). Most recently, Eckart Förster has argued for Goethe's centrality to this developmental picture in *The Twenty-Five Years of Philosophy* (Cambridge: Harvard University Press, 2012). See also

Dalia Nassar, *The Romantic Absolute: Being and Knowing in Early German Romantic Philosophy, 1795–1804* (Chicago: University of Chicago Press, 2014); and Elizabeth Millán-Zaibert, *Friedrich Schlegel and the Emergence of Romantic Philosophy* (Albany: SUNY Press, 2007).

2. It is not coincidental, of course, that the question of being is raised in a post-Kantian context even as "old" metaphysical questions had been reraised in scientific context. My argument here tends to show that the Romantics combined questions of general metaphysics (ontology) stemming from systemic issues in Kant and post-Kantianism with questions of special metaphysics (cosmological, pneumatological, and theological questions) inherited from their eclectic readings of the metaphysical pretensions (or lack thereof) in contemporary science. The holism and organic strains of these efforts are treated in Kristian Köchy, *Ganzheit und Wissenschaft: Das historische Fallbeispiel der romantischen Naturforschung* (Würzburg: Königshausen und Neumann, 1997).

3. See Frank, "Die Philosophie des sogenannten magischen Idealismus," in *Auswege aus dem deutschen Idealismus* (Frankfurt a.M.: Suhrkamp, 2007), 26–66, which presents Novalis's philosophy as suggesting regulative ideas for alteration of the body and world.

4. Dieter Henrich finds that the notion of a "ground" is a necessary conclusion from the "*Form* der wissenden Selbstbeziehung" for Hölderlin (*Der Grund im Bewußtsein: Untersuchungen zu Hölderlins Denken (1794–1795)* [Stuttgart: Klett-Cotta, 1992], 426). This thesis is a combination, according to Henrich, of Jacobi's contention that all chains of reason presuppose a grounding of that chain with the similar thesis that *Dasein* presupposes an infinite as its proper ground. This "ground" itself remains, as Manfred Frank has argued repeatedly, the source of uncertainty, un-groundedness even, especially in the version of this argument given by Novalis. Their arguments are two sides of the same coin: being is conceived first by the judgmental process, and is a necessary (but not necessarily knowable) putative "origin" of that process. The necessity of that origin speaks to the need for a metaphysical "ground," but not to the need to *know* that ground, thus orienting consciousness in the world (Henrich) or making consciousness always an "inverted" and unfounded search for orientation (Frank). My argument in what follows suggests that this focus on self-consciousness as the leading problem leads to a nonexplanatory focus on subjectivity in Romanticism. Even though this early cluster of texts also explains the mystical or totality-oriented parts of Romantic metaphysics, its neglects the centrality of philosophical *method* in the movement, and thus enmires Romanticism in questions about consciousness as being (or as not-being). Novalis might have put it this way: given (n) world, what combination of function and location makes (n) the function of that

world? In finding this function—called "organ"—the idea was progressively to constitute and (if necessary) change that world. The exploration of the "nature" of that world, i.e., its "empirical constitution," could not at any point be abstracted from this methodological question. See Nassar, *Romantic Absolute*, for a comparison of this epistemological reading with the "ontological" reading of Frederick Beiser.

5. Not unlike the always persistent but always deferred and deferring "testing" articulated by Avital Ronell in her *The Test Drive* (Urbana: University of Illinois Press, 2005), 13.

6. Cf. Paul Guyer, *Kant and the Claims of Knowledge* (Cambridge: Cambridge University Press, 1987), 5.

7. In this point I agree with Nassar, *Romantic Absolute*, who holds that the alternative "being/knowing" is always a "both/and"-equation in German Romanticism.

8. See Winfried Menninghaus, "Die frühromantische Theorie von Zeichen und Metapher," *German Quarterly* 61, no. 1 (1989): 48–58, here 50.

4. THE TRAGIC TASK: DIALECTICAL ORGANS AND THE METAPHYSICS OF JUDGMENT (HÖLDERLIN)

1. Friedrich Hölderlin, Sämtliche Werke (Große Stuttgarter Ausgabe), ed. Friedrich Beißner (Stuttgart: Kohlhammer, 1957), 6.1.217. Hereafter cited as GSA; subsequent citations are in the form 6.1.217, referring to volume (6.1), and page number; cf. Heinse's nearly identical notes at Hölderlin, *GSA* 6.2.76, 78.The position between Germany's mythical past and France's radical present is an excellent indicator of the way in which Hölderlin's Germanic patriotism was inspired precisely by the Jacobin clubs, on which see Pierre Bertaux, *Hölderlin und die Französische Revolution* (Frankfurt a.M.: Suhrkamp, 1969), 20. See also Günther Mieth, *Friedrich Hölderlin: Dichter der bürgerlich-demokratischen Revolution* (Berlin: Rütten und Loening, 1978); and Gerhard Kurz *Mittelbarkeit und Vereinigung: Zum Verhältnis von Poesie, Reflexion und Revolution bei Hölderlin* (Stuttgart: Metzler, 1975).

2. Hölderlin, *GSA* 6.1.215.

3. Andreas Gailus points to Kant's understanding of the Germans as theatrical spectators of the Revolution, and the importance of this event for Kant's moral philosophy of history. See Gailus, *Passions of the Sign: Revolution and Language in Goethe, Kant and Kleist* (Baltimore: Johns Hopkins University Press, 2006), 59. But where sympathy and enthusiasm justify a comparison to Aristotle, Hölderlin develops a metaphysical counterpart to the Revolution that is not meant to circumvent its violence, a point upon which Kant and Schiller agree. Mieth casts this distinction as a resolution of the tension

between following Georg Forster into revolution or Schiller into reform (Mieth, *Friedrich Hölderlin*, 27).

4. Although Soemmerring was the Gontard family doctor and had diagnosed Hölderlin with foundworm a few months earlier. From Soemmerring's notes: "2. Mai 1796 Hölderlin——*Vermes*" (Hölderlin, *GSA* 6.2.72).

5. Mieth suggests in a footnote that Hölderlin's orientation in revolutionary questions was shifted precisely during the Bad Driburg trip from Georg Forster to Heinse (Mieth, *Friedrich Hölderlin*, 211).

6. Hölderlin's admiration was requited ambivalently (see Hölderlin, *GSA* 6.2.113).

7. See Panajotis Kondylis, *Die Enstehung der Dialektik: Eine Analyse der geistigen Entwicklung von Hölderlin, Schelling und Hegel bis 1802* (Stuttgart: Klett-Cotta, 1979).

8. Hölderlin, *GSA* 3.81, with reference to Heraclitus.

9. See Kant, *AA* 28, 18–19; cf. Baumgarten *Metaphysica*, §124. Hölderlin would have read the formulation in the *CPR* B291–92.

10. Hölderlin, *GSA* 4.1249.

11. Which is more than a straightforward "aestheticization"—cf. Charles Larmore, "Hölderlin and Novalis," in *The Cambridge Companion to German Idealism*, ed. Karl Ameriks, 141–61 (Cambridge: Cambridge University Press, 2000), 149; Kondylis, *Enstehung*, 261.

12. Hyperion's closing words (*nächstens mehr* . . .) make very clear that systemic closure is far from achieved in the novel. See Hölderlin, *GSA* 3.160. On the transition from *Hyperion* to *Empedocles*, see Mieth, *Friedrich Hölderlin*, 59. The plan for the tragedy was also laid out in an excised letter from Hyperion to Notara.

13. Brief an Johann Gottfried Ebel, 10. Januar 1797, Hölderlin, *GSA* 6.1.230 (my emphases).

14. Hereafter cited as Kurz, *Mittelbarkeit*, 2; my emphases.

15. Hölderlin, GSA 6.1.203.

16. The sense that Schiller has not filled in this doctrine is likely attached to passages like the following. Schiller argues famously that two "drives" (the intellectual and the sensual) must be put into a *Wechselverhältnis* (with explicit reference to Fichte's notion of *Wechselbestimmung* through the *Spieltrieb*). The parallax between the self and the world as points of human focus, however, means that this reconciliation is always a corruptible presentation. Art— which is the highest form of that reconciliation—provides a regulative picture of the cognitive peace Schiller aims at, but resolves nothing at the level of reality. See Friedrich Schiller, *Sämtliche Werke*, vol. 5 (Munich: Hanser, 1962), 612.

17. Hölderlin to Schiller, 4.9.1795, in Hölderlin, *GSA* 6.1.181.

18. Hölderlin, *GSA* 6.1.329.
19. Or the "priestesses" of nature (ibid.).
20. Ibid.
21. For the involved publication history of this fragment, for which the *terminus ante quem* is 20 April 1795, according to statistical analyses of Hölderlin's orthography, see Dieter Henrich, "Hölderlin on Judgment and Being: A Study in the History of the Origins of Idealism," in Henrich, *The Course of Remembrance and Other Essays on Hölderlin*, ed. Eckart Förster (Stanford, CA: Stanford University Press, 1997), 71–90. This study serves to place the writing of the fragment quite early and forms the basis for the "primacy" argument between Hölderlin and Novalis, historiographically. For the Novalis standpoint, see Manfred Frank, *The Philosophical Foundations of Early German Romanticism*, trans. Elizabeth Millàn-Zaibert (Albany: SUNY Press, 2004), 151–77.
22. The Fichtean background—which ultimately goes back to Platner—of this suggestive etymology has been examined by Violetta Waibel, *Hölderlin und Fichte, 1794–1800* (Paderborn: Schöningh, 2000), 140–63.
23. Of course, judgment is also *unifying*, since its structure is such that it brings together two separate factors (subject and predicate) in various possible categorical ways. Kant's definition of judgment as the unification of concept and intuition provides the general context for this determination. Fichte's notion of judgment as *interaction* between concept and intuition was, as we shall see, also key for Hölderlin's elaboration of the notion presented in *Urtheil und Seyn*.
24. Hölderlin, *GSA* 6.1.216.
25. I stress this contradiction here because I want to avoid emphasizing either judgment or being to the detriment of the other for Hölderlin's method. "*Urtheil und Seyn*" is so short and conceptually unelaborated as to have been a strange choice for the amount of attention it has received. Readers agree that the fragment contains a critique of Fichte—which is clear—and a gesture toward ontology singular in the post-Kantian atmosphere. Manfred Frank has contextualized the fragment among the antifoundationalist efforts of Niethammer in particular in Jena—see *Philosophical Foundations*, 125ff. Frederick Beiser has argued against any absolute primacy of Being over thinking in the fragment, calling Hölderlin's work an attempt at a "transcendental deduction" of something like Spinoza's substance (Beiser, *German Idealism: The Struggle against Subjectivism*, 1781–1801 [Cambridge: Harvard University Press, 2002], 391). By placing a real contradiction between judgment and being, Hölderlin puts them both—including their own complex structures—in a single frame. It is that frame which interests me here, and which is, I think, the lightest way of treading on this overinterpreted page and a half.

26. Spinoza's famous example is of a trader intuitively understanding the law of proportions, "seeing" that 2:3 as 4:6. The number 6 is arrived at not by the logical steps (4 is twice 2; twice three is . . .), but by "intuition" or immediate presentation. Spinoza ultimately argues that this kind of knowledge is reflective of the love which binds the unity of nature or god together. See *Ethica* I prop. 42 scholium, and V more generally. *Intuere* is Latin for "observe, look at, admire"—thus the presentational element of this form of knowledge, and thus Kant's translation in his German works, *anschauen*.

27. Kant, *CPR* B 311ff.

28. See, e.g., Kant, *CPR* B 305–7.

29. See Kant, AA V, 400.

30. *Locus classicus*: Kant, *CPR* B176/A137–B187/A147.

31. His use of the term *Wechselwirkung* (and sometimes *Wechselverhältnis*) for various oppositional relationships within the generative *Ich* is not irrelevant, of course. See Violetta Waibel's "Wechselbestimmung: Zum Verhältnis von Hölderlin, Schiller und Fichte in Jena," in *Fichte und die Romantik: Hölderlin, Schelling Hegel und späte Wissenschaftslehre*, ed. Wolfgang Schrader (Amsterdam: Rodolpi, 1997), 43–71.

32. The later use of the term is in his 1797 *Versuch einer neuen Darstellung der Wissenschaftslehre*, by which time Hölderlin, Schelling, and Novalis were all using the term independently (Johann Gottlieb Fichte, "Neue Darstellung der Wissenschaftslehre," in Johann Gottlieb Fichte, *Schriften zur Wissenschaftslehre: Werke*, vol. 1, ed. Wilhelm G. Jacobs [Frankfurt a.M.: Klassiker, 1997], 205–17).

33. Reinhold's popularizing letters on the critical philosophy were published just as the Pantheism Controversy began to pick up steam in 1786. The classic history is given by Dieter Henrich, *Between Kant and Hegel* (Cambridge: Harvard University Press, 2003).

34. Reinhold's doctrinal sentence reads: "In consciousness representation is differentiated by the subject from both subject and object and then related to both." Johann Gottlieb Fichte, *Rezension Aenesidemus*, in *Gesamtausgabe der Bayerischen Akademie der Wissenschaften*, vol. 2, *Werke 1793–95*, ed. Reinhard Lauth et al. (Stuttgart–Bad Canstatt: frommann-holzboog, 1965), 43n12. Hereafter cited as *RezAen*.

35. Indeed, most of the above reading comes from the key passage from *RezAen*, 49–60, esp. 56–57, here 57.

36. Fichte, *RezAen*, 57.

37. John Neubauer, "Intellektuelle, intellektuale und ästhetische Anschauung: Zur Entstehung der romantischen Kunstauffassung," *Deutsche Vierteljahrsschrift für Literaturwissenschaft und Geistesgeschichte* 46, no. 1 (1972): 294–319, argues that the erstwhile Romantic spelling "intellektuale Anschauung" speaks to a Latin provenance that stems from Spinoza, reintegrating the

intellectual love of God as an ontological (and affective) condition into the epistemological problematic given by Kant.

38. Both efforts ran aground on Jacobi's generalized attack on reason-based systems. Indeed, Jacobi coined the term "nihilism" during the "atheism controversy" which led to Fichte's dismissal from the university in Jena in 1800. See Werner Röhr, ed., *Appellation an das Publikum: Dokumente zum Atheismusstreit, Jena 1798/99* (Leipzig: Reclam, 1991). See also Waibel, *Hölderlin*, 233–87, on the intersubjectivity of the *I* in Fichte, and Hölderlin's response to it.

39. See Violetta Waibel's essay, "'With respect to the antinomies, Fichte has a remarkable idea': Three Answers to Kant and Fichte: Hardenberg, Hölderlin, Hegel," in *Fichte, German Idealism and Early Romanticism*, ed. Daniel Breazeale and Tom Rockmore (Amsterdam: Rodopi, 2010), 301–27; see also Waibel, *Hölderlin*, 205–31; cf. Helmut Müller-Sievers, *Self-Generation: Biology, Philosophy, and Literature around 1800* (Stanford, CA: Stanford University Press, 1997), 65–90.

40. Hölderlin, *GSA* 4.1.251–52.

41. Ibid., 3.236.

42. Ibid., 4.1.255–56.

43. Ibid., 4.1.248.

44. Winfried Menninghaus has written of Hölderlin's *Procedure* as a theory of *depiction* that parallels the "semio-ontology" of the Jena Romantics, seeing in the exteriorization of the "organ" the outer sphere that allows ontology to be marginally dominated by semiotics, viz. Winfried Menninghaus, *Unendliche Verdopplung: Die frühromantische Grundlegung der Kunsttheorie im Begriff absoluter Selbstreflexion* (Frankfurt a.M.: Suhrkamp, 1987), 110–14. Menninghaus writes correctly of the opposed tendencies in Hölderlin's thinking, to allow "being" to appear as the beyond of cognition, and to negate all such metaphysical beyonds in favor of concrete epistemic and poetological necessities. It seems to me, however, that both tendencies are united in the problematic of the organ, and thus that this type of thought is meant to intervene in, not merely obliquely point out, the contradictions it casts as inherent in life.

45. Hölderlin, *GSA* 4.1.248–49.

46. The preceding paragraph is based on ibid.

47. Ibid., 4.1.249; emphasis in original.

48. See Waibel, *Hölderlin*, 220–28.

49. Hölderlin, *GSA* 4.1.249–50; emphasis in original; cf. ibid., 4.1.259.

50. See Waibel, *Hölderlin*, 278–85; see also Ulrich Gaier, *Krumme Regel: Novalis's "Konstruktionslehre des schaffenden Geistes"* (Tübingen: Max Niemeyer, 1970).

51. The Heideggerian notion of *poeisis* suggests itself, especially given the phrase "Welt in der Welt." Yet I think we should avoid reading too quickly forward to Heidegger's attempts to stand in the "clearing" of being, to produce or manifest beings' Being through alteration of our subjectivity. It is precisely Hölderlin whom Heidegger uses to point to what little hope there is, in the technological world-age, for such a project. See Martin Heidegger, "The Question Concerning Technology," in *Martin Heidegger: Basic Writings from "Being and Time" (1927) to "The Task of Thinking" (1964)*, ed. David Farrell Krell (San Francisco: Harper, 1977), 307–43, esp. 340ff. And yet Hölderlin's organology seems to run in a different direction. It constructs and builds an order within a greater order, and suggests with that gesture the possibility of intervening in the greater order.

52. Hölderlin, *GSA* 4.1.266.

53. See Frederick C. Beiser, *Diotima's Children: German Aesthetic Rationalism from Leibniz to Lessing* (Oxford: Oxford University Press, 2009), 196–244.

54. This is a very early expression of the dynamism which makes Lessing so relevant for the next generation, in such writings as *Die Erziehung des Menschengeschlechts*, where a progression in the forms of reason is defended. As we saw above, Jacobi's questionable presentation of Lessing's pantheism strengthened this impression. See also Wulf Koepke, "Der späte Lessing und die junge Generation," *Humanität und Dialog: Beiheft zum Lessing-Jahrbuch* (Detroit: Wayne State University Press, 1982), 211–22.

55. Gotthold Ephraim Lessing, *Werke*, vol. 4, ed. Herbert G. Göpfert (Munich: Hanser, 1973), 163; emphasis in original.

56. Hölderlin, *GSA* 4.1.266.

57. Ibid., 4.1.267.

58. The foregoing paragraph is taken from ibid., 4.1.267–69, quotation on 269.

59. See Waibel's "With respect to the antinomies, Fichte has a remarkable idea."

60. Hölderlin thus anticipates Hegel's generalization of the antinomies in the *Encyclopedia*; cf. Werke 8, 127–28.

61. Uvo Hölscher's groundbreaking work in this area showed that Hölderlin had access to most of what is now gathered in the Presocratic fragments edited by Diels and Kranz; see Uvo Hölscher, *Empedokles und Hölderlin*, ed. Gerhard Kurz (Eggingen: Edition Isele, 2001), 11–22.

62. Hölscher's claim that the debate about *cosmogony* (generation of the universe) was actually a *cosmology* (presentation of the lawful universe) remains important. See Hölscher, *Empedokles*, 36.

63. In a famous letter to Böhlendorff of 4 December 1801, Hölderlin describes the addressee's tragic work as "truly modern," an exit from the kingdom of the living (descent into the *aorgic*) (Hölderlin, *GSA* 6.1.426).

64. See Dennis J. Schmidt, *German and Other Greeks: Tragedy and Ethical Life* (Bloomington: Indiana University Press, 2001), 140.

65. See ibid., 144ff.

66. On Hölderlin's radicality, see Miethe, *Friedrich Hölderlin*, 75, 86; the metaphysical side of this politics is treated in Kurz, *Mittelbarkeit*, 198, who argues that the struggle between the general and the particular is epochal. Jochen Schmidt tracks a shift from revolution to history's transcendent laws across the drafts, in Friedrich Hölderlin, *Sämtliche Werke und Briefe in drei Bänden*, vol. 2, *Hyperion/Empedokles/Aufsätze/Übersetzungen*, ed. Jochen Schmidt and Katharina Grätz (Frankfurt a.M.: Suhrkamp, 1994), 1112.

67. Kurz compares this element of Hölderlin's thought to Hegel's Owl of Minerva in *Mittelbarkeit*, 12–13.

68. Cf. Mieth, *Friedrich Hölderlin*, 50.

69. Gabriel Trop's recent reading of necessity is exemplary: Gabriel Trop, "Modal Revolutions: Friedrich Hölderlin and the Task of Poetry," *MLN* 128, no. 3 (2013): 580–610. Trop's point is that tragic and philosophical necessity coincide is essential to understanding the *Empedocles*: "Specifically, for Hölderlin, the main movement in tragic thinking undertakes a modal shift from the possible to the necessary: the attempt to bind and connect the world of the sensible in such a way that the sensible itself is endowed with intelligibility, or such that the possible becomes resignified as necessary without losing its identity as possibility" (587).

70. Kurz, *Mittelbarkeit*, 12; emphasis in original.

71. Hölderlin, *GSA* 4.1.153–54.

72. Ibid., 4.1.152.

73. Ibid., 4.1.154–55.

74. See Schmidt, *Germans*, 154ff.; and Véronique M. Fotì, *Epochal Discordance: Hölderlin's Philosophy of Tragedy* (Albany: SUNY Press, 2006), 46.

75. See, e.g., Friedrich Schiller, *Sämtliche Werke*, vol. 5, *Philosophische und vermischte Schriften* (Munich: Hanser, 1962), 360.

76. Cf. ibid., 364.

77. Ibid., 362.

78. Schiller, *Sämtliche Werke*, 2:819; emphasis in original.

79. Ibid.

80. Hölderlin, *GSA* 4.1.167–68.

81. Ibid., 4.1.164.

82. Ibid., 4.1.168.

83. Hölderlin, *Empedocles*, 188; Hölderlin, *GSA* 4.1.141.

84. Hölderlin, *GSA* 4.1.4–5, 11.
85. Hölderlin, *Empedocles*, 40. Hölderlin, *GSA* 4.1.5.
86. Hölderlin, *GSA* 4.1.7.
87. Ibid., 4.1.7–8.
88. Ibid., 4.1.8–9.
89. Ibid., 4.1.7.

5. ELECTRIC AND IDEAL ORGANS: SCHELLING AND THE PROGRAM OF ORGANOLOGY

1. *Epigraph*: Schelling, *HKA* 5.106.
2. Cf. Thomas Bach and Olaf Breidbach, *Naturphilosophie nach Schelling* (Stuttgart–Bad Canstatt: frommann holzboog, 2005).
3. Here I follow Panajotis Kondylis, *Die Enstehung der Dialektik: Eine Analyse der geistigen Entwicklung von Hölderlin, Schelling und Hegel bis 1802* (Stuttgart: Klett-Cotta, 1979), 558ff.
4. This is emphasized repeatedly by Frederick C. Beiser, German Idealism: The Struggle against Subjectivism, 1781–1801 (Cambridge: Harvard University Press, 2002). See, e.g., 480, 497–98, 501, 506, and 556.
5. Accounts of that conflict can be found in Dieter Henrich, *Der Grund im Bewußtsein: Untersuchungen zu Hölderlins Denken (1794–1795)* (Stuttgart: Klett-Cotta, 1992), 127–35; and Kondylis, *Enstehung*, 540–51.
6. I thus do not follow the interpretive schema that proposes a "materialist" or "realist" moment in Schelling's writings on *Naturphilosophie*, for which see Iain H. Grant, *Philosophies of Nature after Schelling* (New York: Continuum, 2006); and Dalia Nassar, "From a Philosophy of Self to a Philosophy of Nature: Goethe and the Development of Schelling's Naturphilosophie," *Archiv für Geschichte der Philosophie* 92, no. 3 (2010): 304–21.
7. F. W. J. Schelling, *First Outline of a System of the Philosophy of Nature*, trans. Keith R. Peterson (Albany: SUNY Press, 2004), 48; Schelling, *HKA* 7.112.
8. Richards argues that Schelling maintained a "dynamic evolution" that saw general ideal types in optimal but real development. See Robert John Richards, *The Romantic Conception of Life: Science and Philosophy in the Age of Goethe* (Chicago: University of Chicago Press, 2002), 289–307.
9. Schelling, *HKA* 7.114ff.
10. Schelling, *Outline*, 52; Schelling, *HKA* 7.115.
11. Schelling understands Kant's *a priori* as having not to do with the content of propositions but with the manner of our judgment. See Beiser, *German Idealism*, 524ff. and especially 527; cf. Schelling, *HKA* 8.33, in which the manipulation of nature through experiment is simultaneously a manipula-

tion of knowledge, one which can, in principle, *result* in the a priori, rather than merely deriving from it.

12. See Schelling, *HKA* 7.102–3.
13. Kant, AA V, 373–74.
14. See Beiser, *German Idealism*, 486ff.; and Dalia Nassar, "From a Philosophy of Self to a Philosophy of Nature: Goethe and the Development of Schelling's Naturphilosophie," *Archiv für Geschichte der Philosophie* 92, no. 3 (2010): 304–21.
15. Kant, AA V, 220.
16. Schelling, *HKA* 6.204.
17. The term "substrate" seems to derive here less from Kant (for whom the "substrate" is the supersensible ground, which we must consider regulatively) than from Aristotle, whom Schelling quotes to describe the substrate-concept as "in se teres et rotundum" (in itself smooth and rounded) (Schelling, *HKA* 6.205–6). The citation appears to be from Aristotle, *De generatione*, beginning of section 4, and has clear resonances in Schelling's adaptation ("in sich selbst *Ganzes* und *Beschlossenes*") with classical aesthetic theory.
18. Schelling, *HKA* 6.205; emphasis in original.
19. Ibid., 1.6.206; emphasis in original.
20. Ibid., 1.6.207; emphasis in original.
21. Ibid., 1.6.208–9; emphasis in original. See 209ff., where Schelling defines organs as the result of the individual generation-process combined with the specific form of matter involved. He then disputes the possibility that organs and their structure can give rise to function—rather than the other way around.
22. Cf. ibid., 1.6.256–57; ibid., 1.7.271.
23. Ibid., 1.7.73.
24. Ibid., 1.7.210.
25. Herder, *Ideen*, 94.
26. Ibid., 82; emphasis in original.
27. Ibid., 87–88.
28. Ibid., 88–89.
29. Herder ties organs as the determinant factor in identification of species to the "directionality" of the instinct of the animal, something we will see in the subsequent discourse again (ibid., 99).
30. Cf. ibid., 104–5.
31. Ibid., 88.
32. In *Of the World Soul*, Schelling had given this speech iconic status (Schelling, *HKA* 6.253).
33. For the circumstances of the speech, see Kielmeyer, *Über die Verhältnisse*, 12–13.
34. Larson, "Vital Forces," 242.

35. See Kielmeyer, *Über die Verhältnisse*, 33–35. The force of reproduction is also more *concentrated* in those creatures with sensibility, allowing for their more complex structure (which will include sensibility) to be reproduced. Thus compensation occurs at the level of the species as well as phylogenetically. See pp. 63–67 of the introduction, and 35–37 in the text for Kielmeyer's (possibly first) formulation of the biogenetic law ("ontogeny recapitulates phylogeny") in terms of force-relations.

36. See Carl Friedrich Kielmeyer, *Ueber die Verhältnisse der Organischen Kräfte untereinander in der Reihe der verschiedenen Organisationen, die Geseze und Folgen dieser Verhältnisse, eingeführt von Kai Torsten Kanz* (Marburg an der Lahn: Basilisken-Presse, 1993), 4; cf. Timothy Lenoir, The Strategy of Life Teleology and Mechanics in Nineteenth-Century German Biology (Chicago: University of Chicago Press, 1989), 37–35; and James L. Larson, L. "Vital Forces: Regulative Principles or Constitutive Agents? A Strategy in German Physiology, 1786–1802," Isis 70, no. 2 (1979): 241–43.

37. Cf. Herder, *Ideen*, 94.

38. Kielmeyer, *Über die Verhältnisse*, 37.

39. Ibid., 42.

40. Ibid., 43.

41. Schelling, *HKA* 7.115.

42. See Johann Friedrich Blumenbach, *Institutiones physiologicae* (Göttingen: Dieterich, 1786), 31–38; and Johann Friedrich Blumenbach, *Anfangsgründe der Physiologie*, trans. and ed. Joseph Eyerel (Vienna: Wappler, 1789), 26–31. Larson notes that "the order of enumeration is also the order of successive formation" ("Vital Forces," 238).

43. Blumenbach, *Anfangsgründe* §47, p. 28; Blumenbach, *Institutiones*, §47, p. 34.

44. Cited from Schelling, Friedrich Wilhelm Joseph von Schelling, *Schellings Werke: Auswahl in drei Bänden*, vol. 1, ed. Otto Weiß (Leipzig: Eckardt, 1907), 470.

45. The problem of quality is one of Schelling's principal reasons for rejecting purely mechanistic physics, already in the *Ideas* in 1797. See Beiser, *German Idealism*, 525ff.; and Kondylis, *Entstehung*, 565.

46. See Schelling, *HKA* 8.39. See also Benjamin Specht, *Physik als Kunst: Die Poetisierung der Elektrizität um 1800* (Berlin: de Gruyter, 2010), 103.

47. For the experimental and theoretical background of the Galvanism debate, which integrated theories of electricity with problems in the developing field of experimental physiology, see Francesco Moiso, "Theorien des Galvanismus," in Friedrich Wilhelm Joseph Schelling, *Historisch-kritische Ausgabe, Reihe 1. Werke*, vol. 5/9, *Ergänzungsband: Wissenschaftshistorischer Bericht zu Schellings naturphilosophischen Schriften, 1797–1800*, ed. Hans

Michael Baumgartener (Stuttgart: frommann-holzboog, 1994), 320–75; hereafter cited as Schelling, *HKA* 5/9.320–75.

48. For the claim that Galvani discovered but did not prove the existence of biolectricity, see K. E. Rothschuh, "Von der Idee bis zum Nachweis der tierischen Elektrizität," *Sudhoffs Archiv für Geschichte der Medizin und der Naturwissenschaften* 44, no. 1 (1960): 25–44.

49. See K. E. Rothschuh, "Alexander von Humboldt und die Physiologie seiner Zeit," *Sudhoffs Archiv für Geschichte der Medizin und der Naturwissenschaften* 43, no. 2 (1959): 97–113, esp. 101–2.

50. See Francesco Moiso, "Magnetismus, Elektrizität, Galvanismus," in Schelling, *HKA* 5/9.165–375, here 330.

51. Alexander von Humboldt, *Über die gereizte Muskel- und Nervenfaser nebst Vermuthungen über den chemischen Process des Lebens in der Thier- und Pflanzenwelt*, vol. 1 (Berlin: Posen, 1797), 1.

52. Ibid., 1:235; cf. Moiso, "Magnetismus, Elektrizität, Galvanismus," 334.

53. Johann Wilhelm Ritter, *Beweis, daß ein beständiger Galvanismus den Lebenprocess im Thierreich begleite, nebst neuen Versuchen und Bemerkungen über den Galvanismus* (Weimar: Verlag des Industrie-Comptoirs, 1798). On Ritter's experiments and metaphorical discourse, see Jocelyn Holland's excellent *German Romanticism and Science: The Procreative Poetics of Goethe, Novalis, and Ritter* (New York: Routledge, 2009), 141ff.

54. Moiso, "Magnetismus, Elektrizität, Galvanismus," 366.

55. Ibid., 371–72.

56. Schelling, *HKA* 6.244ff., where Schelling locates electric impulse as "awakened in the nerves themselves."

57. See Beiser, *German Idealism*, 544–48, for an overview.

58. Cf. Schelling, *HKA* 6.193ff.

59. Ibid., 193; emphasis in original.

60. Haller's definition uses *pars*; *Commentarii Societatis Regiae Scientiarum Gottingensis*, vol. 2 (Göttingen, 1753), 116.

61. Schelling, *HKA* 7.129.

62. Ibid., 1.6.194; emphasis in original; cf. Larson, "Vital Forces," 238.

63. Schelling, *HKA* 6.196; emphasis in original.

64. Ibid., 6.225.

65. I agree with Kondylis, *Enstehung*, that there is a progression from unity/harmony in the *Ideen*, to a duality in the *Weltseele*, and on to an explicit dialectical triplicity in the *Entwurf*. See Kondylis, *Entstehung*, 558–96.

66. Schelling, *Outline*, 70; Schelling, *HKA* 7.133; emphasis in original.

67. Schelling, *Outline*, 58–59; Schelling, *HKA* 7.123.

68. Schelling, *HKA* 7.207ff.

69. Ibid., 7.180ff.

70. Ibid., 7.205ff.

71. Schelling puts this point this way: "*The third body in the galvanic chain is namely only necessary so that the opposition between the other two can be maintained*" (ibid., 7.185; emphasis in original).

72. Ibid., 7.188ff.

73. Ibid., 7.190–91.

74. That such a thing is possible is included in the definition of the *specific organ* Schellings gives: "Under the *specificity* of the excitability of an organ I think nothing more than that the receptivity of this organ for a stimulus is determined by the dynamical quality of factors out of which the organ is constructed" (Schelling, *Outline*, 127). Thus if sensibility is involved, the field of potential is established as sensible, and the organ is the embodied locus of that potential.

75. Schelling, *HKA* 5.106.

76. Kant, *CPR* B334/A278.

77. Schelling, *HKA* 5.107.

78. Cited here from Friedrich Wilhelm Joseph von Schelling, *Schellings Werke* 1:438.

79. Schelling, *HKA* 5.81. As Schelling goes on to point out, *intuition* is for him only possible on the basis of real opposition within cognition, on the basis of an original contradiction in self-consciousness. Because self-consciousness has an intuitive register, its deepest form provides the *hidden organ* Schelling posits to answer Kant.

80. This is perhaps most clearly expressed in the famous introduction to the *Entwurf* (written after the completion of the text, and with Goethe's collaboration): "Every experiment is a question put to nature, which it is then forced to answer. But every question contains a hidden judgment *a priori*; every experiment that *is* an experiment is a *prophecy*; experimenting itself is a production of appearances" (Schelling, *HKA* 8.33).

81. Ibid., 3.68–69.

82. Manfred Frank, *The Philosophical Foundations of Early German Romanticism* (Albany: SUNY Press, 2004), 23–39.

83. Schelling, *HKA* 9.25.

84. Kant, *CPR* B1.

85. On the break between Schelling and Fichte, see Beiser, *German Idealism*, 491–506. By 1801, Schelling would claim the parity of subject and object in the absolute, something Fichte could not abide. After his move from Jena in 1800, Fichte sometimes uses "organ" in the Romantic sense (Johann Gottlieb Fichte, *Sämmtliche Werke*, vol. 1 [Berlin: Veit, 1845/1846], 214).

86. This is an excellent example of the Romantic approach to Kant. Where Kant had *also* sought for internally valid forms of Reason (as the faculty that

reflects on and unifies the judgments of the understanding)—and even derived the branches of special metaphysics from them—Schelling treats the mereological complex of judgment as continuous with Reason, bracketing Kant's commitment to content-problems in the *Transcendental Aesthetic* (division of outer and inner sense) and the *Transcendental Analytic* (causation and object-formation).

87. Schelling, *HKA* 9.40; emphasis in original; cf. 41.

88. Ibid., 9.1.60.

89. "Recursive immediacy" is my gloss on Fichte's statement that "the I posits itself as positing." The point is that knowledge of objects and knowledge of the self must be not only combined but also immediately identified as the self. The minimum requirement for self-consciousness thus looks something like, "I know that I am the one thinking of that object and producing the thought of that object."

90. This summary is taken from a few comparable passages in Fichte's writings in 1797/98. See Fichte, *Versuch einer neuen Darstellung der Wissenschaftlehre (1797/98)*, ed. Peter Baumanns (Hamburg: Meiner, 1984); and the *Wissenschaftslehre nova methodo 1798/99*, ed. Erich Fuchs (Hamburg: Meiner, 1994).

91. Schelling, *System des transzendentalen Idealismus*, 60–61.

92. Ibid., 60.

93. Ibid., 328.

94. Dieter Jähnig, *Schelling: Die Kunst in der Philosophie*, vol. 1, *Schellings Begründung von Natur und Geschichte* (Pfüllingen: Neske, 1966), 11.

95. Schelling, *HKA* 9.325.

96. Ibid., 9.306–11.

97. Ibid., 9.312. This, it seems to me, is where Schelling departs from Fichte.

98. Ibid., 9.320.

99. Ibid., 9.322.

100. Manfred Frank, *Der unendliche Mangel an Sein: Schellings Hegelkritik und die Anfänge der Marxschen Dialektik* (Munich: Fink, 1992), see 319–61.

6. UNIVERSAL ORGANS: NOVALIS'S ROMANTIC ORGANOLOGY

1. I cite Novalis from the *Historisch-kritische Ausgabe* (*Novalis Schriften: Die Werke Friedrich von Hardenbergs*, ed. Paul Kluckhohn and Richard Samuel, 3rd ed. [1960; Stuttgart: Kohlhammer, 1977]) in the form: *HKA* 3.391. I cite the *Allgemeines Brouillon* according to entry numbers for generality, as *AB* 657.

2. Nelly Tsouyopoulos has proposed that Brown's doctrines were a Kuhnian-style revolution, ending a normal state of medical science (still Galenic in practice) and uniting theory and therapy. See Tsouyopoulos,

Asklepios und die Philosophen: Paradigmawechsel im 19. Jahrhundert, ed. Claudia Wiesemann et al. (Stuttgart–Bad Canstatt: frommann-holzboog, 2008). On the latter point, Thomas Broman agrees in *The Transformation of German Academic Medicine, 1750–1820* (Cambridge: Cambridge University Press, 1996), 128, 151.

3. John Brown, *The Elements of Medicine, or a Translation of the Elementa Medicinae Brunonis* (London: Johnson, 1788), 27.

4. What follows is treated in detail in Nelly Tsouyopoulos, "The Influence of John Brown's Ideas in Germany," *Medical History*, Supplement no. 8 (1988): 63–74.

5. His work was also influential in its own right, raising his home university and hospital setting in Bamberg to a leading medical center in the early years of the nineteenth century. See Broman, *Transformation*, 149ff.

6. Ibid., 150.

7. The humoral model was no longer accepted at the end of the eighteenth century, but medical practice still focused on emetics and purgatives to balance and eliminate fluids in the sick body. Röschlaub's colleague Adalbert Friedrich Marcus did away with this practice in the name of Brown's doctrine at the hospital in Bamberg, with some success. See Broman, *Transformation*, 154.

8. Johann Gottlieb Fichte, *Grundlage der gesammten Wissenschaftslehre (1794)* (Hamburg: Meiner, 1997), 46.

9. Brown, *Elements*, 59.

10. The classic study of Novalis's uneven relationship to Brown's doctrines and the notion of health is John Neubauer, *Bifocal Vision: Novalis' Philosophy of Nature and Disease* (Chapel Hill: University of North Carolina, 1971).

11. Novalis, *HKA* 3.704.

12. See William Arctander O'Brien, *Novalis: Signs of Revolution* (Durham: Duke University Press, 1995), for the history and an attempt to overcome especially Tieck's editorial construal of Novalis as a mystic.

13. Johann Wolfgang von Goethe, *Goethes Werke, Weimarer Sophienausgabe* (Weimar: Böhlau, 1887), sec. 1 vol. 23, p. 131; subsequent citations are in the form Goethe, WA 1.23.131, and refer to section, volume, and page number. Chapter 8 below concludes with an interpretation of this passage.

14. On the philosophical importance of Novalis's time in Freiberg, see Jocelyn Holland, "From Romantic Tools to Technics: Heideggerian Questions in Novalis' Antropology," *Configurations* 18, no. 3 (2010): 291–307.

15. Novalis, *HKA* 3.252.

16. Novalis was not alone in this designation—August Wilhelm Schlegel also calls the encyclopedia an "organ." August Wilhelm von Schlegel, *Kritische Ausgabe: August Wilhelm Schlegel—Vorlesungen von 1798-1827:*

Vorlesungen über Enzyklopädie der Wissenschaften (1803), vol. 3, ed. Frank Jolles, Edith Höltenschmidt, and Ernst Behler (Paderborn: Schöningh, 2006). See the treatment by Carlos Spoerhase, "Poetik des Plans: Über die Form des Systems um 1800," in "Komplexität und Einfachheit," ed. Albrecht Koschorke (forthcoming).

17. Schlegel, *KFSA* 2.257. Note that Schlegel here means "productive imagination," of the ability to form objects out of intuitions in the first place. Thus the organ, filtered through this Kantian doctrine and its centrality for Fichte, becomes the divine ability of the human to make a world—on which more in chapter 7 below.

18. Cf. Friedrich Schleiermacher, *Über die Religion: Reden and die Gebildeten unter Ihren Verächtern* (Hamburg: Meiner, 1958), 30; hereafter cited as *Reden*.

19. Ibid., 31–32.

20. Cf. ibid., 147.

21. Ibid., 131.

22. Cf. ibid., 36. This separates religion from metaphysics, which is the expression of one of two basic drives—consumptive and resignative—that make up human nature for Schleiermacher; cf. ibid., 28–29.

23. Kant, *CPR* B75.

24. *Reden*, 41.

25. Fichte, *Sämmtliche Werke*, 1:319.

26. That process goes a long way to explaining the Romantic interest in and specific reading of Spinoza. As Herder before him, Schleiermacher prizes Spinoza's "religious" sense (cf. Schleiermacher, *Reden*, 31). The interest is not in dogmatic metaphysics, but in the expansion of the instruments of the broadly conceived philosophical and religious senses.

27. Schleiermacher, *Reden*, 82.

28. Ibid., 107.

29. Ibid., 73–74.

30. E.g., ibid., 47, 54.

31. Ibid., 52; my emphasis.

32. "The magical circle of dominating opinions and epidemic feelings surrounds and encircles everything" (ibid., 54).

33. Ibid., 71–72.

34. Novalis, *HKA* 3.476. Novalis thinks of Fichte's philosophy in the same metaphorical terms, as an experiment in generation. Cf. Novalis, *HKA* 3.477; Jürgen Daiber, *"Experimentalphysik des Geistes"—Novalis und das romantische Experiment* (Göttingen: Vandenhoeck and Ruprecht, 2001).

35. In his most complete theoretical statement, the *Logological Fragments* and *Poëticismen*, Novalis writes that the world is like "ein Niederschlag aus der Menschennatur . . . so ist die Götterwelt eine Sublimation—Beyde geschehen

uno actu—Kein plastisches Praecipitat, ohne geistiges Sublimat" (Novalis, *HKA* 2.531).

36. Ibid., 2.554; cf. 2.556.

37. See Johann Gottfried Herder, *Werke in zehn Bänden*, vol. 4, *Schriften zu Philosophie, Literatur, Kunst und Altertum*, ed. Jürgen Brummack and Martin Bollacher (Frankfurt a.M.: Klassiker 1994), 786.

38. See Arthur O. Lovejoy, The Great Chain of Being: A Study of the History of an Idea (1936; Cambridge: Harvard University Press, 1964). On the "infinity" of the universe, see Alexander Koyré, *From the Closed World to the Infinite Universe* (Baltimore: Johns Hopkins University Press, 1968); cf. Hans Blumenberg, *Die Genesis der kopernikanischen Welt* (Frankfurt a.M.: Suhrkamp, 1975).

39. Kant, AA II, 387.

40. Following Schlegel and Schleiermacher, here the efficacious organ is the imagination (Novalis, *HKA* 3.401); cf. *AB* 327.

41. Novalis, *HKA* 3.301; cf. Novalis, *HKA* 4.281.

42. Blumenberg judges this line of thought (in the version Schlegel elaborates in his lectures on Transcendental Idealism in 1800) inconsistent: "The idea of the unfinished state of the world is combined, in Schlegel, with the Romantic predilection for organic metaphors. That is a remarkable incompatibility, since the organic system does not have the character of a substrate; that is, it does not tolerate artificial intervention" (Hans Blumenberg, *Genesis of the Copernican World*, 65–66). The oppositional pair organic/mechanical, which appears as the autonomy of a being following its own law (and not God's) versus the economy of motion for outcome in mechanics, is undercut here by the organ, which is precisely conceived as internally developing and nevertheless adaptable to ends. Organology reconciles use and freedom, economy and spontaneity, at the level of metaphysics, undercutting what Blumenberg sees as a necessary shift into metaphorics.

43. Cf. also Hans Blumenberg, *Arbeit am Mythos* (Frankfurt a.M.: Suhrkamp, 1979), 620, and the previous note. "Incompleteness" is connected to "perfectibility," on which see Ernst Behler, *Unendliche Perfektibilität: Europäische Romantik und Französische Revolution* (Munich: Schöningh, 1989).

44. Novalis, *HKA* 2.452–53; Novalis, *HKA* 3.401–3; *AB* 702 claims that nature is simply "more" after it has passed through the philosophical organ.

45. This reading of the fragment shifts the balance from the problems of orientation and what Kant calls "ideas," as the fragment is read by Rodolphe Gasché, to that of judgment. See Rodolphe Gasché, "Foreword: Ideality in Fragmentation," in Friedrich Schlegel, *Philosophical Fragments*, trans. Peter Firchow (Minneapolis: University of Minnesota Press, 1991), vii–xxxi. Gasché is here building on the work of Jean-Luc Nancy and Philippe Lacoue-Lab-

arthe, *The Literary Absolute: The Theory of Literature in German Romanticism* (Albany: State University of New York Press, 1988), which in turn is strongly influenced by Maurice Blanchot, "The Athenaeum," trans. Deborah Esch and Ian Balfour, *Studies in Romanticism* 22, no. 2 (1983): 163–72.

46. See Caroline Welsh, *Hirnhöhlenpoetiken: Theorien zur Warhnehmung in Wissenschaft, Ästhetik und Literatur um 1800* (Freiburg: Rombach, 2003), 205ff., for a brilliant reading of the *Apprentices* partly based on Novalis's knowledge of Soemmerring's *Seelenorgan*, esp. 243–44, where Welsh shows that the Genius—who is in control of all organs—makes new fields of perception available.

47. For the relation of woman to the linguistic process of *Bildung*, see Friedrich Kittler, "Heinrich von Ofterdingen als Nachrichtenfluß," in *Novalis: Wege der Forschung*, ed. Gerhard Schulz (Darmstadt: Wissenschaftliche Buchgesellschaft, 1986), 480–508.

48. Novalis, *HKA* 1.98–99.

49. Ibid., *HKA* 1.103.

50. Jocelyn Holland, *German Romanticism and Science:* The Procreative Poetics of Goethe, Novalis, and Ritter (New York: Routledge, 2009), 109; cf. Chad Wellmon, *Becoming Human: Romantic Anthropology and the Embodiment of Freedom* (University Park: Pennsylvania State University Press, 2010).

51. On this point, see Chengxi Tang, *The Geographic Imagination of Modernity: Geography, Literature, and Philosophy in German Romanticism* (Palo Alto, CA: Stanford University Press, 2008), 146–50.

52. Novalis, *HKA* 1.85.

53. Ibid., 3.293, *AB* 295.

54. Ibid., 3.293.

55. This passage also makes clear how science and poetry share a common basis in the new mythology. Cf. ibid., 2.461. This conception of the poetic emergence of nature itself and its quasi-mystical uses may have been influenced by Karl von Eckartshausen's *Probaseologie*, which contains a section on "organology" as a study of the parts of God's nature.

56. Ibid., 1.83; cf. ibid., 2.448–49.

57. Ibid., 4.262–63.

58. See the excellent treatment of encyclopedia by Chad Wellmon, "Touching Books: Diderot, Novalis and the Encyclopedia of the Future," *Representations* 114, no. 1 (2011): 65–102.

59. See David Wood, "A Scientific Bible: Novalis and the Encyclopedistics of Nature," in *The Book of Nature in Early Modern and Modern History*, ed. Klaas van Berkel and Arjo Vanderjagt (Leuven: Peeters, 2006), 167–79. The projects of encyclopedia and Bible run parallel: the Bible is the *organon* of religious life, and the "New Bible" will become the higher *organon* in just the

sense Kant had excluded. The obvious reference is to Kant's claim that the *Kritik* is a universal method for philosophy, not that philosophy itself. (For the prehistory of the encyclopedic relation to the Bible in the Enlightenment, see Jonathan Sheehan, "From Philology to Fossils: The Biblical Encyclopedia in Early Modern Europe," *Journal of the History of Ideas* 64, no. 1 [2003]: 41–60; and Jonathan Sheehan, *The Enlightenment Bible: Translation, Scholarship, Culture* [Princeton: Princeton University Press, 2005]).

60. On Novalis's notion and practice of construction, see Ulrich Gaier, *Krumme Regel; Novalis' Konstruktionslehre des Schaffenden Geistes und Ihre Tradition* (Tübingen: M. Niemeyer, 1970).

61. Novalis, *HKA* 3.432; *AB* 841.

62. Novalis, *HKA* 4.264.

63. Schleiermacher himself seems to oppose any Bible or organon for religion (Schleiermacher, *Reden*, 68).

64. Kant separates between "illegal" and "legal" forms of these conflicts. Those which are illegal are based on the higher faculties attempting to usurp the truth-tribunal of the philosophical faculty, either through pretension or through governmental manipulation, while those which are legal are arguments about truth's social and technical application and its convenience to historical and governmental circumstance.

65. The religion of reason was first fully articulated in *Religion within the Bounds of Mere Reason*, and it was the censoring of that book which prevented the *Conflict*'s publication until 1798.

66. Kant, AA II, 37; cf. 42, 44, 45 48, 51, and 53.

67. Novalis, *HKA* 3.420–21.

68. Ibid., 3.420.

69. Ibid., *HKA* 2.388.

70. Kant, *CPR* B xxxvi–xxxvii; emphasis in original.

71. Ibid., B xxxv; emphasis in original.

72. Novalis, *HKA* 2.388.

73. See *AB* 69, 90, 92, 176, 282, and 552. Together these passages emphasize that the instruments of the various sciences are made instrumental first by the generality of the encyclopedic effort, which also makes the various sciences themselves organs.

74. Novalis, *HKA* 3.332; *AB* 453.

75. Ulrich Stadler claims that the tool makes both its user and its use also take on the character of mediation, and that this mediation can, for Novalis, raise itself to a "higher" immediate mediation. See Ulrich Stadler, *Die theuren Dinge* (Bern and Munich: Francke, 1980), 150–83. See also Novalis, *HKA* 3.391; *AB* 656–58.

76. Novalis, *HKA* 3.385; *AB* 642.

77. Ibid., 3.466; *AB* 1075.

78. See Novalis, *HKA* 3.266–67 and 297; *AB* 137 and 322. Novalis, *HKA* 3.301; *AB* 338 makes clear that magical idealism is tied to the outside world coming under rational control as well, and this under the heading "METAPHYSICS." For the latter to make sense, organology must be the base discipline. I do not agree with Manfred Frank that this idealism is "regulative." See "Die Philosophie des sogenannten magischen Idealismus," in Manfred Frank, *Auswege aus dem deutschen Idealismus* (Frankfurt a.M.: Suhrkamp, 2007), 27–67. The point is that there is a philosophical basis for this type of second-order possibility of real determination—so far from tarrying in the nonground of subjectivity, Novalis is pushing for a reproducible speculative encounter with the historical real.

79. As Daiber has pointed out, Novalis differs from his friend Ritter on the score of experimentation, because Novalis does not think that mere experiment will answer transcendental questions (*Jürgen Daiber, Experimentalphysik des Geistes: Novalis und das Romantische Experiment* [Göttingen: Vandenhoeck and Ruprecht, 2001], 99–115). Here we can see why: organs are not merely physical locations, nor are they transcendent sources of truth. They are always split between their physical and spiritual determinations— they are always the forms of possible judgment. Physiology and medicine are treated always from this perspective, where organ is both the concrete and physical location of a sense, and the potentially ideal historical condition of magical reality (possibility). See Novalis, *HKA* 3.307–8 (*AB* 370–72); 314–15 (*AB* 399); 317–19 (*AB* 409); 322–23 (*AB* 437); 327–29 (*AB* 446); 331–32 (*AB* 452); 369–71 (*AB* 593).

80. Novalis, *HKA* 2.550.

81. Ibid., 3.665; cf. Leif Weatherby, "The Romantic Circumstance: Novalis between Kittler and Luhmann," SubStance 43 (2014): 46–66.

82. Novalis, *HKA* 2.577.

83. Historical development, which precipitates and sublimates *organs* from the primordially organological cosmos, introduces intertia into their "lower" forms. So Novalis, *HKA* 2.450–51; *Blüthenstaub* 80; similarly at Novalis, *HKA* 3.281; *AB* 235.

84. Similarly at Novalis, *HKA* 3.401; *AB* 694; Novalis, *HKA* 3.316; *AB* 402.

85. So Novalis, *HKA* 3.333; *AB* 457. Critique is thus, through the filter of the forms of judgment in development (organs, or the possibility of experience as the possibility of possibility), the substantial link between Bible, encyclopedia, and the individual disciplines.

86. Novalis, *HKA* 2.587; cf. Jan-Peter Pudelek, *Der Begriff der Technikästhetik und ihr Ursprung in der Poetik des 18. Jahrhunderts* (Würzburg: Königshausen und Neumann, 2000), 203.

87. Novalis, *HKA* 2.553.

88. Novalis mentions the categories in *AB* 820, where he argues that *Kriticism* has provided a new way of looking at the link between subject and object, as one of active appropriation and self-alienation. The upshot is that "we glimpse ourselves in the system *as part*." Humans become organs, partial points of contact between substances in the universe, and often their own instruments (cf. Novalis, *HKA* 3.297; *AB* 321; *HKA* 3.292; *AB* 291; and *HKA* 3.410–11; *AB* 737). For Novalis's anthropology, see Chad Wellmon, "Lyrical Feeling: Novalis' Anthropology of the Senses," *Studies in Romanticism* 47, no. 4 (2008): 453–77; and Florian Roder, *Menschwerdung des Menschen: Der magische Idealismus im Werke des Novalis* (Stuttgart: Mayer, 1997).

89. Novalis, *HKA* 3.335; *AB* 460.

90. Novalis, *HKA* 3.335; *AB* 460.

91. Novalis, *HKA* 3.359, *AB* 540; cf. Novalis, *HKA* 3.302, *AB* 343; Novalis, *HKA* 2.442–43, which uses Schleiermacher's sense of a "mediator" to claim that anything can be the organ-mediator of religiosity.

92. See Helmut Schanze, *Romantik und Aufklärung: Untersuchungen zu Friedrich Schlegel und Novalis* (Nuremberg: Hans Carl, 1976), 114–51; see also John Neubauer, *Symbolismus und symbolische Logik: Die Idee der* ars combinatoria *in der Entwicklung der modernen Dichtung* (Munich: Fink, 1978); and Michael Gamper, *Elektropoetologie: Fiktionen der Elektrizität 1740–1870* (Göttingen: Wallstein, 2009), 103–52.

93. Novalis, *HKA* 3.257.

94. See Friedrich Kittler, "Die Nacht der Substanz," in *Kursbuch Medienkultur Die maßgeblichen Theorien von Brecht bis Baudrillard*, ed. Claus Pias et al. (Stuttgart: Deutsche Verlags-Anstalt, 1999), 507–25; cf. Friedrich Kittler, "Heinrich von Ofterdingen als Nachrichtenfluß," in *Novalis: Wege der Forschung*, ed. Gerhard Schulz, 480–508 (Darmstadt: Wissenschaftliche Buchgesellschaft, 1986), 495; and, comprehensively, Friedrich Kittler, *Discourse Networks 1800/1900*, trans. Michael Metteer and Chris Cullens (Stanford: Standford University Press, 1999).

95. Novalis, *HKA* 3.272.

96. Gotthold Ephraim Lessing, *Werke*, vol. 8 (Munich: Hanser, 1970), 557–60.

97. Fritz Mauthner, *Wörterbuch der Philosophie*, vol. 2 (Leipzig: Meiner, 1923), 81–89. Mauthner simply assumes the identity of sense in Lessing and organ in Hemsterhuis.

98. Hans-Joachim Mähl argues that Hemsterhuis presents the "passive" side of mythological perception for Novalis, while Fichte represents the "active." See Hans-Joachim Mähl, *Die Idee des goldenen Zeitalters im Werke des Novalis: Studien zur Wesensbestimmung der frühromantischen Utopie und zu ihren ideenge-*

schichtlichen Voraussetzungen (1965; Tübingen: Niemeyer, 1994), 266–87. Stadler, *Die theuren Dinge*, has revised this thesis by calling attention to the ambivalence between passivity and activity in Hemsterhuis's own work. See also Dalia Nassar, *The Romantic Absolute: Being and Knowing in Early German Romantic Philosophy* (Chicago: University of Chicago Press, 2013), 39–48.

99. Stadler, *Die theuren Dinge*, 157–64.

100. Mähl, *Die Idee*, 278ff. Novalis's notes on this writing make clear that he is not simply adopting passivity, but is thinking parallel to Hemsterhuis. See Novalis, *HKA* 2.362–63.

101. For a general study of semiotics in the (German) Enlightenment, see David Welbery, *Lessing's Laocoon: Semiotics and Aesthetics in the Age of Reason* (Cambridge: Cambridge University Press, 1984).

102. This summary is taken from Franz Hemsterhuis, *Lettre sur l'homme et ses rapports* (Paris, 1772), 9ff., and esp. 10.

103. This ability has a theoretical history which extends backward to antiquity, and was particularly intensely treated by medieval scholasticism under the heading *synderesis*. See Jan Verplaetse, *Localizing the Moral Sense: Neuroscience and the Search for the Cerebral Seat of Morality, 1800–1930* (New York: Springer, 2009), 1–30, for the pre-nineteenth-century theories.

104. Frans Hemsterhuis, *Oeuvres philosophiques*, ed. L.S.P. Meyboom, print of the Leeuwarden edition 1846–50 (New York: Hildesheim, 1972), 2:138. Cited at Stadler, *Die theuren Dinge*, 164.

105. Other than his extensive excerpting of various works by Hemsterhuis, Novalis mentions Hemsterhuis's *Theory of the Moral Sense* in Novalis, *HKA* 3.420–21; *AB* 782, where we started our investigation of Novalis's differences with Kant's *Conflict*. From that standpoint it becomes clear that what the moral organ makes possible is the production of moral *true dogmata*, experiential sentences that describe the moral world or allow the experience of the world as moral.

106. Novalis, *HKA* 2.368.

107. Ibid., 2.364.

108. Ibid., 2.370.

109. Ibid., 2.372–73; emphasis in original.

110. Ibid., 1.331.

111. Ibid.

112. Ibid.

113. Ibid.,1.331–32.

114. Ibid., 1.332.

115. Ibid.

116. The scholarship on Novalis and Fichte is broad and varied. The old view that the Romantics were Fichteans, which was laid aside during the

twentieth century's editorial production and deepening of source-investigation, has returned as the picture of Fichte as as subjective and totalizing idealist has been called into question. See Bernard Loheide, *Fichte und Novalis: Transcendentalphilosophisches Denken im romantisierenden Diskurs* (Amsterdam: Rodopi, 2000); see also the excellent review article on this literature by Dalia Nassar, "Interpreting Novalis' *Fichte-Studien*," *Deutsche Vierteljahrsschrift für Literaturwissenschaft und Geistesgeschichte* 84, no. 3 (2010): 315–41. My view diverges from that of Manfred Frank, who insists on a "pre-reflexive" moment of self in Novalis. See Manfred Frank, *"Unendliche Annäherung": Die Anfänge der Philosophischen Frühromantik* (Frankfurt a.M.: Suhrkamp, 1997), but also and especially *Selbstgefühl* (Frankfurt a.M.: Suhrkamp, 2002); cf. Frederick Beiser, *German Idealism:* The Struggle against Subjectivism, 1781–1801 (Cambridge: Harvard University Press, 2002), 407–35.

117. See Novalis, *HKA* 3.445; *AB* 921.

118. Novalis, *HKA* 3.335; *AB* 463.

119. The *Foundation of Natural Right* is itself an excellent example of the reactionary/progressive ambivalence in radical metaphysics of the period. For some of the more reactionary moments, see Adrian Daub, *Uncivil Unions: The Metaphysics of Marriage in German Idealism and Romanticism* (Chicago: University of Chicago Press, 2012), 36–71, and further in that book for the Romantic appropriation.

120. Johann Gottlieb Fichte, Johann Gottlieb Fichte's sämmtliche Werke (Berlin: Veit und Comp., 1845–46), 3:39.

121. J. G. Fichte, *Foundations of Natural Right: According to the Principles of the Wissenschaftslehre*, trans. Michael Baur (Cambridge: Cambridge University Press, 2000), 60–61; ibid., 64.

122. Ibid., 70ff., esp. 72.

123. Ibid., 76ff. At page 78 Fichte divides between the product of nature (which is organ) and the *artificial product* which points quasi-organically to an external *telos*. This external goal demotes the artificial product from *organ* to *tool*. The difference between animals and humans is then cast in partially traditional anthropological terms (humans lack specific organs) as that between *formation* (*Bildung*) and infinite *formativeness* (*Bildsamkeit*) (79–80). By attributing this *Bildsamkeit* as intuitive-organic form and content of (other) self to the other individual, the invidivual first realizes her humanity as common humanity. Fichte goes on to discuss important organs which have developed for humans, including the hand and the voice.

124. Ibid., 85.

125. Violetta Waibel has called attention to Novalis's interest in the *organ* in the *Foundation of Natural Right*, connecting it to a passage on the "inner" and "outer" organs in the *Fichte-Studien*. See Waibel, "'Inneres, äußeres

Organ': Das Problem der Gemeinschaft von Seele und Körper in den *Fichte-Studien* Friedrich von Hardenbergs," *Athenäum: Jahrbuch für Romantik* 10 (2000): 159–81.

126. Novalis, *HKA* 1.97.

127. Ibid.

128. See ibid., 2.460–61.

129. Cf. ibid., 1.101–2.

130. So ibid., 2.412/13; ibid., 2.436/37.

131. That is the message of his *Glauben und Liebe*; cf. Terry Pinkard, *German Philosophy, 1760–1840: The Legacy of Idealism* (Cambridge: Cambridge University Press, 2002), 169.

132. Schelling was most caustic in response, composing a satirical poem. It seemed to him that Novalis had gone back on the elements of Enlightenment that were included in the Romantic program. See Richards, *Romantic Conception* (Chicago: University of Chicago Press, 2002), 103ff.; and Pinkard, *German Philosophy*, 165ff.

133. Novalis, *HKA* 3.513; ibid., 3.520.

134. Ibid., 3.522.

135. The first study to compare Novalis and Hegel was that of Theodor Haering, *Novalis als Philosoph* (Stuttgart: Kohlhammer, 1954). The suggestive presence of Hegel in Novalis scholarship persists, up through Stadler, *Die theuren Dinge*, and on to the present. However we take this tendency, Hegel's negative comments on Novalis in the *Vorlesungen über die Ästhetik* do capture part of Novalis's program. Hegel himself accorded Novalis the honor of describing his work as "annihilation" made into an art form, which Novalis more or less confirms at Novalis, *HKA* 3.517.

136. Novalis, *HKA* 3.523.

137. Ibid., 3.521.

138. Cf. Stadler, *Die theuren Dinge*, 183, which confirms this capacity, but then strangely casts its goal as "immediacy" at 184.

7. BETWEEN MYTH AND SCIENCE: NATURPHILOSOPHIE AND THE ENDS OF ORGANOLOGY

1. *Epigraph*: Ernst Cassirer, *Zur Metaphysik der symbolischen Formen, Nachgelassene Manuskripte und Texte*, vol. 1 (Hamburg: Meiner, 1995), 24, cited in Nitzan Lebovic, *The Philosophy of Life and Death: Ludwig Klages and the Rise of a Nazi Biopolitics* (New York: Palgrave Macmillan, 2013), 77.

2. See, for example, Thomas Kuhn, "Conservation of Energy as an Example of Simultaneous Discovery," in Thomas Kuhn, *The Essential Tension: Selected Studies in Scientific Tradition and Change* (Chicago: University of Chicago Press, 1977), 66–104.

3. Ernst Cassirer, *The Problem of Knowledge: Philosophy, Science and History since Hegel* (1950; New Haven: Yale University Press, 1978), 17–19.
4. Schelling, *HKA* 9.328.
5. Ibid., 9.329.
6. Cf. Manfred Frank, *Der kommende Gott: Vorlesungen über die neue Mythologie* (Frankfurt a.M.: Suhrkamp, 1982), 153–88.
7. Lorenz Oken, *Über das Universum als Fortsetzung des Sinnensystems: Ein pythagoräisches Fragment* (Jena: Frommann, 1808), 10.
8. See Timoth Lenoir, "Operationalizing Kant: Manifolds, Models and Mathematics in Helmholtz's Theories of Perception," in *The Kantian Legacy in Nineteenth-Century Science*, ed. Michael Friedman and Alfred Nordmann (Cambridge: MIT Press, 2006), 141–211.
9. Oken, *Über das Universum*, 11.
10. Karl Friedrich Burdach, *Blicke ins Leben*, vol. 1, *Comparative Psychologie, erster Theil* (Leipzig: Voß, Leipzig, 1842), 19.
11. "It was only from around 1830 that the individual 'organism' became a recurrent technical term in various research fields" (Tobias Cheung, "From the Organism of a Body to the Body of an Organism: Occurrence and Meaning of the Word "Organism" from the Seventeenth to the Nineteenth Centuries," *British Journal for the History of Science* 39, no. 3 [2006]: 335).
12. Carus, *Von den Naturreichen, ihrem Leben und ihrer Verwandtschaft: Eine physiologische Abhandlung* (Dresden: Gärtner, 1818), 5; cited in Kristian Köchy, *Ganzheit und Wissenschaft: Das historische Fallbeispiel der romantischen Naturforschung* (Würzburg: Königshausen und Neumann, 1997), 147.
13. See Carl Gustav Carus, *Zwölf Briefe über das Erdleben* (Stuttgart: Balz'sche Buchhandlung, 1841), 17; Köchy, *Ganzheit und Wissenschaft*, 147.
14. Gotthilf Heinrich Schubert, *Spiegel der Natur: Ein Lesebuch zur Belehrung und Unterhaltung* (Erlangen: Palm and Enke, 1854), 442; cited in Köchy, *Ganzheit und Wissenschaft*, 146.
15. Gotthilf Heinrich Schubert, *Ansichten von der Nachtseite der Naturwissenschaft* (Dresden: Arnold, 1808), 4.
16. Ibid., 4–5.
17. This is cast in terms of God's kingdom (see ibid., 8).
18. Görres's early political dedications gave way to a scientific bent in the years 1800–1806, but "the same emancipatory impulse that informed all of Görres' politics in the 1790s now pervaded his aesthetics" (Jon Vanden Heuvel, *A German Life in the Age of Revolution* [Washington, DC: Catholic University of America Press, 2001], 106).
19. On Romantic physiology, see Karl Eduard Rothschuh, *Geschichte der Physiologie* (Berlin: Springer, 1953), 97ff.

20. See Joseph von Görres, *Exposition der Physiologie: Organologie* [Koblenz: Lassaulz, 1805], 160–61). The introduction to the work has an extended polemic against Gall's doctrines, putting a point on the difference between Gall's organology and Romantic organology in method. See chapter 3 of this volume. Görres reasons that life lies centered in the brain, which makes for itself a grasping organ and to which plastic material then attaches itself as more organs. See Görres, *Exposition*, 144–45. See Michael Hagner, *Homo cerebralis: Der Wandel vom Seelenorgan zum Gehirn* (Frankfurt a.M.: Suhrkamp, 2008), 170–80.

21. Goethe to Eichstädt, April 21, 1804, cited in Vanden Heuvel, *German Life*, 117.

22. See Caroline Welsh, *Hirnhöhlenpoetiken: Theorien zur Warhnehmung in Wissenschaft, Ästhetik und Literatur um 1800* (Freiburg: Rombach, 2003), 253–93.

23. Görres even follows Schelling in the "triple" structure of the organ, see *Exposition*, 150ff.

24. Ibid., 4–5.

25. For example, as applied to sexual difference: ibid., 163–64.

26. The "organology of the human body" is based on astral influences. See ibid., 5.

27. Joseph Görres: *Die teutschen Volksbücher*, in Joseph Görres, *Gesammelte Schriften*, vol. 3, *Geistesgeschichtliche und literarische Schriften I (1803–1808)*, 179; cf. Görres, *Exposition*, 143–44.

28. Görres, *Die teutschen Volksbücher*, 179–80.

29. Friedrich Schlegel, *Kritische Friedrich-Schlegel-Ausgabe, Erste Abteilung: Kritische Neuausgabe*, vol. 2 (Paderborn: Schöningh, 1967), p. 315. Subsequent citations are in the form *KFSA* 2.315 and refer to volume and page number.

30. See, e.g., ibid., 2.319.

31. Wit is described as an "organ of cognition" in *Athenäum* 220, on which see Gerhard Neumann, *Ideenparadiese: Untersuchungen zur Aphoristik von Lichtenberg, Novalis, Friedrich Schlegel und Goethe* (Munich: Fink, 1976), 452.

32. Ibid., 1.346–47; emphasis in original.

33. Ibid., 1.256.

34. Ibid., 1.272.

35. Ibid.,1.346–47; emphasis in original.

36. Ibid., 2.314.

37. Ibid., 2.319.

38. Ibid., 2.262.

39. "Intellectual intuition is the categorical imperative in theory" (ibid., 2.176). The formulation suggests just the idealism demanded in the *Speech*: the categorical imperative will be the transcendental arbiter of not just content but also form, now in cognition and not just morals.

40. See ibid., 2.267.

41. See Frank, *Der kommende Gott*, 176ff.

42. Schlegel's suggestion is that the constitution corresponds to reason, legislative power to the understanding, judicial power to the power of judgment, and executive to sensibility (see Schlegel, *KFSA* 7.15). This suggestion remains unelaborated, but the basic Romantic suggestion (that history should be included as part of theoretical philosophy) is given in this thick but more basic objection (see ibid., 7.23; cf. ibid., 7.24).

43. Ibid., 7.17.

44. Schlegel, *KFSA* 7.25; emphasis in original.

45. See Hagner, *Homo cerebralis*, 89 for a differentiation of Gall's doctrine from the popular image of phrenology. The term "phrenology" certainly occurs earlier than 1815, however, for example in the *Phenomenology of Spirit*.

46. Franz Joseph Gall, "Schreiben über seinen bereits geendigten Prodromus über die Verrichtungen des Gehirns der Menschen und der Thiere, an Herrn Jos. Fr. Von Retzer," *Neuer teutscher Merkur* 3 (1798): 311–32.

47. Hagner uses this term because Gall conceives of the brain (for the first time) as a complex series of organs, thus anticipating both localization-theory and the "cerebralization" of the human. The brain becomes an organ in its own right, instead of a kind of medium between soul and organ (the organ of the soul). See Hagner, *Homo cerebralis*, 93–94.

48. Gall, "Verrichtungen," 320–21.

49. Ibid., 322–23.

50. See Hagner, *Homo cerebralis*, 99ff.

51. See ibid., 122.

52. F. W. J. Schelling, *Sämmtliche Werke*, ed. Karl-Friedrich-August Schelling (Stuttgart: Cotta, 1856–61, 1.7.543; emphasis in original.

53. G. F. W. Hegel, *Werke in 20 Bänden*, vol. 3, *Die Phänomenologie des Geistes* (Frankfurt a.M.: Suhrkamp, 1986), 251; hereafter cited as Hegel, *PhdG*, followed by page number.

54. See ibid., 252.

55. Cf. ibid., 251, 254.

56. I use Terry Pinkard's translation-in-progress, http://terrypinkard.weebly.com/phenomenology-of-spirit-page.html; Hegel, *PhdG*, 260; emphasis in original.

57. Hegel, *PhdG*, 261.

58. Ibid., 234.

59. Ibid., 246. This seems to me to be the only literal use of "organ" in the *Phenomenology*.

60. Ibid., 245.

61. See ibid., 259.

62. Here Hegel finally brushes the organ aside (ibid., 247).
63. Ibid., 263.
64. Ibid., 258.
65. See ibid., 259.
66. Ibid., 262; translation modified. Cf. Hegel, Werke 14, 369–70.
67. F. W. J. Schelling, *Philosophical Investigations into the Essence of Human Freedom*, trans. Jeff Love and Johannes Schmidt (Albany: SUNY Press, 2006), 22; Friedrich Wilhelm Joseph Schelling, *Philosophische Untersuchungen über das Wesen der Menschlichen Freiheit und die damit zusammenhängenden Gegenstände*, ed. Thomas Buchheim (Hamburg: Meiner 1997), 24; hereafter cited as *FS*, followed by page number.
68. Andrew Bowie, *Schelling and Modern European Philosophy: An Introduction* (New York: Routledge 1993), 94.
69. Christine Korsgaard, *Creating the Kingdom of Ends* (Cambridge: Cambridge University Press, 1996).
70. See Paul Guyer, *Kant* (London: Routledge, 2006), 307–60.
71. Schelling found an ally in Böhme as he perused the tradition in the spring of 1809. See his readings, which included Augustine, Boethius, and Luther, documented in F. W. J. Schelling, *Philosophische Entwürfe und Tagebücher: 1809–1813, Philosophie der Freiheit und der Weltalter*, ed. Hans Jörg Sandkühler et al. (Hamburg: Meiner, 1994). Documentation of Schelling's reading of "JB" (Böhme) is at 11. See Robert Brown, *The Later Philosophy of Schelling: The Influence of Boehme in the Works of 1809–1815* (Lewisburg: Bucknell University Press, 1977), esp. 114–51; and Paola Mayer, *Jena Romanticism and Its Appropriation of Jakob Böhme: Theosophy, Hagiography, Literature* (Montreal: McGill-Queen's University Press, 1999), 179–222.
72. See Christopher Lauer, *The Suspension of Reason in Hegel and Schelling* (New York: Continuum 2010); on the figure "ground-consequent," see Thomas Buchheim, "Das Prinzip des Grundes und Schellings Weg zur Freiheitsschrift," in *Schellings Weg zur Freiheitsschrift: Legende und Wirklichkeit*, ed. Hans Michael Baumgartner and Wilhelm G. Jacobs (Stuttgart–Bad Canstatt: frommann-holboog, 1996), 223–40.
73. Schelling, *Philosophical Investigations*, 16; *FS*, 17.
74. In the second of the dialogues in *God*, Herder separates between force and ur-force, thus denying the infinity of God identity with that of the attributes. To be sure, the difference between these infinities is already in Spinoza (substance is clearly infinite in a different, more comprehensive sense than any single attribute), but Herder's conclusion to a minimally transcendental God *in which* everything is (panentheism) would have been unacceptable to Spinoza.
75. See Manfred Durner, "Die Naturphilosophie im 18. Jahrhundert und der naturwissenschaftliche Unterricht in Tübingen: Zu den Quellen von

Schellings Naturphilosophie," *Archiv für Geschichte der Philosophie* 73 (1991): 71–103; Herder is treated at 83–84.

76. Ibid., 84.

77. Schelling had written in similar terms as the identity system gave way to his theological interests. So, for example: "When the absolute knows itself it does so through us, so that we are the organs of its self-knowledge." Here as cited in Beiser, *German Idealism*, 594. For commentary on Schelling's development between 1801 and 1809, most helpful is the volume edited by Christian Danz and Jörg Jantzen, *Gott, Natur, Kunst und Geschichte: Schelling zwischen Identitätsphilosophie und Freiheitsschrift* (Göttingen: Vienna University Press, 2011).

78. Cf. Wilhelm Jacobs, "Vom Ursprung des Bösen zum Wesen der menschlichen Freiheit oder Transzendentalphilosophie und Metaphysik" in *Schellings Weg zur Freiheitsschrift: Legende und Wirklichkeit*, ed. Baumgartner and Jacobs (Stuttgart–Bad Canstatt: frommann-holzboog, 1996), 11–28.

79. Schelling, *FS*, 85.

80. Atemporal, primarily, because a "thing-in-itself" on the Kantian model. I do not treat here the view of ethics which Schelling partially adopts and partially critiques from Kant's *Religion within the Limits of Mere Reason*. The best treatment of this complex issue is Richard J. Bernstein, *Radical Evil: A Philosophical Interrogation* (Malden: Blackwell, 2002), 9–46, 76–98. See also Emile Fackenheim's "Kant and Radical Evil," *University of Toronto Quarterly* 23 (1954): 339–53.

81. Schelling, *Philosophical Investigations*, 75; *FS*, 85.

82. See Schelling, *Philosophical Investigations*, 46.

83. Ibid., 74; *FS*, 84.

84. Schelling, *Philosophical Investigations*, 54; *FS*, 61.

85. Otto Friedrich Bollnow, *Zwischen Philosophie und Pädagogik: Vorträge und Aufsätze* (Aachen: Weitz, 1988), 163.

86. Here I am revaluing Eckart Förster's recent negative judgment of this characteristic of Schelling's philosophy. Förster writes: "if intellectual intuition is to be retained as the method of our intuition of nature, that is only possible on the basis of a depotentiation (a suppression or neutralization) of the intuiting subject. The question however remains whether an intellectual intuition in which one abstracts from the intuiting subject can really amount to more than word-play. . . . If we are now to abstract from the producing subject, then there would have to be a unity of being and thought which could exist without appearing as the product of a subject" (Eckart Förster, *The Twenty-Five Years of Philosophy* [Cambridge: Harvard University Press, 2012], 248–49). Intellectual intuition might precipitate in all kinds of products tracing their lineage only indirectly back to a "subject"—Schelling's

method is meant to cope with the breakdown of the line between natural and artificial hidden inside the contradictory synthesis of being and thought.

8. TECHNOLOGIES OF NATURE: GOETHE'S HEGELIAN TRANSFORMATIONS

1. Johann Peter Eckermann, *Goethes Gespräche in den letzten Jahren seines Lebens*, Montag den 2. [?] August 1830, WA (Anhang), 7.320–23, see note 9 below for citational style. Also cited in Toby A. Appel, *The Cuvier-Geoffroy Debate: French Biology in the Decades before Darwin* (Oxford: Oxford University Press, 1987), 1. Goethe's involvement in the debate—this episode included— has been treated comprehensively by Dorothea Kuhn, *Empirische und ideelle Wirklichkeit: Studien über Goethes Kritik des französischen Akademiestreites* (Graz: Böhlau, 1967); for this conversation, see 56–58. On Goethe's ghostly presence in and beyond the texts of his famous conversations in the last decade of his life (in which the episode with Soret is recorded), see Avital Ronell, *Dictations: On Haunted Writing* (Urbana: University of Illinois Press, 2006).

2. Georges Cuvier, "Elegy of Lamarck," *Edinburgh New Philosophical Journal* 20 (January 1836): 14.

3. See Theo Stammen, *Goethe und die politische Welt* (Würzburg: Ergon, 1999); and Ekkehart Krippendorf, *Goethe: Politik gegen den Zeitgeist* (Frankfurt a.M.: Insel, 1999). Both political scientists relate Goethe's politics to his natural-scientific observational methods, as does Astrida Orle Tantillo, *The Will to Create: Goethe's Philosophy of Nature* (Pittsburgh: University of Pittsburgh Press, 2002).

4. See Appel, *Debate*, 11ff. for the rise of the French Academy in zoology; cf. ibid., 106. It is important to note that, while Goethe saw an ally in Geoffroy, he actually removed himself from the fray to point to the importance of philosophy in the natural sciences more generally.

5. Goethe's insistence on the pride of place of his scientific works among his efforts, and the long-standing dismissal of this claim in favor of his poetic works, is well-known. See, for example, Tantillo, *Will*, 1.

6. The claim I make in this chapter is based on the subterranean semantics of the term "organ," and stands in a diagonal relationship to the rich literature on Goethe and form that has developed in recent years. In the tradition of Dieter Henrich's "constellation work," both David Wellbery ("Form und Idee: Skizze eines Begriffsfeldes um 1800," in *Morphologie und Moderne: Goethes "anschauliches Denken" in den Geistes- und Kulturwissenschaften seit 1800*, ed. Jonas Maatsch [Berlin: de Gruyter, 2014],17–43) and Eckart Förster have claimed that Goethe belongs among the Idealists (on which more below)—a claim I agree with and which Goethe's use of the term "organ" tends to substantiate. Wellbery argues that Goethe practices "endogenous form,"

neither oppositionally paired with a substance nor purely constructivist. Eva Geulen, on the other hand, has argued that in both scientific observation and writing, Goethe's morphology is "the site of a massive transformation of the notion of form" (Eva Geulen, "Serialization in Goethe's Morphology," *Compar(a)ison* 29 [2013]: 53–70, here 56), toward a functionalist form that reemerges in European Modernism. I say that my claims are "diagonal" to this debate for two reasons. First, in my attempt to show that the term "organ" appears in Goethe's works through Romantic discourse and its sources, I bracket the question of form. This is because—second—I think that the term "organ" tends to show a strain of Goethe's work that pushes morphology into what I have elsewhere called a kind of "technological freedom" with respect to these different models of form (see my "Das Innere der Natur und ihr Organ"). This line of thinking has benefited from inspiring and informative discussion with Elisabeth Strowick.

7. Recently David Wellbery has claimed a deep affinity between Goethe and Hegel as articulators of Idealism, a point on which I agree: "The Imagination of Freedom: Goethe and Hegel as Contemporaries," in *Goethe's Ghosts: Reading and the Persistence of Literature*, ed. Simon Richter and Richard Block (Rochester: Camden House, 2013), 217–39.

8. Goethe, *Faust*, trans. Walter Kaufman (1961; New York: Anchor, 1990), 145; translation modified. Johann Wolfgang von Goethe, *Goethes Werke: Weimarer Sophienausgabe* (Weimar: Böhlau, 1887), sec. 1, vol. 14 (1887), p. 57; hereafter cited as WA; subsequent citations are in the form 1.14.57 and refer to section, volume, and page number. I will cite the Weimar edition throughout, though I have also consulted and sometimes find it necessary to cite the Frankfurt and Munich editions (cited as FA and MA, respectively), and especially the *Leopoldina-Ausgabe* of Goethe's scientific works.

9. Tantillo, *Will*, 18–19.

10. *Pace* Isaiah Berlin, for example, *The Roots of Romanticism* (Princeton: Princeton University Press, 1999); see the important corrective in Robert E. Norton, "The Myth of the Counter-Enlightenment," *Journal of the History of Ideas* 68 (2007): 635–58.

11. G. F. W. Hegel, *Werke in 20 Bänden*, vols. 8–10, *Enzyklopädie der philosophischen Wissenschaften* (Frankfurt a.M.: Suhrkamp, 1979). I cite as *Enz. I*, §38. Also quoted at WA 1.14.91.

12. Hegel, *Enz. II*, §246, Zusatz.

13. See Sally Sedgwick, *Hegel's Critique of Kant: From Dichotomy to Identity* (Oxford: Oxford University Press, 2012).

14. On this internal split and its consequences, see John McDowell, *Mind and World with a New Introduction* (Cambridge: Harvard University Press, 1996). A similar thesis is defended by Slavoj Žižek in *Tarrying with the*

Negative: Kant, Hegel, and the Critique of Ideology (Durham: Duke University Press, 1993), 18–22.

15. Hegel, *Enz. I*, §52.
16. Ibid., *I*, §48.
17. Cf. Rolf-Peter Horstmann, "Hegel's Phenomenology of Spirit as an Argument for a Monistic Ontology," *Inquiry* 49 (2006): 103–18; and Frederick Beiser, *Hegel* (New York: Routledge, 2005), 51–80.
18. On the concept-dialectic, see Gunnar, "Die aufgeklärte Aufklärung," in *Kant and the Future of the European Enlightenment / Kant und die Zukunft der europäischen Aufklärung*, ed. Heiner F. Klemme (New York: de Gruyter, 2009), 43–68; and Rolf-Peter Horstmann, *G. W. F. Hegel: Eine Einführung (zusammen mit D. Emundts)* (Stuttgart: Reclam, 2002). A "non-metaphysical" but formally clarifying analysis is in Klaus Hartmann, "Hegel: A Non-Metaphysical View," in *Studies in Foundational Philosophy*, by Hartmann (Amsterdam: Rodopi, 1988).
19. Hegel, *Enz I*, §24.
20. Hegel and Goethe's interactions generally fall into three periods. They were first introduced in Jena and Weimar during Goethe's intensive friendship with Schelling, which started in the last years of the 1790s. During this time, they worked together on experiments in the botanical gardens. This has been documented by Eckart Förster, "Die Bedeutung von §§76, 77 der *Kritik der Urteilskraft* für die Entwicklung der nachkantischen Philosophie. Teil 1," in *Kant und der Frühidealismus*, ed. Jürgen Stolzenberg (Hamburg: Meiner, 2007), 59–80. The second period of interaction (for which, see Frederick Burwick, *The Damnation of Newton: Goethe's Color Theory and Romantic Perception* [New York: de Gruyter, 1986], 58–79) occurred during Hegel's time in Nürnberg in 1816–17, when he aided in experiments, this time in the later work on the theory of color, specifically on "entoptic" colors. This collaboration led Hegel to take sides with Goethe in the color controversy, and ushered in what I will call the third period of their interaction, from roughly 1821 until Hegel's death. This period is marked by letters and a few visits in Weimar, and is generally cordial and friendly but distant. The only analysis of this last period of which I am aware is Karl Löwith's short introductory study in his *From Hegel to Nietzsche: Revolution in Nineteenth-Century Thought*, trans. David Green (New York: Anchor, 1967), 2–29.
21. Hegel, *Enz. I*, §24, Zusatz 3.
22. Hegel, *Vorlesungen über die Ästhetik I (VüdÄ I)*, *Werke*, vols. *13–15* (Frankfurt a.M.: Suhrkamp, 1986), 173.
23. Hegel, *Enz. I*, §24, Zusatz 3.
24. See, e.g., *VüdÄ I*, 184, 192, and 203 (the eye as organ not only of sight but of the soul); page 195 makes the metaphor social: individuals are the

expressive organs of family and state, a line of thought that Hegel rarely exploits in the *Philosophy of Right* (with a significant exception, where classes are defined as organs, at 471).

25. Hegel, *VüdÄ I*, 174; my emphasis. Cf. Hegel, *VüdÄ I*, 174.
26. *Enz. II*, §244, Zusatz.
27. WA 4.33.295. Cited by Löwith, *From Hegel to Nietzsche*, 4.
28. Hegel, *Enz. II*, §246, Zusatz.
29. For Hegel's polemics against Newton, see *Enz. II*, §§ 270, 318, 320. Goethe's relationship to Newton is particularly sensitively handled by Tantillo, *Will to Create*. Note that in a relatively temperate moment in the *Doctrine of Colors*, Goethe affords Newton an "organ," albeit the presumably limited one of mathematics (WA 2.4.97; cf. MA 17.829, where mathematics [and dialectics] is construed as an "organ of the inner higher sense").
30. Categorical change or its lack is what holds back revolutions (Hegel, *Enz. II*, §246, Zusatz).
31. Ibid.
32. Ibid.
33. WA 1.3.105.
34. Cf. Richards, *The Romantic Conception of Life: Science and Philosophy in the Age of Goethe* (Chicago: University of Chicago Press, 2002), 430–31.
35. On Goethe's psychology and the sublimation which may affect his relation to Romanticism, see Eckart Goebel, *Beyond Discontent* (New York: Bloomsbury, 2012), 1–46.
36. Johannes Grave has warned against identification-games in criticism of the novel. See *Der "ideale Kunstkörper": Johann Wolfgang Goethe als Sammler von Druckgraphiken und Zeichnungen* (Göttingen: Vandenhoek und Ruprecht, 2006), 168; see also his "Ideal and History: Johann Wolfgang Goethe's Collection of Prints and Drawings," *Artibus et Historiae* 27, no. 53 (2006): 175–86.
37. WA 1.47.162. Earlier mentions of the periodical are at pp. 146 and 152.
38. On the tradition of this interpretation, see Simon Richter, *Laocoon's Body and the Aesthetics of Pain: Winckelmann, Lessing, Herder, Moritz, Goethe* (Detroit: Wayne State Unversity Press, 1992); and also his excellent introduction to *Literature of Weimar Classicism*, ed. Richter (Rochester, NY: Camden House, 2005), 3–45.
39. Christopher Young and Thomas Gloning, *A History of the German Language through Texts* (London: Routledge, 2004), 248–49, suggest that the term would still have appeared to the German reading public as a *Lehnwort*, not as German, in 1790.
40. An early use of organ further points to its derivation from the Herder-Moritz-Goethe connection. In his famous letter to Herder of 17 May 1787 on

his *aperçu* about the *ur-plant*, he says that he recognizes the "true Proteus" in "that organ of the plant that we usually speak of as the leaf" (WA 1.32.44).

41. See H. B. Nisbet, "Herder, Goethe, and the Natural 'Type,'" in Publications of the English Goethe Society, vol. 37, Papers Read before the Society 1966–67, ed. Elizabeth M. Wilkinson et al., 81–119 (Leeds: Maney and Son, 1967), 105. Richards, *Romantic Conception*, 416, points out that this overlap between form and development puts Goethe between preformationism and epigeneticism in biology, with the "transcendental leaf" open to epigenetic developments, but forming the encapsulation common in preformationist theories.

42. Peter Hanns Reill has related this methodology to Goethe's general procedure. See his "Bildung, Urtyp and Polarity: Goethe and Eighteenth-Century Physiology," *Goethe Yearbook* 3 (1986): 139–48.

43. Johann Wolfgang von Goethe, *The Metamophosis of Plants*, ed. Gordon Miller (Cambridge: MIT Press, 2009), 5–6; WA 2.6.25–26.

44. Goethe, *Metamorphosis*, 80; WA 2.6.71; cf. ibid., 2.6.91.

45. Transcribed in WA 1.32.314–15.

46. See Simon Richter, *Laocoon's Body and the Aesthetics of Pain: Winckelmann, Lessing, Herder, Moritz, Goethe* (Detroit: Wayne State University Press, 1992).

47. The spectacular breakdown is given in the "Tabelle," for which see WA 1.47.338.

48. Ibid., 1.47.19.

49. Ibid., 1.47.80.

50. So in the introduction to the *Propyläen* (ibid., 1.47.13).

51. Ibid., 1.47.171.

52. Ibid.

53. Ibid., 1.47.174.

54. Sounding more like Schelling than any other contemporary philosopher. See Wolf von Engelhardt, *Goethes Weltansichten: Auch eine Biographie* (Weimar: Hermann Böhlau Nachfolger, 2007), 217ff.; and Jeremy Adler, "The Aesthetics of Magnetism: Science, Philosophy, and Poetry in the Dialogue between Goethe and Schelling," in *The Third Culture: Literature and Science*, ed. Elinor S. Schaffer (New York: de Gruyter, 1998), 66–103.

55. WA 2.6.25–26.

56. Which he began to gather late in the second decade of the nineteenth century, and published starting in 1820. See Kuhn, *Empirische und ideelle Wirklichkeit*, 42ff.

57. E.g., Hegel, *Enz. I*, §24, Zusatz 3.

58. WA 2.11.349.

59. Ibid., 2.6.5.

60. But then, neither is anyone else in the younger generation. I do not see the point of insistences on Goethe's self-assessment as a "hartnäckiger Realist," given the overwhelming evidence pointing in a more subtle direction, as in the quotation here. See, for example, Robert Richards, "Nature Is the Poetry of Mind, or How Schelling Solved Goethe's Kantian Problems," in *The Kantian Legacy in Nineteenth-Century Science*, ed. Michael Friedman and Alfred Nordmann (Cambridge: MIT Press, 2006), 27–51.

61. See Kant, *CPR* A 104.

62. Which involves the "heightening of a spiritual faculty" (WA 2.11.128–29).

63. Ibid. 2.11.162.

64. *Locus classicus* for talk of "higher experience" is Goethe's controversial essay "The Experiment as Mediator between Subject and Object," composed in 1792 and then also included in the *Morphologische Hefte*. For a Kantian reading, see von Engelhardt, *Goethes Weltansichten*, 167–87.

65. *De anima* 423b/424a. I will cite the Bekker numbers for convenience. Translations are from *The Basic Works of Aristotle*, ed. Richard Mckeon (New York: Random House, 1941), 535–607. Transliterations are mine.

66. *De anima* 424a–b.

67. Ibid., 429a/19.

68. WA 2.11.59.

69. "In every case the mind which is actively thinking is the objects which it thinks" (*De anima* 431b).

70. WA 2.1.ix–x.

71. For Goethe's systematic position against Newton in this regard, see Hartmut R. Schönherr, *Einheit und Werden: Goethes Newton-Polemik als systematische Konsequenz seiner Naturkonzeption* (Würzburg: Königshausen und Neumann, 1993); and Burwick, *Damnation*.

72. WA 2.1.384.

73. Spinoza's notion of "attributes"—as infinite in their kind—which do not divide *natura* (which is absolutely infinite) might be the source or model of this type of division. See *Ethics* I, Defs. IV and VI, also Prop. X. See also Engelhardt, *Goethes Weltansichten*, 157–67; and David Bell, *Spinoza in Germany from 1670 to the Age of Goethe* (London: Institute of Germanic Studies, 1984).

74. WA 2.1.323–24; my emphasis.

75. Cf. Wellbery, "The Imagination of Freedom."

76. Cf. Förster, "Die Bedeutung," 59–80; and Förster, "Goethe and the 'Auge des Geistes,'" *Deutsche Vierteljahrsschrift für Literaturwissenschaft und Geistesgeschichte* 75, no. 1 (2001): 87–101, with an explicit comparison to Hegel's "idea" at 98, and the notion that "new sense-organs" will develop as a result of this doctrine (98).

77. Cf. MA 8.30–31, on which *Romantic Conception*, 491.
78. Cf. WA (Anhang), 3.290–91.
79. *Locus classicus*: Kant, *CPR* B176/A137–B187/A147.
80. The results of this engagement are treated thoroughly by Eckart Förster, "Die Bedeutung." Förster claims that Goethe doesn't defend "intellectual intuition" but rather "intuitive understanding." Although I think Förster is right that Goethe and his contemporaries were not pushing for the conceptual creation of objects in intellectual intuition, we will see below that Goethe nevertheless did mean to introduce a kind of conceptual intervention in the object-world through this line of thinking.
81. See Eckart Förster, *The Twenty-Five Years of Philosophy: A Systematic Reconstruction*, trans. Brady Bowman (Cambridge: Harvard University Press, 2012), 250–76, "The Method of Intuitive Understanding."
82. Kant, *CPR* B92/A67ff.
83. A number of readings of Goethe's speculative work emphasize this practical element. See, for example, Förster, "Die Bedeutung"; Gunnar Hindrichs, "Goethe's Notion of an Intuitive Power of Judgment," *Goethe Yearbook* 18 (2011): 51–67; and Chad Wellmon, "Goethe's Morphology of Knowledge, or the Overgrowth of Nomenclature," *Goethe Yearbook* 17 (2010): 153–77. These readings emphasize that Goethe's philosophical work is practice-oriented, indeed suggest that he introduces a focus on operation into idealism.
84. Cf. Kant, *CPR* B93/A68.
85. Cf. Paul Guyer, "The Arguments of the Critique," in *The Cambridge Companion to Kant's "Critique of Pure Reason,"* ed. Guyer (Cambridge: Cambridge University Press, 2010), 140.
86. WA 2.11.158.
87. See, on this issue, Hindrichs, "Goethe's Notion," 62.
88. For Brady Bowman, following Rolf-Peter Horstmann, this doubling over forms a requirement for "subjectivity" in Hegel, which is found only where a relation between two forms mirror each other, thus forming an identity which is the identity of each with itself *and* of the other simultaneously. See Brady Bowman, "Goethean Morphology, Hegelian Science: Affinities and Transformations," *Goethe Yearbook* 18 (2011): 166–67.
89. See ibid., 165, where Bowman accepts perhaps too much of the Hegelian charge against Goethe.
90. Hegel, *Enz. II*, §246.
91. Goethe had critiqued Hegel, more and less convincingly, since their first work together in Jena. But until this period there is no evidence that he did so with the intention of his remarks reaching Hegel.

92. The conversation demonstrates the consistency of the issues in the dialogue between Goethe and Hegel: the categorical, nature, and *history*. See WA (Anhang), 6.180–81.

93. The conversation's entirety may be found at WA (Anhang), 6.179–81.

94. See WA (Anhang) 6.180.

95. WA 4.44.88.

96. See Kuhn, *Empirische und ideelle Wirklichkeit*, 48. Goethe's account at WA 4.44.146.

97. August de Candolle, *Organographie végétale* (Paris: Deterville, 1827), vii–viii; my emphasis.

98. MA 18.2.534. The term *composition* was key in the larger debate, but Goethe is likely *not* disagreeing with the sense given in Geoffroy de St.-Hilaire's *unité de composition*, since this is meant to point to the philosophical underpinnings of zoology in analogy, a point with which Goethe agrees.

99. As Robert Richards has argued, Goethe was closer than previously realized to Darwin's evolutionary concept, precisely because he allowed for external circumstances to influence formation (see Richards, *Romantic Conception*, 486).

100. See Kuhn, *Empirische und ideelle Wirklichkeit*, 48–63.

101. Hegel, *Enz. II*, § 249.

102. Cf. Jonathan Crary, *Techniques of the Observer: On Vision and Modernity in the Nineteenth Century* (Cambridge: MIT Press, 1992), 97–100; and Hartmut Boehme, "The Metaphysics of Phenomena: Telescope and Microscope in the Works of Goethe, Leeuwenhoek and Hooke," in *Collection, Laboratory, Theater: Scenes of Knowledge in the 17th Century* (Berlin: de Gruyter, 2005), 355–94.

103. For Goethe's more general final efforts in biology, see Kuhn, *Empirische und ideelle Wirklichkeit*, 48–63.

104. See ibid., 155. On externalities and deformities in the evolutionary picture and "another" classicism," see Astrida Tantillo, "Goethe's "Classical" Science," in *The Literature of Weimar Classicism*, ed. Simon Richter (Rochester, NY: Camden House, 2005), 323–47.

105. Goethe to W. von Humboldt (FA 11, 549). This notion had been introduced into the academy-debate by Geoffroy himself, who asked in his own *Principes de philosophie zoologique* if human and animal organs could not be compared: is the human *hand* not the form from which the degradations of the paw and so forth derive? See Kuhn, *Empirische und ideelle Wirklichkeit*, 97–98.

106. WA 2.13.114.

107. MA 17.917; September 1830–dating of *Maximen* at MA 17.1245.

108. See A. L. Peck, introduction to Aristotle, *The Parts of Animals*, ed. Peck; *The Movement of Animals and the Progression of Animals*, ed. E. S. Forster (London: Heinemann, 1937), 9ff.

109. Karl Schlechta's *Goethe in seinem Verhältnisse zu Aristoteles: Ein Versuch* (Frankfurt a.M.: Klostermann, 1938) deals comprehensively with Goethe's readings in Aristotle. He mentions this episode at 81–82 as an example of Goethe's agreement with Aristotle on the "law of compensation" or the "balance of organs" in the organism. Schlechta's only notice of Goethe's having read the *De anima* comes at 28, with reference to the historical part of the *Doctrine of Colors*. My suggestion that there is a closer reading and philological relationship to this text below will also show that Goethe's "agreement" with Aristotle extends beyond fascination with the same "problems." For Goethe's engagement in 1827/8 with the Aristotelian "*problemata*," see 53–62.

110. WA 2.13.114. Undated according to the MA.

111. As well as Galen and Hippocrates (MA 18.2.529).

112. Galen, *On the Usefulness of the Parts of the Body*, trans. Margaret Tallmadge May (Ithaca: Cornell University Press, 1968), 71. The passage from the *De partibus* is at 4.10.687a20–21, *organon ti pro organon*. See May's commentary on the translation tradition at the same location. Claudii Galeni, *Opera Omnia*, vol. 3, ed. Carolus Gottlob Kühn (1822), 8–9. My transcription and transliteration are from the edition Goethe used.

113. *De anima* 432a. Aristotle also uses this phrase in the politics, where he argues (1253b) that slaves are "instruments for other instruments" in the sense that their use is productive rather than merely to be consumed, as in other kinds of property (beds, tables). Organology tends to elide the category of unproductive property altogether. Cf. Wendy Brown, *Undoing the Demos: Neoliberalism's Stealth Revolutions* (New York: Zone, 2015), 88ff.

114. WA 2.13.114; MA 18.2.529; April 1829. I am not aware of a previous mention of this philological connection to Galen in the literature.

115. This clarifies the simultaneous "triviality" and "obscurity" which Seibicke finds in the phrase. See Wilfried Seibicke, *Technik: Versuch einer Geschichte der Wortfamilie um techne in Deutschland vom 16. Jahrhundert bis etwa 1830* (Düsseldorf: VDI, 1968), 242.

116. Cf. Kuhn, *Empirische und ideelle Wirklichkeit*, 155, rejecting Wilhelm Dilthey, *Weltanschauung und Analyse des Menschen seit Renaissance und Reformation* (*Gesammelte Schriften II*), ed. Georg Misch (Göttingen: Vandenhoek und Ruprecht, 1991), 391–416.

117. WA 1.23.131.

118. Cf. Nicholas Boyle, *Goethe: The Poet and His Age*, vol. 2, *Revolution and Renunciation* (Oxford: Oxford University Press, 2000), 416–17.

119. See Rüdiger Campe, "Kafkas Institutionenroman. Der Proceß, Das Schloß," in *Gesetz: Ironie*, ed. Campe and Michael Niehaus (Krottenmühl: Synchron, 2004), 197–208.

120. Lukàcs sets *Wilhelm Meister* as the paradigm of the novel of the individual, presenting the compromise with one's own life that Romanticism can only "poeticize" while struggling fruitlessly against, trying to escape the "transcendental homelessness" that makes the novel necessary in the first place (Georg Lukàcs, *The Theory of the Novel: A Historico-Philosophical Essay on the Forms of Great Epic Literature*, trans. Anna Bostock [Cambridge: MIT Press, 1971], 132–44).

121. Thus the novel gives rise to a new subject, through its own capacity as bourgeois medium. For Wilhelm's theater as medium, see Gerhard Neumann, "'Mannigfache Wege gehen die Menschen': Romananfänge bei Goethe und Novalis," in *Goethe und das Zeitalter der Romantik*, ed. Walter Hinderer (Würzburg: Königshausen und Neumann, 2002), 1–91.

122. Schlegel, *KFSA* 1.2.130.

9. INSTEAD OF AN EPILOGUE: COMMUNIST ORGANS, OR TECHNOLOGY AND ORGANOLOGY

1. Samuel Butler, *Erewhon, or, Over the Range* (London: Trübner and Co., 1872), 218.

2. Gilles Deleuze and Félix Guattari, *Anti-Oedipus: Capitalism and Schizophrenia*, trans. Robert Hurley et al. (Minneapolis: University of Minnesota, 1983), 284; emphasis in original.

3. The other famous source for the notion, and the origin of the phrase, is Antonin Artaud. See Deleuze and Guattari, *Anti-Oedipus*, 8–9.

4. Deleuze and Guattari, *Anti-Oedipus*, 10.

5. Karl Marx and Friedrich Engels, *Collected Works*, vol. 24, *Marx and Engels, 1874–1883* (1989) (New York: Lawrence and Wishart, 1975–2004), 467–68.

6. There has been considerable controversy recently over Marx's concept of nature, with the point of contention in Alfred Schmidt, *The Concept of Nature in Marx*, trans. Ben Fowkes (London: NLB, 1971), who claims that Marx wanted to separate a (salutary) dominance over nature from the dominance of human over human (12–13). Cf. Anselm Rabinbach, *The Human Motor: Energy, Fatigue, and the Origins of Modernity* (Berkeley: University of California Press, 1990), 45–84; and John Bellamy Foster, *Marx's Ecology: Materialism and Nature* (New York: Month Review, 2000); both argue for Marx's philosophical materialism extending beyond his social doctrines.

7. Karl Marx and Friedrich Engels, *Collected Works*, vol. 35, *Capital* (1996) (New York: Lawrence and Wishart, 1975–2004), 375–76; *Marx Engels Werke*, vol. 23, *Das Kapital* (Berlin: Dietz, 1962), 392n89; my emphases; hereafter cited as Marx, *Kapital*, followed by page number.

8. *Marx Engels Werke*, vol. 30 (Berlin: Dietz, 1964), 578; hereafter cited as *MEW*, followed by volume and page numbers.

9. Charles Darwin, *On the Origin of Species by Means of Natural Selection, or The Preservation of the Favoured Races in the Struggle for Life*, 3rd ed. (London: John Murray, 1861), 85.

10. Charles Darwin, *On the Origin of Species by Means of Natural Selection, or The Preservation of the Favoured Races in the Struggle for Life*, 4th ed. (London: John Murray, 1866), 94.

11. See Robert Richards, *The Meaning of Evolution: The Morphological Construction and Ideological Reconstruction of Darwin's Theory* (Chicago: University of Chicago Press, 1992), 63–91.

12. Marx/Engels, *Capital*, 346, citing Darwin, *Origin*, 1st ed. (1859), 149, from which I have replaced the citation here.

13. Darwin, *Origin of Species*, 4th ed., 170.

14. Rabinbach, *Human Motor*, argues for Marx's adoption of terminology from scientific popularizers like Ludwig Büchner (the brother of the poet Georg). If he adopts the terms, however, he appropriates them for the complexity of his own system, in which organic and mechanical models serve only as examples.

15. Marx anticipates the meaning of this term that did not become common until around 1900, namely to designate the material artifacts created to aid human efforts. In addition, he sometimes anticipates the sociotechnological thought of the twentieth century by merging the earlier meaning ("scientific knowledge of production") with this later meaning. See Günther Ropohl, "Karl Marx und die Technik," in *Die technikhistorische Forschung in Deutschland von 1800 bis zur Gegenwart*, ed. Wolfgang König and Helmuth Schneider (Kassel: Kassel University, 2007), 63–85; Leo Marx, "'Technology': The Emergence of a Hazardous Concept," *Social Research* 64, no. 3 (1997): 561–77; and Guido Frison, "Technical and Technological Innovation in Marx," *History and Technology: An International Journal* 6, no. 4 (2008): 299–324. See also Wilfried Seibicke, *Technik: Versuch einer Geschichte der Wortfamilie um* techne *in Deutschland vom 16. Jahrhundert bis etwa 1830* (Düsseldorf: VDI, 1968).

16. *MEW* 40.516.

17. Ibid., 40.515–16.

18. The classic terms of the contemporary debate were set out by Louis Althusser, and they concern the role of the "human" in Marx. The sharp distinction made by Althusser can be set against Kolakowski's view that there is only a resetting of terms (see Leszek Kolakowski, *Main Currents in Marxism*, trans. P. S. Falla [New York: Norton, 2005], 146–50).

19. By shifting the object of critique from local areas (like religion) to the very form of any possible set of propositions (that is, by "secularizing"

critique, in Marx's language), Marx sets his discourse literally against the entire constitution of the human-produced world—my rephrasing of the argument at Warren Breckman, *Marx, the Young Hegelians, and the Origins of Radical Social Theory* (Cambridge: Cambridge University Press, 1999), 292ff.

20. Kolakowski describes this as the "Fichteanization" of Hegel; see Kolakowski, *Main Currents*, 43–80.

21. Max Horkheimer, *Gesammelte Schriften*, vol. 4, *Schriften 1936–1941*, ed. Alfred Schmidt (Frankfurt a.M.: Fischer, 1988), 174.

22. Walter Benjamin, *Gesammelte Schriften*, vol. 1, pt. 2, ed. Rolf Tiedemann and Hermann Schweppenhäuser (Frankfurt a.M.: Suhrkamp, 1980), 499–500.

23. Max Horkheimer and Theodor Adorno, *Dialektik der Aufklärung: Philosophische Fragmente* (Frankfurt a.M.: Fischer, 1988), 151.

24. Where the term "organ" occurs in elsewhere in the *Dialektik*, it is usually subordinated to the logic of pure (or bad) instrumentality. Needless to say, this logic usually overlooks the Romantic sensibility about organicity.

25. Horkheimer, "Traditionelle und kritische Theorie," 175.

26. See Karl Marx and Friedrich Engels, *Werke, Ergänzungsband, I. Teil* (Berlin: Dietz, 1968), 546; cited as Marx, *Manuskripte*, 546. ,

27. Ludwig Feuerbach, *Kleine philosophische Schriften (1842–1845)* (Leipzig: Meiner, 1950), 150, cf. 164.

28. See John Toews, *Hegelianism: The Path toward Dialectical Humanism, 1805–1841.* (Cambridge: Cambridge University Press, 1980), 327; cf. Breckman, *Dethroning*, 199.

29. The human and natural sciences should thus become one; see Marx, *Manuskripte*, 543.

30. Ibid., 541; emphasis in original.

31. See ibid., 540.

32. But also of other humans: the section contains a long analysis of the "communism of women" in non-Hegelian socialist experimental societies (see ibid., 534ff.).

33. Ibid., 540; emphasis in original.

34. On the development of the term *ideology*, see Emmet Kennedy, "Ideology from Destutt de Tracy to Marx," *Journal of the History of Ideas* 40, no. 3 (1979): 353–68. As Kennedy notes, the original *idéologues* in post-Revolutionary Paris sometimes used the epigenetic metaphor with respect to the generation of ideas from impressions. For Marx, ideology is not a set of illegitimate ideas, but a set of illegitimate ideas about the origins of ideas.

35. Marx, *Kapital*, 193.

36. Ibid., 198.

37. Ibid., 189; translation modified; ibid., 194.

38. Ibid., 193.

39. On this interchangeability in the light of Victorian anxieties and disabilities studies, see Tamara Ketabgian, *The Lives of Machines: The Industrial Imaginary in Victorian England* (Ann Arbor: University of Michigan, 2011), 17–44.

40. Cf. Kostas Axelos, *Alienation, Praxis, & Techne in the Thought of Karl Marx*, trans. Ronald Bruzina (Austin: University of Texas Press, 1976), 77–78.

41. Marx, *Capital*, 189–90; Marx, *Kapital*, 194–95, my emphasis.

42. Marx, *Kapital*, 195.

43. Ibid., 200.

44. Marx, *Capital*, 205; Marx, *Kapital*, 209.

45. *MEW* 42.600, where again Goethe's *Faust* is invoked.

46. Thus anticipating the notion of society as a machine for such twentieth-century thinkers as Lewis Mumford and Jacques Ellul.

47. Recently, some scholars have attempted to return to Engels and rescue him from this judgment, with varied results. The most significant of these attempts is that of Helena Sheehan, *Marxism and the Philosophy of Science: A Critical History* (Atlantic Highlands, NJ: Humanities, 1993), 21–67.

48. *MEW* 20.445.

49. Ibid., 20.446.

50. See the intertwined emergence of automatic labor and the discipline of political economy: Maxine Berg, *The Machinery Question and the Making of Political Economy, 1815–1848* (Cambridge: Cambridge University Press, 1980).

51. See David Ricardo, *On the Principles of Political Economy and Taxation* (London: Murray, 1821), 466–83.

52. Marx, *Capital*, 444–45; Marx, *Kapital*, 464–65, translation modified.

53. Marx, *Capital*, 423; Marx, *Kapital*, 442, translation modified.

54. Marx, *Kapital*, 440.

55. Marx has several sources for the notion of this inversion, of which the earliest seems to be Ferguson: "It may even be doubted, whether the measure of national capacity increases with the advancement of arts. Many mechanical arts, indeed, require no capacity; they succeed best under a total suppression of sentiment and reason; and ignorance is the mother of industry as well as of superstition. Reflection and fancy are subject to err; but a habit of moving the hand, or the foot, is independent of either. Manufactures, accordingly, prosper most, where the mind is least consulted, and where the workshop may, without any great effort of imagination, be considered as an engine, the parts of which are men" (cited in *MEW* 4.147–48). Similar remarks in the passages that follow, at 151, 153, 155, and 157. In these passages, Marx is working with Ure's and Charles Babbage's ideas, and beginning to argue that machinofacture is distinct in both its consequences for living work and its effects in the

relations of production from the division of labor analyzed by the Classical political economists. The machine-driven factory is thus a materialization of social domination, as, e.g., 153.

56. Andrew Ure, *Philosophy of Manufacture; or, an Exposition of the Scientific, Moral, and Commercial Economy of the Factory System* (London: Clowes and Sons, 1835), 2.

57. Ibid., 13–14.

58. Marx, *Kapital*, 393.

59. Ibid., 398.

60. Ibid., 400.

61. Here we see "the immediate technical basis of large industry" (ibid., 403, and again at 406).

62. Ibid., 408.

63. Marx, *Capital*, 384–85; Marx, *Kapital*, 402, translation modified.

64. Marx, *Capital*, 389; Marx, *Kapital*, 407, translation modified.

65. *MEW* 25.897–919.

66. Karl Marx and Friedrich Engels, Karl Marx, Friedrich Engels Gesamtausgabe (MEGA) (Berlin: Dietz, 1972–), 2.3.6.2059.

67. Ibid., 2.3.6.2060–61.

68. Michel Foucault, *"Society Must Be Defended": Lectures at the Collège de France, 1975–76* (New York: Picador, 2003), 246.

69. Émile Durkheim, *The Division of Labor in Society*, trans. W. D. Halls (New York: Free Press, 1997).

70. Ernst Kapp, *Grundlinien einer Philosophie der Technik: Zur Entstehungsgeschichte der Cultur aus neuen Gesichtspunkten* (Westermann: Braunschweig, 1877).

71. Georg Simmel, *The Metropolis and Mental Life: The Sociology of Georg Simmel* (New York: Free Press, 1976).

72. Heidegger thinks of the organ as a "tool built into its user," developed from an ability (*Fähigkeit*) and not merely incidentally supplying a capacity (*Fertigkeit*). But the transcendental moment housed in the organ is not, in organology, shifted to the organism, as it is in vitalism (which Heidegger is also trying to avoid), but is treated in isolation from and as generative of forms like "machines" and "bodies." Organology, as I have been arguing, obviated vitalism. See Martin Heidegger, *The Fundamental Concepts of Metaphysics: World, Finitude, Solitude*, trans. William Mcneill and Nicholas Walker (Bloomington: Indiana University Press, 1995), 219; translation modified.

73. Humberto R. Maturana and Francisco J. Varela, *Autopoiesis and Cognition: The Realization of the Living* (London: Reidel, 1980).

74. Jacques Lacan, *The Seminar of Jacques Lacan*, book 11, *The Four Fundamental Concepts of Psychoanalysis*, trans. Alan Sheridan (New York: Norton, 1978), 187.

75. Ibid., 196 ff.

76. Cf. Avital Ronell's analysis of Freud's (and Hegel's) conceptions of "dual organs" with respect to the telephone as a synecdoche of technology in general in *The Telephone Book: Technology, Schizophrenia, Electric Speech* (Lincoln: University of Nebraska Press, 1991), 102–6.

77. Slavoj Žižek , *Organs without Bodies: On Deleuze and Consequences* (New York: Routledge, 2003).

78. Blumenberg, *Geistesgeschichte*, 13; emphasis in original.

BIBLIOGRAPHY

Adelung, Johann Christoph. *Verusch eines vollständig grammatisch-kritischen Wörterbuches der hochdeutschen Mundart mit beständiger Vergleichung der übrigen Mundarten, besonder aber der Oberdeutschen. Dritter Theil, von L— Scha-* . Leipzig: Breitkopf, 1777.
Adler, Jeremy. "The Aesthetics of Magnetism: Science, Philosophy, and Poetry in the Dialogue between Goethe and Schelling." In *The Third Culture: Literature and Science*, edited by Elinor S. Schaffer, 66–103. New York: de Gruyter, 1998.
Alt, Peter-André. *Schiller: Eine Biographie*. Vol.1, 1759–1791. Munich: Beck, 2000.
Ameriks, Karl. *Kant's Theory of Mind: An Analysis of the Paralogisms of Pure Reason*. Oxford: Oxford University Press, 2000.
Andriopoulos, Stefan. *Possessed: Hypnotic Crimes, Corporate Fiction, and the Invention of Cinema*. Translated by Peter Jansen and Stefan Andriopoulos. Chicago: University of Chicago, 2008.
Appel, Toby A. *The Cuvier-Geoffroy Debate: French Biology in the Decades before Darwin*. New York: Oxford University Press, 1987.
Aristotle. *Basic Works*. Edited by Richard Peter McKeon. New York: Random House, 2001.
———. *The Basic Works of Aristotle*. Edited by Richard McKeon. New York: Random House, 1941.
———. *Generation of Animals*. Translated by A. L. Peck. Cambridge: Harvard University Press, 2000.
———. *On the Soul; Parva Naturalia; On Breath*. Cambridge: Harvard University Press; London: Heinemann, 1935.
———. *Parts of Animals. With an English Translation by A. L. Peck. Movement of Animals. Progression of Animals. With an English Translation by E. S. Forster.* Translated by Edward Seymour Forster and Arthur Leslie Peck. London: Heinemann; Cambridge: Harvard University Press, 1937.
———. *Problems II: Books XXII–XXXVIII*. Translated by W. S. Hett. Edited by H. Rackham. London: Heinemann, 1965.

Axelos, Kostas. *Alienation, Praxis, & Techne in the Thought of Karl Marx.* Translated by Ronald Bruzina. Austin: University of Texas Press, 1976.
Bach, Thomas, and Olaf Breidbach. *Naturphilosophie nach Schelling.* Stuttgart: frommann-holzboog, 2005.
Barber, W. H. *Leibniz in France, from Arnauld to Voltaire: A Study in French Reactions to Leibnizianism, 1670–1760.* Oxford: Clarendon, 1955.
Beck, Lewis White. *Early German Philosophy; Kant and His Predecessors.* Cambridge: Belknap Press of Harvard University Press, 1969.
Behler, Ernst. "The Origins of the Romantic Literary Theory." *Colloquium Germanica* 2 (1968): 109–26.
———. *Unendliche Perfektibilität: Europäische Romantik und Französische Revolution.* Munich: Schöningh, 1989.
Beiser, Frederick C. *Diotima's Children: German Aesthetic Rationalism from Leibniz to Lessing.* Oxford: Oxford University Press, 2009.
———. *German Idealism: The Struggle against Subjectivism, 1781–1801.* Cambridge: Harvard University Press, 2002.
———. *Hegel.* New York: Routledge, 2005.
———. *The Romantic Imperative: The Concept of Early German Romanticism.* Cambridge: Harvard University Press, 2003.
Bell, David. *Spinoza in Germany from 1670 to the Age of Goethe.* London: Institute of Germanic Studies, University of London, 1984.
Benjamin, Walter. *Gesammelte Schriften.* Edited by Rolf Tiedemann. Frankfurt a.M.: Suhrkamp, 1972–99.
Berg, Maxine. *The Machinery Question and the Making of Political Economy, 1815–1848.* Cambridge: Cambridge University Press, 1980.
Berlin, Isaiah. *The Roots of Romanticism.* Edited by Henry Hardy. Princeton, NJ: Princeton University Press, 1999.
Bernstein, Richard J. *Radical Evil: A Philosophical Interrogation.* Malden, MA: Blackwell, 2002.
Bertaux, Pierre. *Hölderlin und die Französische Revolution.* Frankfurt a.M.: Suhrkamp, 1969.
Blanchot, Maurice. "The Athenaeum." Translated by Deborah Esch and Ian Balfour. *Studies in Romanticism* 22, no. 2 (1983): 163–72.
Blumenbach, Johann Friedrich. *Anfangsgründe der Physiologie.* Translated and edited by Joseph Eyerel. Vienna: Christ. Frid. Wappler, 1789.
———. *Institutiones Physicologicae.* Göttingen: Dieterich, 1786.
———. *Über den Bildungstrieb.* Göttingen: Dieterich, 1791.
Blumenberg, Hans. *Arbeit am Mythos.* Frankfurt a.M.: Suhrkamp, 1979.
———. *Die Genesis der Kopernikanischen Welt.* Frankfurt a.M.: Suhrkamp, 1975.
———. *The Genesis of the Copernican World.* Cambridge: MIT, 1989.

———. *Die Lesbarkeit der Welt*. Frankfurt a.M.: Suhrkamp, 2000.
———. *Geistesgeschichte der Technik: Mit einem Radiovortrag auf CD*. Translated by Alexander Schmitz. Edited by Bernd Stiegler. Frankfurt a.M.: Suhrkamp, 2009.
———. *Lebenszeit und Weltzeit*. Frankfurt a.M.: Suhrkamp, 1986.
———. *The Legitimacy of the Modern Age*. Translated by Robert M. Wallace. Cambridge: MIT Press, 1983.
———. *Paradigmen zu einer Metaphorologie*. Frankfurt a.M.: Suhrkamp, 1998.
———. *Shipwreck with Spectator: Paradigm of a Metaphor for Existence*. Translated by Stephen Rendall. Cambridge: MIT Press, 1997.
———. *Zu den Sachen und Zurück*. Edited by Manfred Sommer. Frankfurt a.M.: Suhrkamp, 2002.
Bodmer, Johann Jakob, and Johann Jakob Breitinger. *Schriften zur Literatur*. Edited by Volker Meid. Stuttgart: Reclam, 1980.
Boehme, Hartmut. "The Metaphysics of Phenomena: Telescope and Microscope in the Works of Goethe, Leeuwenhoek and Hooke." In *Collection, Laboratory, Theater: Scenes of Knowledge in the 17th Century*, edited by Helmar Schramm, Ludger Schwarte, and Jan Lazardzig, 355–94. Berlin: de Gruyter, 2005.
Bollnow, Otto Friedrich. *Zwischen Philosophie und Pädagogik: Vorträge und Aufsätze*. Aachen: Weitz, 1988.
Bos, A. P. *The Soul and Its Instrumental Body: A Reinterpretation of Aristotle's Philosophy of Living Nature*. Leiden: Brill, 2003.
Bowie, Andrew. *Schelling and Modern European Philosophy: An Introduction*. London: Routledge, 1993.
Bowman, Brady. "Goethean Morphology, Hegelian Science: Affinities and Transformations." *Goethe Yearbook* 18, no. 1 (2011): 159–81.
Boyle, Nicholas. *Goethe: The Poet and His Age*. Oxford: Oxford University Press, 2000.
Brandis, Johann Dietrich. *Versuch über die Lebenskraft*. Hannover: Verlag der Hahn'schen Buchhandlung, 1795.
Breckman, Warren. *Marx, the Young Hegelians, and the Origins of Radical Social Theory: Dethroning the Self*. Cambridge: Cambridge University Press, 1999.
Broman, Thomas Hoyt. *The Transformation of German Academic Medicine, 1750–1820*. Cambridge: Cambridge University Press, 1996.
Brown, John. *The Elements of Medicine, Or, A Translation of the Elementa Medicinae Brunonis. With Large Notes, Illustrations, and Comments*. London: J. Johnson, 1788.
Brown, Robert F. *The Later Philosophy of Schelling: The Influence of Boehme on the Works of 1809–1815*. Lewisburg: Bucknell University Press, 1977.
Brown, Wendy. *Undoing the Demos: Neoliberalism's Stealth Revolutions*. New York: Zone, 2015.

Brunner, Otto, et al., eds. *Geschichtliche Grundbegriffe: Historisches Lexikon zur politisch-sozialen Sprache in Deutschland*. Stuttgart: Klett-Cotta, 1997.

Buchheim, Thomas. "Das Prinzip des Grundes und Schellings Weg zur Freiheitsschrift." In *Schellings Weg zur Freiheitsschrift: Legende und Wirklichkeit*, edited by Hans Michael Baumgartner and Wilhelm G. Jacobs, 223–40. Stuttgart—Bad Cannstatt: frommann-holzboog, 1996.

Burdach, Karl Friedrich. *Blicke ins Leben*. Leipzig: Voss, 1842.

Burwick, Frederick. *The Damnation of Newton: Goethe's Color Theory and Romantic Perception*. New York: de Gruyter, 1986.

Butler, Samuel. *Erewhon; or Over the Range*. London: Trübner &, 1872.

Campe, Johann Heinrich. *Wörterbuch zur Erklärung und Verdeutschung der unserer Sprache aufgedrungenen fremden Ausdrücke: Ein Ergänzungsband zu Adelung's und Campe's Wörterbüchern*. Braunschweig: Schulbuchhandlung, 1801.

———. *Wörterbuch zur Erklärung und Verdeutschung der unserer Sprache aufgedrungenen fremden Ausdrücke: Ein Ergänzungsband zu Adelung's und Campe's Wörterbüchern*. Braunschweig: Schulbuchhandlung, 1813.

Campe, Rüdiger. "Kafkas Institutionenroman: *Der Proceß, Das Schloß*." In *Gesetz: Ironie*, edited by Rüdiger Campe and Michael Niehaus, 197–208. Krottenmühl: Synchron, 2004.

Candolle, August de. *Organographie végétale*. Paris: Deterville, 1827.

Canguilhem, Georges. *Knowledge of Life*. Translated by Paola Marrati. Edited by Todd Meyers and Stefanos Geroulanos. New York: Fordham University Press, 2008.

Carus, Carl Gustav. *Von den Naturreichen, ihrem Leben und ihrer Verwandtschaft: Eine Physiolog. Abh*. Dresden: Gaertner, 1818.

———. *Zwölf Briefe über das Erdleben*. 1841. Edited by Ekkehard Meffert. Stuttgart: Freies Geistesleben, 1986.

Cassirer, Ernst. *Kant's Life and Thought*. New Haven: Yale University Press, 1981.

———. *Nachgelassene Manuskripte und Texte*. Edited by John Michael Krois, Oswald Schwemmer, and Klaus Christian Köhnke. Hamburg: Meiner, 1995.

———. *The Philosophy of the Enlightenment*. Translated by Peter Gay. Princeton, NJ: Princeton University Press, 1971.

———. *The Problem of Knowledge; Philosophy, Science, and History since Hegel*. New Haven: Yale University Press, 1950.Cheung, Tobias. "From the Organism of a Body to the Body of an Organism: Occurrence and Meaning of the Word 'Organism' from the Seventeenth to the Nineteenth Centuries." *British Journal for the History of Science* 39, no. 3 (2006): 319–39.

———. "Organismen: Agenten zwischen Innen- und Außenwelten, 1780–1860." Transcript. Bielef, 2014.

Commentarii Societatis Regiae Scientiarum Gottingensis. Vol. 2. Göttingen, 1753.

Crary, Jonathan. *Techniques of the Observer: On Vision and Modernity in the Nineteenth Century*. Cambridge: MIT Press, 1992.
Cunningham, Andrew. *The Anatomist Anatomis'd: An Experimental Discipline in Enlightenment Europe*. Farnham, Surrey, England: Ashgate, 2010.
Cuvier, Georges. "Elegy of Lamarck." *Edinburgh New Philosophical Journal* 20 (1836):1–22.
Daiber, Jürgen. *Experimentalphysik des Geistes: Novalis und das romantische Experiment*. Göttingen: Vandenhoeck and Ruprecht, 2001.
Danz, Christian, Jörg Jantzen, and Karl Baier, eds. *Gott, Natur, Kunst und Geschichte: Schelling zwischen Identitätsphilosophie und Freiheitsschrift*. Göttingen: Vienna University Press, 2011.
Darwin, Charles. *On the Origin of Species by Means of Natural Selection, or The Preservation of the Favoured Races in the Struggle for Life*. 3rd ed. London: John Murray, 1861.
———. *On the Origin of Species by Means of Natural Selection, or The Preservation of the Favoured Races in the Struggle for Life*. 4th ed. London: John Murray, 1866.
Daub, Adrian. *Uncivil Unions: The Metaphysics of Marriage in German Idealism and Romanticism*. Chicago: University of Chicago Press, 2012.
Deleuze, Gilles, and Félix Guattari. *Anti-Oedipus: Capitalism and Schizophrenia*. Minneapolis: University of Minnesota Press, 1983.
de Man, Paul. *Aesthetic Ideology*. Edited by Andrzej Warminski. Minneapolis: University of Minnesota, 1996.
Descartes, René. *A Discourse on Method*. Translated by Ian MacLean. Oxford: Oxford University Press, 2006.
———. *The World and Other Writings*. Translated by Stephen Gaukroger. Cambridge: Cambridge University Press, 2004.
Dick, Manfred. "Der Dichter und der Naturforscher: Samuel Thomas Soemmerring und Wilhelm Heinse." In *Samuel Thomas Soemmerring und die Gelehrten der Goethezeit: Beiträge eines Symposions in Mainz vom 19. bis 21. Mai 1983*, edited by Gunter Mann and Franz Dumont, 203–29. Stuttgart: Fischer, 1985.
Diderot, Denis, and Jean le Rond D'Alembert. *Encyclopédie, ou dictionnaire raisonné des sciences, des arts et des métiers, etc.* University of Chicago: ARTFL Encyclopédie Project (Spring 2011 edition). Edited by Robert Morrissey. http://encyclopedie.uchicago.edu/.
Dilthey, Wilhelm. *Gesammelte Schriften*. Edited by Georg Misch. Göttingen: Vandenhoek und Ruprecht, 1991.
Duchesneau, François. *La physiologie des lumières: Empirisme, modèles et théories*. Boston: Nijhoff, 1982.
Durkheim, Émile. *The Division of Labor in Society*. Translated by W. D. Halls. New York: Free Press, 1997.

Durner, Manfred. "Die Naturphilosophie im 18. Jahrhundert und der Naturwissenschaftliche Unterricht in Tübingen. Zu den Quellen von Schellings Naturphilosophie." *Archiv für Geschichte der Philosophie* 73, no. 1 (1991): 71–103.

Eckartshausen, Karl von. *Probaseologie oder praktischer Theil der Zahlenlehre der Natur: ein Schlüssel zu den Hieroglyphen der Natur.* Leipzig: Gräff, 1795.

Engelhardt, Wolf von. *Goethes Weltansichten: Auch eine Biographie.* Weimar: Böhlau, 2007.

Euler, Werner. "Die Suche nach dem Seelenorgan: Kants philosophische Analyse einer anatomischen Entdeckung Soemmerrings." *Kant-Studien* 93, no. 4 (2002): 453–80.

Fackenheim, Emile. "Kant and Radical Evil." *University of Toronto Quarterly* 23 (1954): 339–53.

Feuerbach, Ludwig. *Kleine philosophische Schriften (1842–1845).* Edited by Max Gustav Lange. Leipzig: Meiner, 1950.

Feuereisen, Karl Gottlieb. *Pflanzen-Organologie oder Etwas aus dem Pflanzenreiche: insonderheit die sonderbaren Würkunge des Nahrungssaftes in den Gewächsen.* Hannover: Pockwitz, 1780.

Fichte, Johann Gottlieb. *Fichte: Foundations of Natural Right.* Translated by Michael Baur. Edited by Frederick Neuhouser. Cambridge: Cambridge University Press, 2000.

———. *Grundlage der gesammten Wissenschaftslehre (1794).* Hamburg: Meiner, 1997.

———. *J.-G.-Fichte-Gesamtausgabe der Bayerischen Akademie der Wissenschaften.* Edited by Reinhard Lauth. Stuttgart–Bad Cannstatt: frommann-holzboog, 1965.

———. *Johann Gottlieb Fichte's sämmtliche Werke.* Berlin: Veit und Comp., 1845–46.

———. *Versuch einer neuen Darstellung der Wissenschaftslehre (1797/98).* Edited by Peter Baumanns. Hamburg: Meiner, 1984.

———. *Wissenschaftslehre nova methodo 1798/99.* Edited by Erich Fuchs. Hamburg: Meiner, 1994.

Förster, Eckart. "Die Bedeutung von §§76, 77 der *Kritik der Urteilskraft* für die Entwicklung der nachkantischen Philosophie. Teil 1." In *Kant und der Frühidealismus*, edited by Jürgen Stolzenberg, 59–80. Hamburg: Meiner, 2007.

———. "Goethe and the 'Auge des Geistes.'" *Deutsche Vierteljahrsschrift für Literaturwissenschaft und Geistesgeschichte* 75, no. 1 (2001): 87–101.

———. *The Twenty-Five Years of Philosophy.* Cambridge: Harvard University Press, 2012.

Forster, Michael N. *After Herder: Philosophy of Language in the German Tradition.* Oxford: Oxford University Press, 2010.

Foster, John Bellamy. *Marx's Ecology: Materialism and Nature.* New York: Monthly Review Press, 2000.
Fóti, Véronique Marion. *Epochal Discordance: Hölderlin's Philosophy of Tragedy.* Albany: SUNY Press, 2006.
Foucault, Michel. *The Foucault Reader.* Edited by Paul Rabinow. New York: Pantheon, 1984.
———. *The Order of Things: An Archaeology of the Human Sciences.* New York: Pantheon, 1971.
———. *Society Must Be Defended: Lectures at the Collège de France, 1975–76.* Edited by Mauro Bertani, Alessandro Fontana, and François Ewald. Translated by David Macey. New York: Picador, 2003.
Frank, Manfred. *Auswege aus dem deutschen Idealismus.* Translated by Elizabeth Millán-Zaibert. Frankfurt a.M.: Suhrkamp, 2007.
———. *Der kommende Gott: Vorlesungen über die neue Mythologie.* Frankfurt a.M.: Suhrkamp, 1982.
———. *Der unendliche Mangel an Sein: Schellings Hegelkritik und die Anfänge der Marxschen Dialektik.* Munich: Fink, 1992.
———. *The Philosophical Foundations of Early German Romanticism.* Albany: SUNY Press, 2004.
———. *Selbstgefühl: Eine historisch-systematische Erkundung.* Frankfurt a.M.: Suhrkamp, 2002.
———. *"Unendliche Annäherung": Die Anfänge der philosophischen Frühromantik.* Frankfurt a.M.: Suhrkamp, 1997.
Franks, Paul W. *All or Nothing: Systematicity, Transcendental Arguments, and Skepticism in German Idealism.* Cambridge: Harvard University Press, 2005.
Frey, Christiane. "The Art of Observing the Small: On the Borders of the subvisibilia (from Hooke to Brockes)." *Monatshefte* 105, no. 3 (2013): 376–88.
Friedman, Michael. *Kant and the Exact Sciences.* Cambridge: Harvard University Press, 1992.
Frison, Guido. "Technical and Technological Innovation in Marx." *History and Technology* 6, no. 4 (1988): 299–324.
Fuchs, Thomas. *The Mechanization of the Heart: Harvey and Descartes.* Translated by Marjorie Grene. Rochester, NY: University of Rochester Press, 2001.
Gaier, Ulrich. *Krumme Regel: Novalis' Konstruktionslehre des schaffenden Geistes und ihre Tradition.* Tübingen: Niemeyer, 1970.
Gailus, Andreas. *Passions of the Sign: Revolution and Language in Kant, Goethe, and Kleist.* Baltimore: Johns Hopkins University Press, 2006.
Galen. *Claudii Galeni opera omnia.* Edited by Karl Gottlob Kühn. Lipsiae: Prostat in Officina Libraria Car. Cnoblochii, 1822.

———. *On the Usefulness of the Parts of the Body: Peri chreias morian de usu partium*. Translated by Margaret Tallmadge May. Ithaca, NY: Cornell University Press, 1968.

Galenus. *On Antecedent Causes*. Edited by Robert J. Hankinson. Cambridge: Cambridge University Press, 1998.

Gall, Franz Joseph. "Schreiben über seinen bereits geendigten Prodromus über die Verrichtungen des Gehirns der Menschen und der Their, an Herrn Jos. Fr. Von Retzer." *Neuer teutscher Merkur* 3 (1798): 311–32.

Gamper, Michael. *Elektropoetologie: Fiktionen der Elektrizität, 1740–1870*. Göttingen: Wallstein, 2009.

Gasché, Rodolphe. "Ideality in Fragmentation." Foreword to *Philosophical Fragments*, by Friedrich von Schlegel. Translated by Peter Firchow. Minneapolis: University of Minnesota Press, 1991.

Gehler, Johann Samuel Traugott. *Physikalisches Wörterbuch oder Versuch: Erklärung der vornehmsten Begriffe und Kunstwörter der Naturlehre mit kurzen Nachrichten von der Geschichte der Erfindungen und Beschreibungen der Werkzeuge begleitet in alphabetischer Ordnung. Dritter Theil von Liq bis Sed.* Schwickert: Leipzig, 1790.

Geulen, Eva. "Serialization in Goethe's Morphology." *Compar(a)ison* 29 (2013): 53–70.

Gigante, Denise. *Life: Organic Form and Romanticism*. New Haven: Yale University Press, 2009.

Gloning, Thomas, and Christopher Young. *A History of the German Language through Texts*. London: Routledge, 2004.

Goebel, Eckart. *Beyond Discontent: 'Sublimation' from Goethe to Lacan*. Translated by James C. Wagner. New York: Bloomsbury, 2012.

Goethe, Johann Wolfgang von. *Faust*. Translated by Walter Arnold Kaufmann. 1961. New York: Anchor, 1990.

———. *Sämtliche Werke. Briefe, Tagebücher und Gespräche*. Edited by Dieter Borchmeyer et al. Frankfurt am Main: Deutscher Klassiker Verlag, 1985 ff.

———. *Sämtliche Werke nach Epochen seines Schaffens*. Edited by Karl Richter et al. Munich: Hanser, 1985–.

———. *Goethes Werke: Weimarer Sophienausgabe*. Edited by Gustav Von Loeper. Weimar: Böhlau, 1887.

———. *The Metamorphosis of Plants*. Edited by Gordon L. Miller. Cambridge: MIT Press, 2009.

Görres, Joseph von. *Exposition der Physiologie: Organologie*. Koblenz: Lassaulx, 1805.

———. *Gesammelte Schriften*. Vol. 3, *Geistesgeschichtliche und literarische Schriften I (1803–1808)*. Cologne: Gilde, 1926.

Grant, Iain H. *Philosophies of Nature after Schelling*. New York: Continuum, 2006.

Grave, Johannes. *Der "ideale Kunstkörper": Johann Wolfgang Goethe als Sammler von Druckgraphiken und Zeichnungen*. Göttingen: Vandenhoeck and Ruprecht, 2006.

———. "Ideal and History: Johann Wolfgang Goethe's Collection of Prints and Drawings." *Artibus et Historiae* 27, no. 53 (2006): 175–86.

Guyer, Paul. *The Cambridge Companion to Kant's "Critique of Pure Reason."* Cambridge: Cambridge University Press, 2010.

———. *Kant*. London: Routledge, 2006.

———. *Kant and the Claims of Knowledge*. Cambridge: Cambridge University Press, 1987.

Haering, Theodor. *Theodor Haering: Novalis als Philosoph*. Stuttgart: Kohlhammer, 1954.

Hagner, Michael. *Homo cerebralis: Der Wandel vom Seelenorgan zum Gehirn*. Frankfurt a.M.: Suhrkamp, 2008.

Haller, Albrecht von. *Primae lineae physiologiae*. Göttingen: van Rossum, 1758.

———. *Primeae lineae physiologiae in usum praelectionum academicarum*. Göttingen: Acad. Bibl., 1751.

———. *Sur la formation du coeur dan le poulet*. Lausanne: Bousquet, 1758.

———. *Umriss der Geschäfte des körperlichen Lebens*. Berlin: Haude und Spener, 1770.

Hamacher, Werner. *Premises: Essays on Philosophy and Literature from Kant to Celan*. Stanford: Stanford, 1999.

Hamann, Johann Georg. *Writings on Philosophy and Language*. Edited by Kenneth Haynes. Cambridge: Cambridge University Press, 2007.

Hankinson, R. J. *The Cambridge Companion to Galen*. Cambridge: Cambridge University Press, 2008.

———. "Philosophy of Nature." In *The Cambridge Companion to Galen*, 210–42. Cambridge: Cambridge University Press, 2008.

Hartmann, Klaus. *Studies in Foundational Philosophy*. Amsterdam: Rodopi, 1988.

Hegel, Georg Wilhelm Friedrich. *Phenomenology of Spirit*. Translated by Terry Pinkard. http://terrypinkard.weebly.com/phenomenology-of-spirit-page.html.

———. *Werke in zwanzig Bänden*. Edited by E. Moldenhauer and K. M. Michel. Frankfurt a.M: Suhrkamp, 1969–71.

Heidegger, Martin. *Basic Writings: From "Being and Time" (1927) to "The Task of Thinking" (1964)*. Edited by David Farrell Krell. New York: Harper and Row, 1977.

———. *The Fundamental Concepts of Metaphysics: World, Finitude, Solitude*.

Translated by William Mcneill and Nicholas Walker. Bloomington: Indiana University Press, 1995.

Heimsoeth, Heinz. "Zur Herkunft und Entwicklung von Kants Kategorientafel." *Kant Studien* 54, no. 1–4 (1963): 376–403.

Heinz, Marion, ed. *Herder und die Philosophie des deutschen Idealismus*. Amsterdam: Rodopi, 1997.

———. *Sensualistischer Idealismus: Untersuchungen zur Erkenntnistheorie des jungen Herder (1763–1778)*. Hamburg: Meiner, 1994.

Helfer, Martha. *The Retreat of Representation: The Concept of* Darstellung *in German Critical Discourse*. Albany: SUNY Press, 1996.

Hemsterhuis, Frans. *Lettre sur l'homme et ses rapports avec le commentaire inédit de Diderot*. Edited by Denis Diderot and Georges May. New Haven: Yale University Press, 1964.

———. *Oeuvres philosophiques*. Edited by L. S. P. Meyboom. Hildesheim: Olms, 1972.

Henrich, Dieter. *Between Kant and Hegel: Lectures on German Idealism*. Edited by David S. Pacini. Cambridge: Harvard University Press, 2008.

———. *Der Grund im Bewußtsein: Untersuchungen zu Hölderlins Denken (1794–1795)*. Stuttgart: Klett-Cotta, 1992.

———. "Fichte's Original Insight." Translated by David Lachterman. *Contemporary German Philosophy* 1 (1982): 15–53.

———. "Hölderlin on Judgment and Being: A Study in the History of the Origins of Idealism." In *The Course of Remembrance and Other Essays on Hölderlin*, edited by Eckart Förster, 71–90. Stanford, CA: Stanford University Press, 1997.

Herder, Johann Gottfried. *Werke in zehn Bänden*. Edited by Jürgen Brummack and Martin Bollacher. Frankfurt a.M.: Klassiker Verlag, 1985–.

Heuvel, Jon Vanden. *A German Life in the Age of Revolution*. Washington, DC: Catholic University of America Press, 2001.

Hindrichs, Gunnar. "Die aufgeklärte Aufklärung." In *Kant and the Future of the European Enlightenment / Kant und die Zukunft der europäischen Aufklärung*, edited by Heiner F. Klemme, 43–68. New York: de Gruyter, 2009.

———. "Goethe's Notion of an Intuitive Power of Judgment." *Goethe Yearbook* 18, no. 1 (2011): 51–65.

Hölderlin, Friedrich. *The Death of Empedocles: A Mourning-Play*. Translated by David Farrell Krell. Albany: SUNY Press, 2008.

———. *Sämtliche Werke (Große Stuttgarter Ausgabe)*. Edited by Friedrich Beißner. Stuttgart: Kohlhammer, 1957.

———. *Sämtliche Werke und Briefe*. Edited by Jochen Schmidt and Katharina Grätz. Frankfurt a.M.: Suhrkamp, 1994.

Holland, Jocelyn. "From Romantic Tools to Technics: Heideggerian Questions in Novalis's Anthropology." *Configurations* 18, no. 3 (2010): 291–307.

———. *German Romanticism and Science: The Procreative Poetics of Goethe, Novalis, and Ritter.* New York: Routledge, 2009.

Hölscher, Uvo. *Empedokles und Hölderlin*. Edited by Gerhard Kurz. Eggingen: Isele, 2001.

Hörl, Erich. *Die technologische Bedingung*. Frankfurt a.M.: Suhrkamp, 2011.

Horkheimer, Max. *Gesammelte Schriften*. Edited by Alfred Schmidt and Gunzelin Schmid Noerr. Frankfurt a.M.: Fischer, 1985–96.

Horkheimer, Max, and Theodor W. Adorno. *Dialektik der Aufklärung: Philosophische Fragmente*. 1969. Frankfurt a.M.: Fischer, 1988.

Horstmann, Rolf-Peter. "Hegel's Phenomenology of Spirit as an Argument for a Monistic Ontology." *Inquiry* 49 (2006): 103–18.

Horstmann, Rolf-Peter, and Dina Emundts. *Georg Wilhelm Friedrich Hegel: Eine Einführung*. Stuttgart: Reclam, 2002.

Humboldt, Alexander von. *Ansichten der Natur*. Darmstadt: Wissenschaftliche Buchgesellschaft, 1987.

———. *Aphorismen aus der chemischen Physiologie der Pflanzen*. Translated by Gotthelf Fischer von Waldheim. Weimar: Voß und Compagnie, 1794.

———. *Florae fribergensis specimen: Plantas cryptogamicas praesertim subterraneas exhibens. Accedunt aphorismi ex doctrina physiologiae chemicae plantarum*. Berlin: Rottmann, 1793.

———. "Life-Force, or the Genius of Rhodes." Translated by Leif Weatherby. *Yearbook of Comparative Literature* 58 (2012): 163–68.

———. *Über die gereizte Muskel- und Nervenfaser nebst Vermuthungen über den chemischen Process des Lebens in der Thier- und Pflanzenwelt*. Vol. 1. Berlin: Posen, 1797.

Israel, Jonathan I. *Radical Enlightenment: Philosophy and the Making of Modernity, 1650–1750*. Oxford: Oxford University Press, 2001.

Jacobi, Friedrich Heinrich. *David Hume über den Glauben oder Idealismus und Realismus*. Breslau: Loewe, 1787.

———. *Werke: Gesamtausgabe*. Edited by Klaus Hammacher. Hamburg: Meiner, 1998.

Jacobs, Wilhelm G. "Vom Ursprung des Bösen zum Wesen der menschlichen Freiheit oder Transzendentalphilosophie und Metaphysik." In *Schellings Weg zur Freiheitsschrift: Legende und Wirklichkeit*, edited by Hans Michael Baumgartner and Jacobs, 11–28. Stuttgart–Bad Cannstatt: frommann-holzboog, 1996.

Jähnig, Dieter. *Schelling: Die Kunst in der Philosophie*. Pfullingen: Neske, 1966.

Jauß, H. R., ed. *Nachahmung und Illusion: Kolloquium Gießen Juni 1963, Vorlagen und Verhandlungen*. Munich: Eidos, 1964.

Jolley, Nicholas. *The Cambridge Companion to Leibniz*. Cambridge: Cambridge University Press, 1995.
Jütte, Robert. *Geschichte der Sinne: Von der Antike bis zum Cyberspace*. Munich: Beck, 2000.
Kant, Immanuel. *Critique of Pure Reason*. Translated by Paul Guyer. Cambridge: Cambridge University Press, 1998.
———. *Kant's gesammelte Schriften*. Berlin: Reimer, 1900–.
———. *Notes and Fragments: Logic, Metaphysics, Moral Philosophy, Aesthetics*. Edited by Paul Guyer. New York: Cambridge University Press, 2005.
Kant, Immanuel. *Werkausgabe*. Ed. Wilhelm Weischedel. Frankfurt a.M.: Suhrkamp, 1977.
Kapp, Ernst. *Grundlinien einer Philosophie der Technik: Zur Entstehungsgeschichte der Cultur aus neuen Gesichtspunkten*. Braunschweig: Westermann, 1877.
Kennedy, Emmet. "'Ideology' from Destutt De Tracy to Marx." *Journal of the History of Ideas* 40, no. 3 (1979): 353.
Ketabgian, Tamara Siroone. *The Lives of Machines: The Industrial Imaginary in Victorian Literature and Culture*. Ann Arbor: University of Michigan Press, 2011.
Kielmeyer, Carl Friedrich. *Ueber die Verhältnisse der organischen Kräfte untereinander in der Reihe der verschiedenen Organisationen, die Gesetze und Folgen dieser Verhältnisse*. Introduction by Kai Torsten Kanz. Marburg an der Lahn: Basilisken-Presse, 1993.
Kittler, Friedrich. "Die Nacht der Substanz." In *Kursbuch Medienkultur die maßgeblichen Theorien von Brecht bis Baudrillard*, edited by Claus Pias, Britta Neitzel, Oliver Fahle, Lorenz Engell, and Joseph Vogl, 507–25. Stuttgart: Deutsche Verlags-Anstalt, 1999.
———. *Discourse Networks 1800/1900*. Translated by Michael Metteer and Chris Cullens. Stanford, CA: Stanford University Press, 1999.
———. "Heinrich von Ofterdingen als Nachrichtenfluß." In *Novalis: Wege der Forschung*, edited by Gerhard Schulz, 480–508. Darmstadt: Wissenschaftliche Buchgesellschaft, 1986.
Köchy, Kristian. *Ganzheit und Wissenschaft: Das historische Fallbeispiel der romantischen Naturforschung*. Würzburg: Königshausen und Neumann, 1997.
Koepke, Wulf. "Der späte Lessing und die junge Generation." In *Humanität und Dialog. Beiheft zum Lessing-Jahrbuch*, 211–22. Detroit: Wayne State University Press, 1982.
Kołakowski, Leszek. *Main Currents of Marxism: The Founders, the Golden Age, the Breakdown*. Translated by P. S. Falla. New York: Norton, 2005.
Kondylis, Panajotis. *Die Enstehung der Dialektik: Eine Analyse der geistigen Entwicklung von Hölderlin, Schelling und Hegel bis 1802*. Stuttgart: Klett-Cotta, 1979.

Korsgaard, Christine M. *Creating the Kingdom of Ends*. Cambridge: Cambridge University Press, 1996.

Koschorke, Albrecht. *Körperströme und Schriftverkehr: Mediologie des 18. Jahrhunderts*. Munich: Fink, 2003.

———. "Physiological Self-Regulation: The Eighteenth-Century Modernization of the Human Body." *MLN* 123, no. 3 (2008): 469–84.

Koyré, Alexandre. *From the Closed World to the Infinite Universe*. Baltimore: Johns Hopkins University Press, 1968.

Koyré, Alexandre, and I. Bernard Cohen. "The Case of the Missing Tanquam: Leibniz, Newton & Clarke." *Isis* 52, no. 4 (1961): 555–66.

Krippendorf, Ekkehart. *Goethe: Politik gegen den Zeitgeist*. Frankfurt a.M.: Insel, 1999.

Kuhn, Dorothea. *Empirische und ideelle Wirklichkeit: Studien über Goethes Kritik des französischen Akademiestreites*. Graz: Böhlau, 1967.

Kuhn, Thomas S. *The Essential Tension: Selected Studies in Scientific Tradition and Change*. Chicago: University of Chicago Press, 1977.

Kurz, Gerhard. *Mittelbarkeit und Vereinigung: Zum Verhältnis von Poesie, Reflexion und Revolution bei Hölderlin*. Stuttgart: Metzler, 1975.

Lacan, Jacques. *The Four Fundamental Concepts of Psycho-analysis*. Translated by Alan Sheridan. New York: Norton, 1978.

Lacoue-Labarthe, Philippe, and Jean-Luc Nancy. *The Literary Absolute: The Theory of Literature in German Romanticism*. Albany: SUNY Press, 1988.

Lambert, Johann Heinrich. *Neues Organon, oder Gedanken über die Erforschung und Bezeichnung des Wahren und dessen Unterscheidung vom Irrtum und Schein*. Leipzig: Wendler, 1764.

Larmore, Charles. "Hölderlin and Novalis." In *The Cambridge Companion to German Idealism*, edited by Karl Ameriks, 141–61. Cambridge: Cambridge University Press, 2000.

Larson, James L. "Vital Forces: Regulative Principles or Constitutive Agents? A Strategy in German Physiology, 1786–1802." *Isis* 70, no. 2 (1979).

Lauer, Christopher. *The Suspension of Reason in Hegel and Schelling*. London: Continuum, 2010.

Lebovic, Nitzan. *The Philosophy of Life and Death: Ludwig Klages and the Rise of a Nazi Biopolitics*. New York: Palgrave Macmillan, 2013.

Leibniz, Gottfried Wilhelm. *New Essays on Human Understanding*. Translated and edited by Peter Remnant and Jonathan Bennett. New York and Cambridge: Cambridge University Press, 1981.

———. *Philosophical Essays*. Edited by Roger Ariew and Daniel Garber. Indianapolis: Hackett, 1989.

———. *Philosophical Texts*. Edited by Richard Francks and R. S. Woolhouse. New York: Oxford University Press, 1998.

———. *Sämtliche Schriften und Briefe. IV: 6, 1695–1697*. Edited by Friedrich Breiderbeck et al. Berlin: de Gruyter, 2008.

———. *Theodicee, das ist, Versuch von der Güte Gottes, Freyheit des Menschen, und vom Ursprung des Bösen*. Translated by Johann Christoph Gottsched. Hannover and Leipzig: Verlag der försterischen Erben, 1763.

Lenoir, Timothy. "Göttingen School." Studies in the History of Biology 5 (1981): 11–205.

———. "Operationalizing Kant: Manifolds, Models and Mathematics in Helmholtz's Theories of Perception." In *The Kantian Legacy in Nineteenth-Century Science*, edited by Michael Friedman and Alfred Nordmann, 141–211. Cambridge: MIT Press, 2006.

———. *The Strategy of Life Teleology and Mechanics in Nineteenth-Century German Biology*. Chicago: University of Chicago Press, 1989.

Lessing, Gotthold Ephraim. *Gotthold Ephraim Lessing: Werke*. Edited by Herbert G. Göpfert. Munich: Hanser, 1973.

Lindberg, David C., Mary Jo Nye, Roy Porter, and Theodore M. Porter, eds. *The Cambridge History of Science*. Cambridge: Cambridge University Press, 2003.

Loheide, Bernard. *Fichte und Novalis: Transcendentalphilosophisches Denken im romantisierenden Diskurs*. Amsterdam: Rodopi, 2000.

Longuenesse, Béatrice. *Kant on the Human Standpoint*. Cambridge: Cambridge University Press, 2005.

Lovejoy, Arthur O. *The Great Chain of Being: A Study of the History of an Idea*. 1936. Cambridge: Harvard University Press, 1964.

———. "Schiller and the Genesis of German Romanticism." *MLN* 35, no. 1 (1920): 1–10.

Löw, Reinhard. *Philosophie des Lebendigen: Der Begriff des Organischen bei Kant, Sein Grund und seine Aktualität*. Frankfurt a.M.: Suhrkamp, 1980.

Löwith, Karl. *From Hegel to Nietzsche: The Revolution in Nineteenth-Century Thought*. New York: Anchor, 1967.

Lukács, Georg. *The Theory of the Novel: A Historico-Philosophical Essay on the Forms of Great Epic Literature*. Cambridge: MIT Press, 1971.

Mähl, Hans-Joachim. *Die Idee des goldenen Zeitalters im Werke des Novalis: Studien zur Wesensbestimmung der frühromantischen Utopie und zu ihren ideengeschichtlichen Voraussetzungen*. 1965. Tübingen: Niemeyer, 1994.

Mann, Gunter, Franz Dumont, and Gabriele Wenzel-Naos, eds. *Samuel Thomas Soemmerring und die Gelehrten der Goethezeit: Beiträge eines Symposions in Mainz vom 19. bis 21. Mai 1983*. Stuttgart: Fischer, 1985.

Marx, Karl, and Friedrich Engels. *Karl Marx, Frederick Engels: Collected Works*. New York: International, 1975–2004.

———. *Karl Marx, Friedrich Engels Gesamtausgabe (MEGA)*. Berlin: Dietz, 1972–.
———. *Karl Marx, Friedrich Engels: Werke*. Berlin: Dietz, 1956–90.
Marx, Leo. "Technology: The Emergence of a Hazardous Concept." *Technology and Culture* 51, no. 3 (2010): 561–77.
Matala de Mazza, Ethel. *Der verfaßte Körper: Zum Projekt einer organischen Gemeinschaft in der politischen Romantik*. Freiburg: Rombach, 1999.
Maturana, Humberto R., and Francisco J. Varela. *Autopoiesis and Cognition: The Realization of the Living*. Dordrecht, Holland: Reidel, 1980.
Mauthner, Fritz. *Wörterbuch der Philosophie: Neue Beiträge zu einer Kritik der Sprache*. Leipzig: Meiner, 1923.
Mayer, Paola. *Jena Romanticism and Its Appropriation of Jakob Böhme: Theosophy, Hagiography, Literature*. Montreal: McGill-Queen's University Press, 1999.
McDowell, John. *Mind and World: With a New Introduction*. Cambridge: Harvard University Press, 1996.
McLaughlin, Peter. "Soemmerring und Kant: Über das Organ der Seele und den Streit der Facultäten." In *Samuel Thomas Soemmerring und die Gelehrten der Goethezeit:Beiträge eines Symposions in Mainz vom 19. bis 21. Mai 1983*, edited by Gunter Mann et al., 191–203. Stuttgart: Fischer, 1985.
Meinecke, Friedrich. *Historism: The Rise of a New Historical Outlook*. London: Routledge and Kegan Paul, 1972.
Menninghaus, Winfried. "Die frühromantische Theorie von Zeichen und Metapher." *German Quarterly* 61, 1 (1989): 48–58.
———. *Unendliche Verdopplung: Die frühromantische Grundlegung der Kunsttheorie im Begriff absoluter Selbstreflexion*. Frankfurt a.M.: Suhrkamp, 1987.
Mensch, Jennifer. *Kant's Organicism: Epigenesis and the Development of Critical Philosophy*. Chicago: University of Chicago Press, 2013.
Mieth, Günter. *Friedrich Hölderlin: Dichter der bürgerlich-demokratischen Revolution*. Berlin: Rütten and Loening, 1978.
Millán-Zaibert, Elizabeth. *Friedrich Schlegel and the Emergence of Romantic Philosophy*. Albany: SUNY Press, 2007.
Mitchell, Robert. *Experimental Life: Vitalism in Romantic Science and Literature*. Baltimore: Johns Hopkins University Press, 2013.
Moiso, Francesco. "Magnetismus, Elektrizität, Galvanismus." In *Historisch-kritische Ausgabe by Friedrich Wilhelm Joseph von Schelling. Ergänzungsband zu Werke Band 5 bis 9, Wissenschaftshistorischer Bericht zu Schellings Naturphilosophen Schriften 1797–1800*, 320–75. Stuttgart: frommann-holzboog, 1994.
Müller-Sievers, Helmut. *Self-Generation: Biology, Philosophy, and Literature around 1800*. Stanford, CA: Stanford University Press, 1997.
Nadler, Steven M. *Causation in Early Modern Philosophy: Cartesianism, Occasionalism, and Preestablished Harmony*. University Park: Pennsylvania State University Press, 1993.

Nassar, Dalia. "From a Philosophy of Self to a Philosophy of Nature: Goethe and the Development of Schelling's *Naturphilosophe*." *Archiv für Geschichte der Philosophie* 92, no. 3 (2010): 304–21.

———. "Interpreting Novalis' *Fichte-Studien*." *Deutsche Vierteljahrsschrift für Literaturwissenschaft und Geistesgeschichte* 84, no. 3 (2010): 315–41.

———. *The Romantic Absolute: Being and Knowing in Early German Romantic Philosophy*. Chicago: University of Chicago Press, 2013.

Neubauer, John. *Bifocal Vision: Novalis' Philosophy of Nature and Disease*. Chapel Hill: University of North Carolina Press, 1971.

———. "Intellektuelle, intelktuale und ästhetische Anschauung. Zur Entstehung der romantischen Kunstauffassung," *Deutsche Vierteljahrsschrift für Literaturwissenschaft und Geistesgeschichte* 46, no. 1 (1972): 294–319.

———. *Symbolismus und symbolische Logik: Die Idee der Ars Combinatoria in der Entwicklung der modernen Dichtung*. Munich: Fink, 1978.

Neumann, Gerhard. "'Mannigfache Wege gehen die Menschen': Romananfänge bei Goethe und Novalis." In *Goethe und das Zeitalter der Romantik*, edited by Walter Hinderer, 71–91. Würzburg: Königshausen und Neumann, 2002.

———. "Romantische Aufklärung. Zu E.T.A. Hoffmanns Wissenschaftspoetik," in *Aufklärung als Form. Beiträge zu einem historischen und aktuellen Problem*, edited by Helmut Schmiedt and Helmut J. Schneider, 106–148. Würzburg: Königshausen und Neumann, 1997.

Newton, Isaac. *Opticks, or, a Treatise of the Reflections, Refractions, Inflections and Colours of Light*. London: Innys, 1730.

Nisbet, H. B. "Herder, Goethe, and the Natural 'Type.'" In *Publications of the English Goethe Society vol. XXXVII: Papers Read before the Society 1966–67*, edited by Elizabeth M. Wilkinson et al., 81–119. Leeds: Maney and Son, 1967.

Norton, Robert E. "The Myth of the Counter-Enlightenment." *Journal of the History of Ideas* 68 (2007): 635–58.

Novalis. *Notes for a Romantic Encyclopedia: Das allgemeine Brouillon*. Edited and translated by David Wood. Albany: SUNY Press, 2007.

———. *Novalis Schriften: Die Werke Friedrich von Hardenbergs*. Edited by Paul Kluckhohn and Richard H. Samuel. 1960. Stuttgart: Kohlhammer, 1977.

Nyhart, Lynn K. *Biology Takes Form: Animal Morphology and the German Universities, 1800-1900*. Chicago: University of Chicago Press, 1995.

O'Brien, William Arctander. *Novalis, Signs of Revolution*. Durham: Duke University Press, 1995.

Oken, Lorenz. *Über das Universum als Fortsetzung des Sinnensystems: Ein Pythagoräisches Fragment von Oken*. Jena: Frommann, 1808.

Otis, Laura. *Müller's Lab*. Oxford: Oxford University Press, 2007.

Pinkard, Terry P. *German Philosophy, 1760–1860: The Legacy of Idealism*. Cambridge: Cambridge University Press, 2002.

Platner, Ernst. *Anthropologie für Aerzte und Weltweise*. Leipzig: Dyck, 1772.

———. *Neue Anthropologie für Aerzte und Weltweise: Mit Besonderer Rücksicht auf Physiologie, Pathologie, Moralphilosophie und Aesthetik*. Leipzig: Crusius, 1790.

Pudelek, Jan-Peter. *Der Begriff der Technikästhetik und ihr Ursprung in der Poetik des 18.Jahrhunderts*. Würzburg: Königshausen und Neumann, 2000.

Rabinbach, Anselm. *The Human Motor: Energy, Fatigue, and the Origins of Modernity*. Berkeley: University of California Press, 1990.

Reil, Johann Christian. *Gesammelte kleine physiologische Schriften*. Vienna: Doll, 1811.

———. *Von der Lebenskraft*. Leipzig: Johann Ambrosius Barth, 1910.

Reill, Peter Hanns. "*Bildung*, *Urtyp* and Polarity: Goethe and Eighteenth-Century Physiology." *Goethe Yearbook* 3, no. 1 (1986): 139–48.

———. *Vitalizing Nature in the Enlightenment*. Berkeley: University of California Press, 2005.

Ricardo, David. *On the Principles of Political Economy and Taxation*. London: John Murray, 1821.

Richards, Robert John. *The Meaning of Evolution: The Morphological Construction and Ideological Reconstruction of Darwin's Theory*. Chicago and London: University of Chicago Press, 1992.

———. "Nature Is the Poetry of Mind, or How Schelling Solved Goethe's Kantian Problems." In *The Kantian Legacy in Nineteenth-Century Science*, edited by Michael Friedman and Alfred Nordmann, 27–51. Cambridge: MIT Press, 2006.

———. *The Romantic Conception of Life: Science and Philosophy in the Age of Goethe*. Chicago: University of Chicago Press, 2002.

Richter, Simon. *Laocoon's Body and the Aesthetics of Pain: Winckelmann, Lessing, Herder, Moritz, Goethe*. Detroit: Wayne State University Press, 1992.

———, ed. *The Literature of Weimar Classicism*. Rochester, NY: Camden House, 2005.

———. "Medizinischer und ästhetischer Diskurs im 18. Jahrhundert: Herder und Haller über Reiz." *Lessing-Yearbook* 25 (1993): 83–97.

Ritter, Johann Wilhelm. *Beweis, dass ein Beständiger Galvanismus den Lebensprocess in dem Thierreich begleite: Nebst neuen Versuchen und Bemerkungen über den Galvanismus*. Weimar: Verlag des Industrie-Comptoirs, 1798.

Robinet, Jean-Baptiste-René. *De la nature (IV)*. Amsterdam: van Harrevelt, 1761.

Roder, Florian. *Menschwerdung des Menschen: Der magische Idealismus im Werk des Novalis*. Stuttgart: Mayer, 1997.

Roe, Shirley A. *Matter, Life, and Generation: Eighteenth-Century Embryology and the Haller-Wolff Debate*. Cambridge: Cambridge University Press, 1981.

Röhr, Werner, ed. *Appellation an Das Publikum . . . : Dokumente zum Atheismusstreit um Fichte, Forberg, Niethammer; Jena 1789/99.* Leipzig: Reclam, 1991.

Ronell, Avital. *Dictations: On Haunted Writing.* Bloomington: Indiana University Press, 1986.

———. *The Telephone Book: Technology, Schizophrenia, Electric Speech* (Lincoln: University of Nebraska Press, 1991),

———. *The Test Drive.* Urbana: University of Illinois Press, 2005.

Ropohl, Günther. "Karl Marx und die Technik." In *Die technikhistorische Forschung in Deutschland von 1800 bis zur Gegenwart*, edited by Wolfgang König and Helmuth Schneider, 63–85. Kassel: Kassel University Press, 2007.

Rothschuh, Karl. "Alexander von Humboldt und die Physiologie seiner Zeit." *Sudhoffs Archiv für Geschichte der Medizin und der Naturwissenschaften* 43, no. 2 (1959): 97–113.

———. *Geschichte der Physiologie.* 1953. Berlin: Springer, 2012.

———. "Von der Idee bis zum Nachweis der tierischen Elektrizität." *Sudhoffs Archiv für Geschichte der Medizin und der Naturwissenschaften* 44, no. 1 (1960): 25–44.

Saller, Reinhard. *Schöne Ökonomie: Die poetische Reflexiion der Ökonomie in frühromantischer Literatur.* Würzburg: Königshausen und Neumann, 2007.

Schanze, Helmut. *Romantik und Aufklärung: Untersuchungen zu Friedrich Schlegel und Novalis.* Nürnberg: Carl, 1966.

Schelling, Friedrich Wilhelm Joseph von. *First Outline of a System of the Philosophy of Nature.* Translated by Keith R. Peterson. Albany: SUNY Press, 2004.

———. *Historisch-kritische Ausgabe.* Edited by Jörg Jantzen, Thomas Buchheim, Jochem Hennigfeld, Wilhelm G. Jacobs, and Siegbert Peetz. Stuttgart: frommann-holzboog, 1976–.

———. *Philosophical Investigations into the Essence of Human Freedom.* Translated by Jeff Love and Johannes Schmidt. Albany: SUNY Press, 2006.

———. *Philosophische Entwürfe und Tagebücher: 1809–1813, Philosophie der Freiheit und der Weltalter.* Edited by Lothar Knatz, Hans Jörg Sandkühler, and Martiv Schraven. Hamburg: Meiner 1994.

———. *Philosophische Untersuchungen über das Wesen der menschlichen Freiheit und die damit zusammenhängenden Gegenstände.* Edited by Thomas Buchheim. Hamburg: Meiner, 1997.

———. *Sämmtliche Werke.* Edited by Karl-Friedrich-August Schelling. Stuttgart: Cotta, 1856–61.

———. *Schellings Werke: Auswahl in drei Bänden.* Edited by Otto Weiß. Leipzig: Eckardt, 1907.

Schiller, Friedrich. *Sämtliche Werke*. Edited by Gerhard Fricke. Munich: Hanser, 1962.

Schlechta, Karl. *Goethe in seinem Verhältnisse zu Aristoteles: Ein Versuch*. Frankfurt a.M.: Klostermann, 1938.

Schlegel, August Wilhelm von. *Kritische Ausgabe: August Wilhelm Schlegel— Vorlesungen von 1798–1827: Vorlesungen über Enzyklopädie der Wissenschaften* (1803): Vol. 3., edited by Frank Jolles, Edith Höltenschmidt, and Ernst Behler. Paderborn: Schöningh, 2006.

Schlegel, Friedrich von. *Kritische Friedrich-Schlegel-Ausgabe*. Edited by Ernst Behler, Jean Jacques Anstett, and Hans Eichner. Munich: Schöningh, 1958–.

Schleiermacher, Friedrich. *Über die Religion: Reden an die Gebildeten unter ihren Verächtern*. Hamburg: Meiner, 1958.

Schmidt, Alfred. *The Concept of Nature in Marx*. Translated by Ben Fowkes. London: New Left, 1971.

Schmidt, Dennis J. *On Germans & Other Greeks: Tragedy and Ethical Life*. Bloomington: Indiana University Press, 2001.

Schmidt, Siegfried J. *Sprache und Denken als sprachphilosophisches Problem von Locke bis Wittgenstein*. The Hague: Nijhoff, 1968.

Schönherr, Hartmut R. *Einheit und Werden: Goethes Newton-Polemik als systematische Konsequenz seiner Naturkonzeption*. Würzburg: Königshausen and Neumann, 1993.

Schramm, Helmar, Ludger Schwarte, and Jan Lazardzig, eds. *Collection, Laboratory, Theater: Scenes of Knowledge in the 17th Century*. Berlin: de Gruyter, 2005.

Schubert, Gotthilf Heinrich von. *Ansichten von der Nachtseite der Naturwissenschaft*. Dresden: Arnold, 1808.

———. *Spiegel der Natur: Ein Lesebuch zur Belehrung und Unterhaltung*. Erlangen: Palm und Enke, 1854.

Schwann, Theodor. *Mikroskopische Untersuchungen über die Uebereinstimmung in der Struktur und dem Wachstthum der Thiere und Pflanzen*. Berlin: Sander, 1839.

Scribner, F. Scott. *Matters of Spirit: J. G. Fichte and the Technological Imagination*. University Park: Pennsylvania State University Press, 2010.

Sedgwick, Sally S. *Hegel's Critique of Kant: From Dichotomy to Identity*. Oxford: Oxford University Press, 2012.

Seibicke, Wilfried. *Technik: Versuch einer Geschichte der Wortfamilie um Réyvn in Deutschland vom 16. Jahrhundert bis etwa 1830*. Dusseldorf: VDI-Verlag, 1968.

Seyhan, Azade. *Representation and Its Discontents: The Critical Legacy of German Romanticism*. Berkeley: University of California Press, 1992.

Sheehan, Helena. *Marxism and the Philosophy of Science: A Critical History.* Atlantic Highlands, NJ: Humanities Press, 1993.
Sheehan, Jonathan. *The Enlightenment Bible: Translation, Scholarship, Culture.* Princeton, NJ: Princeton University Press, 2005.
———. "From Philology to Fossils: The Biblical Encyclopedia in Early Modern Europe." *Journal of the History of Ideas* 64, no. 1 (2003): 41–60.
Simmel, Georg. *The Metropolis and Mental Life: The Sociology of Georg Simmel.* New York: Free Press, 1976.
Simondon, Gilbert. *Du mode d'existence des objets techniques.* Paris: Aubier, 1989.
Smith, Justin E. H. *Divine Machines: Leibniz and the Sciences of Life.* Princeton, NJ: Princeton University Press, 2011.
———. *The Problem of Animal Generation in Early Modern Philosophy.* Cambridge: Cambridge University Press, 2006.
Soemmerring, Samuel Thomas von. *Über das Organ der Seele.* Königsberg: Friedrich Nicolovius, 1796.
———. *Ueber das Organ der Seele (1796).* Edited by Manfred Wenzel. Basel: Schwabe, 1999.
Specht, Benjamin. *Physik als Kunst: Die Poetisierung der Elektrizität um 1800.* Berlin: de Gruyter, 2010.
Specht, Rainer. *Commercium mentis et corporis: Über Kausalvorstellungen im Cartesianismus.* Stuttgart–Bad Cannstatt: frommann-holzboog, 1966.
Spinoza, Benedictus de. *Ethics.* Edited by G. H. R. Parkinson. Oxford: Oxford University Press, 2000.
Spoerhase, Carlos. "Poetik des Plans: Über die Form des Systems um 1800." Forthcoming.
Stadler, Ulrich. *Die theuren Dinge.* Munich: Francke, 1980.
Stammen, Theo. *Goethe und die politische Welt: Studien.* Würzburg: Ergon, 1999.
Stiegler, Bernard. *For a New Critique of Political Economy.* Cambridge: Polity, 2010.
Tang, Chengxi. *The Geographic Imagination of Modernity: Geography, Literature, and Philosophy in German Romanticism.* Palo Alto: Stanford University Press, 2008.
Tantillo, Astrida Orle. *The Will to Create: Goethe's Philosophy of Nature.* Pittsburgh: University of Pittsburgh Press, 2002.
Taylor, Charles. *Philosophical Arguments.* Cambridge: Harvard University Press, 1995.
Taylor-Alexander, Samuel. "How the Face Became an Organ." Somatosphere. August 11, 2014. http://somatosphere.net/2014/08/how-the-face-became-an-organ.html.
———. *On Face Transplantation: Life and Ethics in Experimental Biomedicine.* Houndmills, Basingstoke: Palgrave Macmillan, 2014.

Toews, John Edward. *Hegelianism: The Path toward Dialectical Humanism, 1805–1841*. Cambridge: Cambridge University Press, 1980.
Tonelli, Giorgio. "Die Umwälzung von 1769 bei Kant." *Kant-Studien* 54, no. 1–4 (1963): 369–75.
———. *Kant's "Critique of Pure Reason" within the Tradition of Modern Logic: A Commentary on Its History*. Edited by David Howard Chandler. Zürich: G. Olms, 1994.
———. "Leibniz on Innate Ideas and the Early Reactions to the Publication of the Nouveaux Essais (1765)." *Journal of the History of Philosophy* 12, no. 4 (1974): 437–54.
Torretti, Roberto. *The Philosophy of Physics*. Cambridge: Cambridge University Press, 1999.
Tresch, John. *The Romantic Machine: Utopian Science and Technology after Napoleon*. Chicago: University of Chicago Press, 2012.
Tresch, John, and Emily I. Dolan. "Toward a New Organology: Instruments of Music and Science." *Osiris* 28 (2013): 278–98.
Trop, Gabriel. "Modal Revolutions: Friedrich Hölderlin and the Task of Poetry." *MLN* 128, no. 3 (2013): 580–610.
Tsouyopoulos, Nelly. *Asklepios und die Philosophen: Paradigmawechsel in der Medizin im 19. Jahrhundert*. Edited by Claudia Wiesemann, Barbara Bröker, and Sabine Rogge. Stuttgart–Bad Cannstatt: frommann-holzboog, 2008.
———. "The Influence of John Brown's Ideas in Germany." *Medical History*, Supplement no. 8 (1988): 63–74.
Ure, Andrew. *Philosophy of Manufacture; or, an Exposition of the Scientific, Moral, and Commercial Economy of the Factory System*. London: Clowes and Sons, 1835.
Vaihinger, Hans. *The Philosophy of "As If": A System of the Theoretical, Practical and Religious Fictions of Mankind*. Translated by Charles K. Ogden. London: Routledge and Kegan Paul, 1965.
Verplaetse, Jan. *Localizing the Moral Sense: Neuroscience and the Search for the Cerebral Seat of Morality, 1800–1930*. New York: Springer, 2009.
Waibel, Violetta. *Hölderlin und Fichte, 1794–1800*. Paderborn: Schöningh, 2000.
———. "'Inneres, äußeres Organ': Das Problem der Gemeinschaft von Seele und Körper in den *Fichte-Studien* Friedrich von Hardenbergs." *Athenäum: Jahrbuch für Romantik* 10 (2000): 159–81.
———. "Wechselbestimmung: Zum Verhältnis von Hölderlin, Schiller und Fichte in Jena." In *Fichte und die Romantik—Hölderlin, Schelling, Hegel und die späte Wissenschaftslehre: "200 Jahre Wissenschaftslehre—Die Philosophie Johann Gottlieb Fichtes"; Tagung der Internationalen J.-G.-Fichte-Gesellschaft*

(*26. September—1. Oktober 1994*) *in Jena*, edited by Wolfgang H. Schrader, 43–71. Amsterdam: Rodopi, 1997.

———. "'With respect to the antinomies, Fichte has a remarkable idea': Three Answers to Kant and Fichte: Hardenberg, Hölderlin, Hegel." In *Fichte, German Idealism, and Early Romanticism*, edited by Daniel Breazeale and Tom Rockmore, 301–27. Amsterdam: Rodopi, 2010.

Weatherby, Leif. *Becoming Human: Romantic Anthropology and the Embodiment of Freedom*. University Park: Pennsylvania State University Press, 2010.

———. "Das Innere der Natur und ihr Organ: Von Albrecht von Haller zu Goethe." *Goethe-Yearbook* 21 (2014): 191–217.

———."Goethe's Morphology of Knowledge, or the Overgrowth of Nomenclature." *Goethe Yearbook* 17 (2010): 153–77.

———. "Lyrical Feeling: Novalis' Anthropology of the Senses." *Studies in Romanticism* 47, no. 4 (2008): 453–77.

———. "The Romantic Circumstance: Novalis between Kittler and Luhmann." *SubStance* 43 (2014): 46–66.

———. "Touching Books: Diderot, Novalis, and the Encyclopedia of the Future." *Representations* 114, no. 1 (2011): 65–102.

Wellbery, David. "Form und Idee: Skizze eines Begriffsfeldes um 1800." In *Morphologie und Moderne: Goethes "anschauliches Denken" in den Geistes- und Kulturwissenschaften seit 1800*, edited by Jonas Maatsch, 17–43. Berlin: de Gruyter, 2014.

———. "The Imagination of Freedom: Goethe and Hegel as Contemporaries." In *Goethe's Ghosts: Reading and the Persistence of Literature*, edited by Simon Richter and Richard Block, 217–39. Rochester: Camden House, 2013.

———. *Lessing's Laocoon: Semiotics and Aesthetics in the Age of Reason*. Cambridge: Cambridge University Press, 1984.Welsh, Caroline. *Hirnhöhlenpoetiken: Theorien zur Warhnehmung in Wissenschaft, Ästhetik und Literatur um 1800*. Freiburg: Rombach, 2003.

Wenzel, Manfred. "Johann Wolfgang von Goethe und Samuel Thomas Soemmerring: Morphologie und Farbenlehre." In *Samuel Thomas Soemmerring und die Gelehrten der Goethezeit: Beiträge eines Symposions in Mainz vom 19. bis 21. Mai 1983*, edited by Gunter Mann et al., 11–35. Stuttgart: Fischer, 1985.

Wessell, Leonard. "Schiller and the Genesis of German Romanticism." *Studies in Romanticism* 10, no. 3 (1971): 176–98.

Wilson, Catherine. *The Invisible World: Early Modern Philosophy and the Invention of the Microscope*. Princeton, NJ: Princeton University Press, 1995.

Winckelmann, Johann Joachim, and Helmut Holtzhauer. *Winckelmanns Werke in einem Band*. Berlin: Aufbau-Verlag, 1969.

Wolf, Jörn Henning. *Der Begriff Organ in der Medizin*. Munich: Fritzsch, 1971.
Wolff, Caspar Friedrich. *Theorie von der Generation*. Berlin: Birnstiel, 1764.
Wolff, Christian. *Erste Philosophie, oder, Ontologie nach wissenschaftlicher Methode Behandelt, in der die Prinzipien der gesamten menschlichen Erkenntnis enthalten sind: 1–78: Lateinisch-deutsch*. Edited and translated by and Dirk Effertz. Hamburg: Meiner, 2005.
Wood, David. "A Scientific Bible: Novalis and the Encyclopedistics of Nature." In *The Book of Nature in Early Modern and Modern History*, edited by Klaas Van Berkel and Arie Johan, 167–79. Leuven, Belgium: Peeters, 2006.
Yolton, John W. *Locke and French Materialism*. Oxford: Clarendon, 1991.
Zammito, John H. *Kant, Herder, and the Birth of Anthropology*. Chicago: University of Chicago Press, 2002.
Zelle, Carsten. "Sinnlichkeit und Therapie: Zur Gleichursprünglichkeit von Ästhetik und Anthropologie um 1750." In *"Vernünftige Ärzte": Hallesche Psychomediziner und die Anfänge der Anthropologie in der deutschsprachigen Frühaufklärung*, edited by Zelle, 6–24. Tübingen: Niemeyer, 2001.
Žižek, Slavoj. *Organs without Bodies: On Deleuze and Consequences*. New York: Routledge, 2003.
———. *Tarrying with the Negative Kant, Hegel, and the Critique of Ideology*. Durham, NC: Duke University Press, 1993.

INDEX

Absolute being, 144
Absolute self-identity, 197
Absolute spirit, 137
Acosmism, 51
Admiration, 155
Adorno, Theodor, 328
Aenesidemus (Schulze), 143
Aesthetics: aestheticization, 25, 42, 207; and anthropology, 32; and being, 40; classical, 13, 84; content, 329; doctrine, 172; drive, 137; in Enlightenment, 110; and intuition, 174, 175, 193, 202, 203; and judgment, 13; and metaphysics, 12, 153; organology, 252; and Romanticism, 32; sense, 136; technicization of, 357n27; unification of opposites, 147
Aetheric current, 182
Aistheterion, 18, 19, 66, 103
Alexis or the Golden Age (Hemsterhuis), 241
Alienation, 222, 400n88
Alogos, 105
Althusser, Louis, 419n18
Amphibolies of Reflexive Concepts (Kant), 194
Analytic of the Understanding (Kant), 194
Anatomical comparison, 308
Anaxagoras, 20, 310
Animal body analogy, 207
Anorganic, 159, 167, 168, 181
Antecedent causality, 4
Anthropology, 110, 247, 311

Anthropology (Platner), 113, 378n18
Anthropology for Doctors and Philosophers (Platner), 117
Anthropology from a Pragmatic Standpoint (Kant), 225
Antigone (Sophocles), 161
Aorgic, 158
Appearance, 179
Apperception, 86, 89
Applied Critique, 235
Applied organology, 348
Ardinghello, or the Happy Isles (Heinse), 132, 160
Aristotle, 2, 4, 13, 16, 17–22, 40, 49, 54, 63, 74, 125, 206, 213, 298, 309, 310, 312, 359nn49–50, 360n62, 389n17, 417n109
Art, 164, 193, 203, 204, 238
Artificial selection, 320
Athenäum (periodical), 31
Äußerung, 293
Automatization process, 339, 421n50

Babbage, Charles, 421n55
Bacon, Francis, 7, 16
Baumgarten, Alexander Gottlieb, 58, 88
Beauty, 205. *See also* Aesthetics
Beckmann, Johann, 210
Behler, Ernst, 31, 358n40
Being: and aesthetics, 40; foreignness of, 127; Hölderlin on, 145–54, 383n25; and intellectual intuition, 135–44; and judgment, 138, 151, 154; reduction to

449

method, 127; representation of, 150, 173; Romantic conceptions of, 125, 126; and Will, 272
Beiser, Frederick, 31, 282, 379n1
Benjamin, Walter, 328
Bernard, Claude, 2
Bible, 224, 226–32, 274, 333, 397n59
Bildung, 312–15
Bildungstrieb, 69
Bioelectricity, 185
Biologized aesthetics, 207
Biology, 44, 45, 207, 256, 280, 309, 319, 323, 362n78
Biopower, 346
Blanchot, Maurice, 31, 397n45
Blumenbach, Johann Friedrich, 64, 68–70, 83, 99, 110, 137, 181, 183, 184
Blumenberg, Hans, 39–44, 57, 65, 351, 352, 362n84, 363n97, 396n42
Böhme, Jakob, 270, 407n71
Bonnet, Charles, 63
Borrowed life, 184
Bos, A. P., 360n59
Bourgeois republicanism, 261
Bowie, Andrew, 270
Bowman, Brady, 415n88
Brandis, Joachim Dietrich, 111, 113
Breckman, Warren, 420n19
The Bride of Messina (Schiller), 165
Broman, Thomas, 209
Brown, John, 34, 186, 208, 225
Büchner, Ludwig, 419n14
Burdach, Karl Friedrich, 255
Butler, Samuel, 316, 317

Campe, Johann Heinrich, 15
Candolle, Augustin Pyramus de, 307
Canguilhem, Georges, 8
Canon of orientation within cognition, 96
Canon of the understanding, 76, 81–83, 88, 96, 133, 139, 196
Capacity, 246, 264, 311, 422n72
Capital (Marx), 324–26, 330, 332, 336, 339–41, 344, 347, 351

Capitalism, 339
Carus, Gustav, 256
Cassirer, Ernst, 15, 38, 251
Categorical function, 286
Categories: generation of, 194, 266, 372n42; Goethe on, 297–306; of modality, 93, 94
Catharticon, 78
Causality: antecedent, 4; effective, 141; efficient, 54, 67; instrumental, 5; mechanical, 67; reciprocal, 177, 178, 180, 183
Cavendish, Henry, 185
Cell theory, 36, 280, 362n79
Cheung, Tobias, 361n66
Child labor, 340
Chomsky, Noam, 376n98
Chorus, 12, 165–67. *See also* Tragic chorus
Christendom or Europe (Novalis), 249
Church doctrine, 115
Clarke, Samuel, 57, 103
Classical tragedy theory, 155
Classicism, 12, 14, 30, 260, 309, 313
Classicism (Hölderlin), 134
Classification, 313. *See also* Categories
Coalition Wars (1792–1802), 132, 161
Cognition: and being, 154; and consciousness, 84, 87, 89, 127, 150, 193, 242; embodied, 255; and judgment, 133, 140, 218; Kant on, 81, 82, 84–91, 109; and knowledge, 126; Schelling on, 187, 195; spatialization of, 213; tools for, 244; transcendental sources of, 195. *See also* Knowledge
Cognitive capacity, 139, 177
The Collector and His Circle (Goethe), 291, 295, 296, 306, 313
Comparative physiology, 180–92
Compensation, 183, 417n109
Complementarity, 311
Complete concept doctrine, 365n9
Conceptuality, 286, 287
The Conflict of the Faculties (Kant), 27, 114, 116, 214, 225, 228

Consciousness: and cognition, 84, 86, 87, 89, 127, 193; and culture, 110; Fichte on, 209; Foucault on, 45; Hegel on, 265; and identity, 40; instance of, 109; and intellectual intuition, 134, 168; and judgment, 146, 147, 150, 152; and nature, 195; objectification of, 174; organic nature of, 102; as purposeful organ, 41; and reason, 104; and self, 204; spectator-consciousness, 169; transcendental, 41. *See also* Self-consciousness
Conservation of force, 55, 365n12
Constellation work, 379n1, 409n6
Contingency, 319
Continuity, 255
Contractibility, 64, 184
Cooperation, 344, 345, 347
Correspondance on Tragedy (Nicolai), 155
Cosmology, 47, 89, 110, 160, 220, 252, 386n62
Critical metaphysics, 260
Critical methodology, 123
Critical Theory, 25, 329
Critique of Practical Reason (Kant), 96
Critique of Pure Reason (Kant), 30, 31, 67, 72, 74, 80, 81, 84, 86, 89, 91, 95, 96, 115, 120, 121, 141, 177, 180, 217, 225, 231, 237, 328
Critique of Teleological Judgment (Kant), 176
Critique of the Power of Judgment (Kant), 52, 67, 68, 83, 93, 96, 97, 140, 141, 157, 159, 160, 177, 178, 203
Culture industry, 329
Cuvier, Georges, 37, 45, 46, 279, 280, 298, 307, 308, 318
Cybernetic instrumentalism, 360n59

Daiber, Jürgen, 399n79
D'Alembert, Jean le Rond, 238
Darwin, Charles, 37, 308, 318–27, 334, 416n99
"*Darwin amongst the Machines*" (Butler), 317

De anima (Aristotle), 13, 20, 21, 300, 309, 310, 417n113
The Death of Empedocles (Hölderlin), 24, 131, 134, 137, 160–70
Deduction, 84. *See also* Logic
De generatione animalium (Aristotle), 63
De historia animalium (Aristotle), 309
Deleuze, Gilles, 317, 349
De partibus animalium (Aristotle), 309
Descartes, René, 1–5, 8, 18, 54, 58, 118, 263
Determinative judgment, 140
Developmental organic model, 156
Dialectical materialism, 337
Dialectical method, 325
Dialogue on Poetry (Schlegel), 10, 226, 259
Dictionary for the Explanation and Germanification of Those Foreign Expressions That Have Been Forced upon Our Language (Campe), 15
Diderot, Denis, 238
Disciplinarity, 39, 45, 251, 352
Discourse on Metaphysics (Leibniz), 59
Discourse on Method (Descartes), 1
Division of labor, 314, 338, 340
Doctrine of Color (Goethe), 301
Dogmatism, 92, 194, 217, 229–31
Dreams of a Spirit-Seer (Kant), 117
Durkheim, Émile, 349
Durner, Manfred, 271

Eckartshausen, Karl von, 356n23
Economic-Philosophical Manuscripts (Marx), 330
Education of the Human Race (Lessing), 228, 273
Effective causality, 141
Efficient causality, 54, 67, 141
Elective Affinities (Goethe), 314
Electricity, 182, 185, 190, 208, 241
Ellul, Jacques, 421n46
Embodiment, 246
Embryogenesis, 66, 206, 207
Empedocles (Hölderlin), 24, 131, 134, 137, 160–70

Empirical realism, 153
Empiricism, 109, 199, 283, 290, 304, 312
Encyclopedia (Diderot), 97
Encyclopedia (Hegel), 308
Encyclopedics, 224–40
Engels, Friedrich, 318, 319, 337, 338, 344, 421n47
Enlightenment, 9, 25, 32, 44, 73, 109, 184, 185. *See also* German Enlightenment
Ensyllogism, 96
Epigenesis, 73–97, 99, 145, 175, 197, 207, 327, 368n48
Epimarchus, 108, 109
Epistemology, 47, 59, 272
Erewhon (Butler), 316, 317
Essay Concerning Human Understanding (Locke), 60
Essay on Human Freedom (Schelling), 29
Essay on the Life Force (Soemmerring), 113
Essence, 117, 137, 366n17
Ethical ramifications of organology, 214
Euclid, 370n19
Evolution, 280, 307–8, 334, 335, 416n99. *See also* Natural selection
Exchange-value, 336
Excitability, 189, 191, 208
Excitation, 208, 209
"The Experiment as Mediator between Subject and Object" (Goethe), 414n64
Exposition of Physiology (Görres), 258

Faculties, 213, 221, 261, 398n64
Feudalism, 250
Feuerbach, Ludwig, 319, 324, 330, 331, 333, 351
Fichte, Johann Gottlieb: on anthropology, 110, 136; categorical systems of, 196; on consciousness, 208; and Criticism, 266; critique of Schelling, 392n85; doctrine of organs, 245; on higher and lower organs, 246; Hölderlin influenced by, 138, 142–44; influence of, 30–32; on intuition, 201, 218; on judgment, 383n23; Kant's influence on, 126; on knowledge, 201, 249, 393n89; on metaphysics, 26–28, 117, 133, 174, 175; Novalis influenced by, 237; on product of nature, 402n123; on representation, 197, 198; Schelling influenced by, 235; on self-consciousness, 84, 87, 140, 209
Fichte-Studies (Novalis), 126
Film, 328, 329
First Outline of a System of the Philosophy of Nature (Schelling), 175
Fischer, Gotthelf, 113
Force: and comparative physiology, 180–92; conservation of, 55, 365n12; force-bound unification, 56; force-continuity, 255; Herder on, 99, 101, 407n74; impulse mechanism, 341; Kant on, 56; Leibniz on, 66; Marx on, 325; model of, 59; plurality of, 120
Formative drive, 64, 69, 70, 106, 137, 181, 184
Förster, Eckart, 379n1, 408n86, 409n6, 415n80
Forster, Georg, 382n3
Foster, John Bellamy, 418n6
Foucault, Michel, 39, 44–46, 256, 346
The Foundation of Natural Right (Fichte), 245, 402n119
Foundations of Natural Law (Fichte), 28
Frank, Manfred, 31, 126, 127, 197, 205, 261, 372n51, 379n1, 380n4, 383n25, 399n78
Frankfurt school, 327, 328
Franklin, Benjamin, 334
Freedom, 25, 147, 150, 202, 236, 272, 350, 351
Freedom Essay (Schelling), 262, 269–73
Freiberg Mining Academy, 211
Freud, Sigmund, 41
Fruitbearing Society, 53
Functionality, 46, 57, 69, 95
Function-problem, 34

Gailus, Andreas, 381n3
Galen, 3–5, 298, 309–12, 356n16

Gall, Franz Joseph, 29, 252, 259, 262–64, 405n20
Galvani, Luigi, 185, 191
Galvanic response, 186, 192, 390n47
Gans, Eduard, 306, 307
Gasché, Rodolphe, 396n45
Genre theory, 133, 145
Geoffroy de St.-Hilaire, Étienne, 46, 279, 280, 298, 307, 318, 321, 409n4, 416n105
German Enlightenment, 5, 23, 53, 110
German Idealism, 48, 124, 149, 171
The German Ideology (Marx), 332
Gigante, Denise, 36
Gland H. *See* Pineal gland
Gloning, Thomas, 412n39
God: existence of, 160; human separation from, 48; knowledge of, 139; man as analogical organ of, 103; power of, 56; self-perception of, 182; and *sensorium*, 102, 103
God (Herder), 99, 110, 213, 219, 220, 407n74
God: Some Conversations (Herder), 271
Goethe, Johann Wolfgang von, 279–315; Aristotelian influences on, 417n109; on categories of organs, 297–306; critiques of, 283; economic model of, 314; Hegel's critique of, 30, 287–90, 411n20, 415n91; on higher experience, 414n64; on Hölderlin, 119, 120; and Idealism, 409n6, 410n7; idealism of, 295; on intellectual intuition, 415n80; on knowledge, 296, 299, 300, 302–3; on morphology, 318; on nature, 254; on Novalis, 249; on organs, 12, 13, 281–82, 292, 293, 295; on reason, 306–12; on representation, 297–306, 348; on taxonomy and idealism, 291–97; and technological metaphysics, 282; on tools of *bildung*, 312–15
Görres, Joseph von, 29, 252, 253, 258, 259, 404n18, 405n20
Gottsched, Johann Christoph, 55, 155
Governance, 250

Grammatical-Critical Dictionary of High-German Speech (Adelung), 62
Grammatology, 239
Grave, Johannes, 412n36
Ground for Empedokles (Hölderlin), 158, 162, 221
Grundrisse (Marx), 344
Guattari, Félix, 317
Guyer, Paul, 81, 89, 373n61

Haering, Theodor, 403n135
Hagner, Michael, 118, 263, 406n47
Haller, Albrecht von, 23, 33–35, 48, 63, 66, 184, 186, 188, 189, 254, 368n48
Hamann, Johann Georg, 375n98
Hardenberg, Friedrich von. *See* Novalis
Harvey, William, 258
Hegel, Georg Wilhelm Friedrich: on absolute spirit, 137; Classical philosophy of, 84; on consciousness, 143, 265, 266; on Goethe, 30, 287–90, 411n20, 415n91; on judgment, 268; on Kant, 282–87; on knowledge, 285, 287, 305, 308; Marx influenced by, 324, 330, 351; on Novalis, 214, 249–50, 403n135; on representation, 304; on Schelling, 171, 262–75
Heidegger, Martin, 43, 349, 386n51, 422n72
Heinrich of Ofterdingen (Novalis), 241, 243
Heinse, Wilhelm, 118, 119, 132, 135, 136, 170
Heinz, Marion, 374n75
Hemsterhuis, Franz, 28, 129, 214, 224, 239, 241, 400n98, 401n105
Henrich, Dieter, 31, 126, 372n50, 379n1, 380n4, 409n6
Herder, Johann Gottfried, 97–107; on analogy of nature, 374n78; cosmology of, 12; on epigenesis, 73, 111, 207; on force, 182, 190, 374n75, 407n74; Kant's critique of, 22–24, 107; Kant's influence on, 58; on language, 375n98; literalization of "organ" by, 71, 73, 293; on metaphys-

ics, 116, 117, 119; on *sensorium*, 66; on Spinoza, 271
Higher experience, 414n64
Higher realism, 233, 311
Hippocrates, 309
Histoire naturelle (Leclerc), 98
Historical materialism, 320
Historiography, 38–46
Hölderlin, Friedrich, 131–70; on art as cognition, 193; on being, 128, 145–54, 283, 383n25; on cognition, 88; on dialectical metaphysics, 118, 150, 350; on genre-theoretical metaphysics of judgment, 145–60, 387n69; on intellectual intuition, 135–44; on intuitive understanding, 154–60; on judgment, 145–54, 176, 383n25; on metaphysics, 8, 24, 25, 29, 32, 110, 111, 119; on mind as animal body, 206; "organ" terminology used by, 124, 269; Platner's influence on, 110, 117, 118; Schelling's critique of, 174; on self-knowledge, 372n50
Holism, 308, 357n29, 380n2
Holland, Jocelyn, 364n114
Hölscher, Uvo, 386nn61–62
Horkheimer, Max, 327, 328, 332, 343, 345
Horstmann, Rolf-Peter, 415n88
Hufeland, Christoph, 225
Human cognition. *See* Cognition
Humanism, 110
Humanization of work, 344
Humboldt, Alexander von, 108, 111, 175, 185, 186, 279
Humboldt, Wilhelm von, 309, 375n98
Hume, David, 74, 85
Humoral pathology, 208, 394n7
Husserl, Edmund, 40, 41, 43
Hylomorphism, 63, 310
Hyperion, 134

Iatropoetics, 258
Idealism: Blumenberg on, 41; of Goethe, 30, 296, 304, 306, 409n6, 410n7; of Hegel, 409n6; and knowledge, 127; of Novalis, 28, 399n78; and realism, 313; and Romanticism, 31, 172; of Schelling, 171, 174, 193, 196, 198, 203, 262, 270, 272
Ideal realism, 10, 260
Ideas (Herder), 98, 106, 115, 181, 193, 213, 220
Ideas for a Philosophy of Nature (Schelling), 175
Identity, 138, 255, 268, 305, 408n77
Ideology, 329, 420n34
Imagination, 11, 13, 229, 261, 304, 395n17
Inaugural Dissertation (Kant), 82
Indifference point, 173
Individualism, 246, 366n17
Industrial Revolution, 319, 326, 342
Infinite judgment, 268
Infinite unification, 146
Innatism, 61
Institute of Social Research, 327
Instrumentalization, 5, 6, 28, 43, 134, 173, 183, 235
Intellectual intuition: Goethe on, 415n80; Hölderlin on, 133–44, 149, 150, 154, 157, 158, 162–64; Kant on, 48, 118, 302; Novalis on, 212; Schelling on, 174, 178, 189, 192–205, 408n86
Intentionality, 41
Interior mould, 64
Intermittence, 42
Interventionist metaphysics, 14
Intuition: Fichte on, 201; Hegel on, 287; Hölderlin on, 154–60; Kant on, 82, 88, 90, 91, 95, 97, 141, 142, 194, 302; Mendelssohn on, 156; and representation, 348; Schelling on, 175–80, 195, 392n79; Schlegel on, 216–19. *See also* Intellectual intuition
Intuitive adequate knowledge, 48, 139
Intuitive judgment, 302
Intuitive understanding, 140, 142, 157, 158, 180
Irony, 31, 239
Irritability, 64, 66, 181–84, 186, 188, 189, 191

Jacobi, Friedrich Heinrich, 51, 126, 154, 217, 373n56, 380n4, 385n38
Jähnig, Dieter, 202
Judgment: genre-theoretical metaphysics of, 145–60; Goethe on, 303, 304; Hegel on, 285; Hölderlin on, 145–54, 383n25; and intellectual intuition, 135–44; Kant on, 85–87, 91–97, 177, 178, 218; mereological nature of, 95; metaphysics of, 170; and reason, 145; Schelling on, 393n86; unifying nature of, 383n23
Judgmentalism, 90
Judgmental knowledge, 199
Judgment and Being (Hölderlin), 126, 142, 144, 146

Kant, Immanuel: biology development role of, 369n67; on canon of the understanding, 81–83; on capacities, 88; categorical systems of, 196, 303; on cognition, 81, 82, 84–91, 109; Critical vision of, 22, 45, 46, 193, 212, 222, 398n59; critiques of, 283; definition of organ, 256, 314; on disciplinary organs, 110–16; division of judgmental cognition into intuition and concept, 26; on epigenesis, 73–97, 195; on faculties, 48, 377n11, 398n64; Fichte influenced by, 126; on force, 56; on freedom, 165; on "general logic," 78; Hegel on, 282–87; on Herder, 22–24, 58, 107, 374n75; on inner sense, 373n62; on intellectual intuition, 48, 118, 302; on intuition, 82, 88, 90, 91, 95, 97, 141, 142, 180, 194, 282, 302; on judgment, 85–87, 91–97, 104, 177, 178, 218, 383n23; on knowledge, 141, 225, 272; and Leibniz, 73, 74, 105, 107, 194; on metaphysics, 9, 13, 19, 20, 22–24, 27, 28; on methodology in metaphysics, 74, 77–78; on moralization of belief, 245; Novalis on, 224–40; on organic object, 68; on *organon*, 80, 120, 229; on organs of brain, 378n17; on "organs of reason," 224; and problem of the empirical, 284; on rationality, 76; on rationalization of will, 269, 274; on "synthetic organon," 286
Kant's Organicism (Mensch), 373n55
Kant-Studies (Novalis), 231
Kapp, Ernst, 329, 349
Kennedy, Emmet, 420n34
Kielmeyer, Carl Friedrich, 24, 111, 174, 175, 181–83, 190
Kittler, Friedrich, 238, 239
Knowledge: absolute, 158; Aristotle on, 300, 360n62; dialectical knowing, 95; disciplinarization of, 39, 251; Fichte on, 201, 249, 393n89; Goethe on, 296, 299, 300, 302–3; Hegel on, 285, 287, 305, 308; Hölderlin on, 174; identification of objects as, 58; intuitive, 139; Kant on, 141, 225, 272; Marx on, 321, 324, 325, 334, 345, 347; as mathematical, 213; Novalis on, 213–14, 231; Schelling on, 195, 197–99; self-justifying, 143; as tool of mind, 20. See also Self-knowledge
Köchy, Kristian, 380n2
Kolakowski, Leszek, 419n18
Korsgaard, Christine, 270
Kühn, Sophie von, 210
Kuhn, Thomas, 37
Kurz, Gerhard, 161

Lacan, Jacques, 349
Lacoue-Labarthe, Philippe, 31, 396n45
Lamarck, Jean-Baptiste, 37, 280
Lambert, Johann Heinrich, 16, 78–81, 239, 370n19, 370n23
Lang, Fritz, 343
Language, 104, 105, 337, 375n98
Lassalle, Ferdinand, 320
Lavater, Johann Kaspar, 265
Lebenskraft, 24
Leclerc, Georges-Louis, 64, 98
Legalism, 229
Leibniz, Gottfried Wilhelm von: on complete concept doctrine, 365n9; on divine preformation of mind, 52–61;

456   Index

doctrine of the intellect, 59; on epigenesis and force, 61–71; and Herder, 73; on "intuitive adequate knowledge," 48; and Kant, 73, 74, 105, 107, 194; on metaphorology of the organ, 92; metaphysics of, 22, 53, 365n8; on Newton's God, 103; on preformation, 207; and Reil, 112; on representations, 366n17
Lenoir, Timothy, 36, 369n67
Lessing, Gotthold Ephraim, 155, 156, 214, 240, 241, 291, 386n54
Letter on the Human and His Relations (Hemsterhuis), 241
Letters on Dogmatism and Criticism (Schelling), 196
"Letters on the Aesthetic Education of Mankind" (Schiller), 108
Life-force, 109
"Life-Force, or the Genius of Rhodes" (Humboldt), 108, 186
Life sciences, 62
Localization theory, 263, 377n16
Locke, John, 60, 85, 105, 207
Logic, 76, 78, 80, 83, 139, 161
Logos, 105
Love, 222
Löw, Reinhard, 360n62
Löwith, Karl, 411n20
Lucinde (Schlegel), 226
Lukàcs, Georg, 418n120

Machine-Fragment, 336
Machinification, 342
Machinofacture, 340, 342, 421n55
Magical idealism, 28, 232–34
Magnetism, 241
Mähl, Hans-Joachim, 241, 400n98
Maimon, Salomon, 126
Malpighi, Marcello, 56
Man, Paul de, 31
Marx, Karl: on communist organs, 327–32; and Darwin, 318–27; on humans as organs, 338–48; on ideology, 420n34; on knowledge, 321, 324, 325, 334, 345, 347; on materialism, 419n15; on mechanical arts, 421n55; on modes of production, 329; on nature, 418n6; on organ's *techne*, 348–52; on political economy, 332–38; on technological development, 172
Materialism, 80, 241, 319, 330, 333, 419n15
Material monism, 308, 309
Mathematics, 215
Mathesis universalis, 44
Matter, 179
Maturana, Humberto, 349
Mauthner, Fritz, 241
Maxims and Reflections (Goethe), 309
Means of production, 333, 335–37, 347
Mechanics, 24, 67, 159, 341, 347
Mediation, 224, 235
Mendelssohn, Moses, 155, 156
Menninghaus, Winfried, 31, 128, 385n44
Mensch, Jennifer, 370n5, 373n55
Mesmer, Franz, 359n46
Meta-Critique (Herder), 104
"Metamorphosis of Plants" (Goethe), 292, 297
Metaphorology, 40, 43, 73, 77, 92, 97, 106, 173
Metaphysics: confrontation between biology and physics, 173; of contradiction, 133; innovation in, 169, 175; of judgment, 24, 149, 152, 153, 155, 176, 305; and knowledge, 95; methodologization of, 127; as "queen of the disciplines," 51
Metropolis (Lang), 343
Microphysics of power, 44
Microscopes, 365n8
Mieth, Günther, 381–82n3
Mimesis, 147, 152, 169, 205, 296–99, 304, 305, 312
Mind-body problem, 2, 12, 374n75
Mittelwerkzeug, 118
Monism, 98, 154, 160, 357n29
Morality, 165, 216, 262–75
Moral technologies, 156

Moritz, Karl Phillip, 294
Morphological Notebooks (Goethe), 294, 297, 305
Morphology, 292, 298, 304, 309, 318
Müller-Sievers, Helmut, 69
Mumford, Lewis, 421n46
Mysticism, 126, 270

Nancy, Jean-Luc, 31, 396n45
Naturalism, 319
Natural selection, 33, 308, 318, 320. *See also* Evolution
Natural technology, 321
Nature: analytical division of, 289; as full of organs, 57; inclusion of reason in, 174; and judgment, 137; rationality of, 289; reveals itself through sense, 301; *techne* of, 323
Naturphilosophie: and consciousness, 109; emergence of, 124, 251–52; and Engels, 338; and Goethe, 289, 290; and Hegel, 305, 308; and metaphysics, 26; and mythological organs, 252–62; and organological biology, 36; and Schelling, 171–74, 177, 181, 182, 187, 189–91, 193, 196–99, 201, 204, 226, 251–75
Necessity, 25, 202, 350
Negative dogmatism, 144
Neologism, 17
Nervengeist, 378n17, 378n19
Neuer teutscher Merkur (periodical), 263
Neues Organon (Lambert), 78
New Anthropology for Doctors and Philosophers (Platner), 117
New Essays on Human Understanding (Locke), 60
New metaphysics, 43, 116–21, 123
New System of the Communication of Substances, and of the Union of Body and Soul (Leibniz), 54–57
Newton, Isaac, 47, 48, 63–65, 103, 127, 206, 236, 284, 289
Nexus effectivus, 97
Nicholas of Cusa, 40
Nicolai, Friedrich, 155

Niethammer, Friedrich Immanuel, 135
Nietzsche, Friedrich, 316
Nihilism, 51, 144, 385n38
Noncontradiction principle, 52
Normativity, 305
Notes for a Romantic Encyclopedia (Novalis), 26
Noumena, 302
Nous, 20, 21, 312
Novalis (Friedrich von Hardenberg), 206–50; on being, 128; on cognition, 84, 88; on cosmological organs, 216–24; critique of Kant, 228; on encyclopedics, 224–40; on freedom and necessity, 350; on function, 380n4; Hegel's critique of, 403n135; on Kant, 224–40, 400n88; on knowledge, 213–14, 231; on metaphysics, 25–29, 31, 32; on moral organs, 240–45; on organology, 232; "organ" terminology used by, 111, 119, 124, 236; on political organ, 212, 248–50; Schelling's critique of, 403n132; on social organs, 245–48
Novum organum scientiarum, 7
Nutritive drive, 181
Nyhart, Lynn, 362n78

Object-consciousness, 202
Objectification, 147, 150, 159, 164, 199, 200, 203
Objectivity, 283
Of the World Soul (Schelling), 175, 177, 187, 188
Oken, Lorenz, 254, 255
The Only Possible Proof (Kant), 79, 93
On Nature (Robinet), 100
"On Perpetual Peace" (Kant), 261
On the Cognition and Sensation of the Human Soul (Herder), 103, 181
"On the Concept of Republicanism" (Schlegel), 261
"On the Different Kinds of Poetry" (Hölderlin), 145
"On the Formative Drive" (Blumenbach), 65, 111

458   Index

On the Formative Imitation of Beauty (Moritz), 294
On the Motion of the Heart (Harvey), 258
On the Organ of the Soul (Soemmerring), 116, 132
"On the Procedure of the Poetic Spirit" (Hölderlin), 145
On the Relations of the Organic Forces amongst Themselves in the Series of the Different Organizations (Kielmeyer), 182
On the Study of Greek Poetry (Schlegel), 11, 260
On the Universe as a Extension of the Sensory System (Oken), 254
Ontosemiology, 31
Opticks (Newton), 47
The Order of Things (Foucault), 44
Ordo inversus, 127
Organ/organs: as appearance's subjective/objective contingency, 238; communist organs, 327–32; as complex mechanisms, 4; as concretion of field of possibility, 124; correlative force of, 68; cosmological function designated by, 48; cosmological organs, 216–24; and developmental aspect of life, 112; as dialectical, 133; dialectical organs, 144, 170, 189; as dividing line from God's *techne*, 62; divine organ, 103; double meaning of, 48; fine organ, 113; as forms of judgment, 232; functions of, 36; for Galenic theoreticians, 5; literalization of, 24, 52, 71, 110, 120, 326; Marx on, 327–38; as mediator between body and spirit, 2, 12, 172; metaphorization of, 6, 53; metaphysical and epistemological, 48; moral organs, 240–45; of motion, 7; mythological organs, 252–62; operating on a continuum of materiality and ideality, 16; of philosophy, 202; as political, 240; political organ, 212, 248–50; primary organs, 98; and question of being linked to question of knowledge, 15, 49; Romanticism's inheritance of term, 107; secondary organs, 98; as separative and binding elements of natural systems, 190; as set of rules for thinking, 16; social organs, 245–48; of the soul, 117, 132, 135, 136; as structure and mediation, 12; transcendentalization of, 189, 268; transplanting, 22–38; as vessel, 356n23. *See also* Organology; Romantic organology
Organe, 23
Organibility, 28, 234
Organicism, 70, 181, 189, 208, 242, 308
Organless knowers, 48
Organographie végétale (Candolle), 307
Organology: crystallization of, 259; dual mission of, 212; and evolutionary biology, 307–8; governance as, 250; as metaphysical *bricolage*, 10; as not biological, 33; Novalis on, 211, 215; and science's objects vs. philosophy's subjects, 302; as study of elements with variable functions, 9; as transcendental technology, 308. *See also* Romantic organology
Organon, 8, 17, 22, 72–81, 92, 115, 127, 200, 229, 311, 359nn49–50
Origin of Species (Darwin), 317, 318, 320

Pantheism, 6, 132, 137, 270, 384n33, 386n54
"The Part Played by Labor in the Becoming-Human of the Ape" (Engels), 337
Passivity, 327, 400n98, 401n100
Perception, 237, 241, 264, 284, 300, 330, 331
Phenomenology, 41
Phenomenology of Spirit (Hegel), 29, 238, 262, 265, 266, 285
Philosophical empiricism, 9
Philosophical faculty, 114
Philosophical Writings (Mendelssohn), 155
Philosophy of art, 201–3
The Philosophy of "As If" (Vaihinger), 6
Philosophy of technology, 350

Phrenology, 29, 262, 263, 265, 267
Physical Dictionary (Gehler), 62
Physics, 289, 323
Physiognomy, 265, 267
Physiological Institutions (Blumenbach), 184
Physiological teleology, 308, 309
Physiology, 2, 5, 233, 355n14
Physis-techne analogy, 18, 19, 24, 43, 57, 63, 101, 113, 134, 139, 142, 161, 163, 202, 204, 205, 238
Pineal gland (gland H), 2, 55, 118
Pittsburgh Hegelians, 371n38
Plan for a German-Loving Fraternity (Leibniz), 52
Plasticity, 112, 347
Platner, Ernst, 110, 116–18, 135, 378n17
Poesie, 138, 202, 204, 261
Poetry, 10–11, 25, 53, 145, 155, 157, 161, 230, 238, 240, 242, 259–60, 397n55
Political economy, 320, 322, 332–38
Political Romanticism, 347–48
Port Royale Logic, 84, 88
Positivism, 251, 255, 269, 377n16
Possibility, 219, 238, 240, 264, 300, 301
Potentia, 221, 264
Pre-established harmony, 54, 59, 85, 367n35
Preformation, 54, 56, 61, 99, 207, 367n41
Prejudgmental unity, 138
Principes de philosophie zoologique (Geoffroy), 416n105
Principles of Economy and Taxation (Ricardo), 338–39
Private property, 331
The Problem of Knowledge (Cassirer), 38
Production economy, 210
Productive imagination, 395n17
Progressive semiosis, 59
Proof, That a Constant Galvanism Accompanies the Life-Process in the Kindgom of Animals (Ritter), 186
Propulsion, 182
Propyläen (periodical), 291, 295
Prosyllogism, 96

Proto-existentialism, 126
Proton aistheterion, 119
Pseudo-Aristotle, 20
Psychology, 42, 89, 105, 285, 373n68
Pudelek, Jan-Peter, 357n27
Pure religious doctrine, 115

Quality, 89, 93–94, 112, 185, 190, 390n45

Rabinbach, Anselm, 419n14
Rational cognition, 187
Rationalism, 9, 23, 289. *See also* Reason
Rational psychology, 89
Rational theology, 89
Realism, 259, 313
Reason: as constructive, 311; Galen on, 311; Goethe on, 306–12; Hegel on, 265, 268, 271–74; and judgment, 145; language as organ of, 104; as organic, 102; Schelling on, 187; Schlegel on, 406n42
Receptivity, 299
Reciprocal causality, 177, 178, 180, 183
Recursive immediacy, 393n89
Reflective judgment, 140, 141, 159, 178
Reflexivity, 298
Reformation, 249
"*Refutation of Idealism*" (Kant), 90
Reil, Johann Christian, 24, 111–13, 148, 377n10, 378n18
Reill, Peter, 63, 413n42
Reinhold, Karl Leonard, 84, 116, 126, 135, 143, 197, 357n29, 384n33
Relations of production, 336, 347
Religion, 115, 217, 218, 220, 222, 229, 261
Religion within the Boundaries of Mere Reason (Kant), 114
Representation: of being, 150, 173; and cognition, 154; Fichte on, 197, 198; Goethe on, 297–306, 348; Hegel on, 266, 286; Hölderlin on, 146, 147; and intuition, 348; Leibniz on, 366n17; Marx on, 323; and metaphysics, 22, 153; and mimesis, 312; Novalis on, 239, 240; Reinhold on, 143; Romantic conceptions of, 125; Schelling on, 183;

self-representation, 226, 299; and transcendental-empirical organ, 44–46
Representation and Its Discontents (Seyhan), 45
Reproduction, 147, 153, 169, 181–83, 190, 390n35
Republican government, 261, 262
Ricardo, David, 320, 338–39
Richards, Robert, 308, 369n67, 388n8, 413n41, 416n99
Ritter, Johann Wilhelm, 111, 186
Robinet, Jean-Baptiste, 100, 102, 258, 374n83
Romantic Encyclopedia (Novalis), 215, 224, 238, 240, 247
Romanticism: and aestheticization of metaphysics, 23; Classicism differentiated, 12; and Idealism, 31, 172; and metaphysics, 14, 42; and organology, 10
Romantic organology: Aristotelian terminological problems, 17–22; Blumenberg's metaphorical organ, 39–44; Foucault's transcendental-empirical organ, 44–46; Herder on, 97–107; and historiography, 38–46; Hölderlin on, 131–70; Jeffersonian strain of, 29, 262; Kant on, 73–97, 110–16; Leibniz's metaphysical organ, 52–71; and metaphysics, 8, 126; Novalis on, 206–50; Schelling on, 171–205; terminology and metaphysics, 1–46; transplanting "organ," 22–38
Ronell, Avital, 423n76
Röschlaub, Andreas, 208
Rothschuh, Karl, 185

Scaliger, Julius, 60
Schelling, Friedrich, 171–205; on cognition, 88; on comparative physiology, 180–92; and Criticism, 266; Fichte's critique of, 235, 392n85; on freedom and necessity, 350; Hegel's critique of, 262–75; on hidden organ, 195, 226; on identity, 408n77; instrumentalization of electricity, 182; on intellectual intuition, 192–205; on intuitive understanding, 175–80, 392n79; on judgment, 84, 388n11, 393n86; on knowledge, 32; and metaphysics, 8, 25, 26, 29; on mythological organs, 252–62; on Novalis, 403n132; "organ" terminology used by, 111, 124; *Poesie* distinguished from art, 203; on romantic metaphysics of morals, 272; on specific organ, 392n74; on substrate, 389n17. *See also Naturphilosophie*
Schematism, 90, 94, 328, 329
Schiller, Friedrich, 11–14, 136, 165–68, 358n39, 382n16
Schlegel, August Wilhelm von, 394n16, 406n42
Schlegel, Caroline, 210, 211
Schlegel, Friedrich: on aesthetic organology, 212, 252, 259, 260, 262; and Classical philosophy, 31; on cognition, 88; on judgment, 84; on metaphysics, 10, 11, 13, 16, 17, 212; on moral organs, 245; "new mythology" of, 212, 216; on productive imagination, 395n17; on task of organology, 239; on tool production, 226; on transcendental imagination, 29
Schleiermacher, Friedrich, 27, 31, 212, 213, 216, 217, 219, 220, 222, 228, 245, 395n26
Schmidt, Alfred, 418n6
Schmidt, Jochen, 387n66
Scholasticism, 401n103
Schubert, Gotthilf Heinrich, 29, 222, 225, 252, 253, 256, 257, 274
Schwann, Theodor, 35, 36, 362nn75–76
Science of Knowledge (Fichte), 138, 142, 143, 218, 245
Secretion, 182
Sedes animae, 119
Seelenorgan (Platner), 170
Seibicke, Wilfried, 417n115
Self, 193, 194, 201, 273, 274
Self-alienation, 400n88
Self-consciousness, 84, 86, 87, 126, 138, 140, 198–200, 202, 203, 265, 372n51, 392n79
Self-constitution, 41, 209
Self-genesis of reason, 105
Self-identification, 138, 161, 162

Index 461

Self-intuited-as-itself vs. self-intuiting-itself, 199
Self-justifying knowledge, 143
Self-knowledge: Aristotle on, 312; divisions of, 198; Goethe on, 312; Herder on, 105; Hölderlin on, 154, 373n50; relationship to knowledge, 231; Schelling on, 408n77
Self-materializing observation, 267
Self-objectification, 164, 200, 201
Self-realization, 201, 268, 313
Self-recognition, 331
Self-reflexivity, 299
Self-representation, 226, 299
Sensation, 155, 186, 218
Senses, 18, 23, 28, 98, 187, 234, 236, 240, 287
Sensibility, 64, 181–84, 190–92
Sensorium, 118, 119, 261
Sensory nerve-pairs, 119
Sensualism, 241
Seyhan, Azade, 45
Shipwreck with Spectator (Blumenberg), 40
Simmel, Georg, 349
Simondon, Gilbert, 8–10, 349
"*Simple Imitation of Nature*" (Goethe), 295
Smith, Adam, 320, 346
Smith, Justin E. H., 359n50
Sociability, 218
Soemmerring, Samuel Thomas, 24, 111, 113, 116–21, 128, 132, 136, 170, 225, 382n4
"*Some Thoughts about the Doctrine of the Skull*" (Schelling), 264
Sophocles, 161
Soret, Frédéric, 279, 280, 307
Soul, 56, 101, 120, 261
Spectatorship, 132, 134, 162, 169
Speeches on Religion for the Educated Amongst Its Despisers (Schleiermacher), 216
Speech on Mythology (Schlegel), 259
Spinoza, Baruch, 11, 31, 48, 103, 133, 137, 139, 217, 260, 271, 384n26, 395n26, 414n73
Spirit/matter axis, 282

Spontaneity, 302
Stadler, Ulrich, 241, 398n75
Stiegler, Bernard, 8–10
Stoicism, 60
Subjectivity, 138
Subject-object mimesis, 25, 305
Subrepted concepts, 120
Substrate, 389n17
Subsumption, 140, 141, 159, 178
Subtle material, 247, 248
Süßmilch, Johann Peter, 104, 105
Swammerdam, Jan, 56
Syllogism, 96
Sympathy, 156, 157
Symphilosophy, 226
Syncretism, 227
Synthetic, 92, 157, 193, 280
Synthetic organon, 286
System of Transcendental Idealism (Schelling), 26, 180, 192

Tabula rasa, 60, 207
Tantillo, Astrida, 281
Taxonomy, 295, 309
Taylor, Charles, 376n100
Techne-physis analogy. *See Physis-techne* analogy
Technicization, 342, 344, 352, 357n27
Technik, 210–12, 239, 240
Technological imagination, 43, 351, 352
Technological metaphysics, 124
Technological possibility, 308
Technological truth, 345
Technological will, 352
Technologized nature, 319
Technomorphism, 18, 43
Teleology, 203, 308, 309
Teleomechanism, 36
"*That There Could be More than Five Senses for the Human*" (Lessing), 240
Theism, 6
Theodicy (Leibniz), 54
Theology, 47, 269
Theory of the Moral Sense (Hemsterhuis), 401n105

Thought: and being, 24; primacy of, 125; thought-determinations, 285, 286; thought-force, 294
Tonelli, Giorgio, 77, 78, 82
Tools: as artifacts of economic development, 322; of *bildung*, 312–15; character of, 342; of cognition, 244; and organs, 329, 333, 342
"*To the Physicist*" (Goethe), 290
Traditional and Critical Theory (Frankfurt school), 327, 329
Tragedy, 134, 156–58, 160–70
Tragic chorus, 12, 165
Transcendental Deduction, 74, 88, 89
"*Transcendental Dialectic*" (Kant), 95
Transcendental-empirical organ, 44–46, 256
Transcendentalism: cognitive capacity, 209; consciousness, 41; idealism, 82, 153, 174, 226, 234, 235; logic, 76, 81–83; morphology, 36, 109, 280; philosophy, 41, 198, 201; realism, 232, 234, 240, 250; technology, 128, 170; unity of apperception, 89, 90
Transference, 41
Transmigration of souls, 56
Treatise on Man (Descartes), 1
Tresh, John, 357n27
Trop, Gabriel, 387n69
Truth, 61, 82, 95, 105, 114, 306, 399n79
Tsouyopoulos, Nelly, 208, 209, 393n2
Turing test, 3
Type-theory, 292

Unanticipated Thoughts Relating to the Use and Embetterment of the German Lanuage (Leibniz), 53
Unconscious activity, 199
Unconscious perceptions, 60
Underdetermination, 267
Understanding, 96, 152. *See also* Canon of the understanding; Cognition
Unity-in-duality, 147
Universals, 13–14, 181, 206–50
Ure, Andrew, 341, 421n55
Ur-theil doctrine, 149

Use-value, 322, 336

Vaihinger, Hans, 6–8
Valéry, Paul, 42
Varela, Francisco, 349
Vascularized composite allografts, 8
Vehikel, 229
Vehikulum, 102
Vermögen (Kant), 27, 106, 120, 213, 231
Verrichtung, 263
Views from the Night-Side of Natural Science (Schubert), 222, 257
Virchow, Rudolf, 35, 36
Vitalist Metaphysics, 24, 422n72
Vita propria, 184
Volta, Alessandro, 185, 186, 191
Von Baer, Karl Ernst, 35, 36

Waibel, Violetta, 402n125
Walsh, John, 185
Wechselwirkung, 384n31
Wellbery, David, 59, 409n6, 410n7
Welsh, Caroline, 397n46
Wessell, Leonard, 358n39
"*What Is Called Enlightenment?*" (Kant), 114
Wilhelm Meister's Years of Apprenticeship (Goethe), 13, 211, 243, 312–14, 418n120
Will, 206, 272–74, 352
Winckelmann, Johann Joachim, 291
Wolf, Jörg Henning, 15, 359n49
Wolff, Caspar Friedrich, 64, 65, 69, 70, 99, 181, 368n48
Wolff, Christian, 51, 58
"*The Work of Art in the Age of Its Technological Reproducibility*" (Benjamin), 328

Yearning, 242
Young, Christopher, 412n39

Zelle, Carsten, 32, 376n4
Žižek, Slavoj, 349, 410n14
Zu den Sachen und Zurück (Blumenberg), 40
Zweckmäßigkeit, 178

forms of living

Stefanos Geroulanos and Todd Meyers, *series editors*

Georges Canguilhem, *Knowledge of Life*. Translated by Stefanos Geroulanos and Daniela Ginsburg. Introduction by Paola Marrati and Todd Meyers.

Henri Atlan, *Selected Writings: On Self-Organization, Philosophy, Bioethics, and Judaism*. Edited and with an Introduction by Stefanos Geroulanos and Todd Meyers.

Catherine Malabou, *The New Wounded: From Neurosis to Brain Damage*. Translated by Steven Miller.

François Delaporte, *Chagas Disease: History of a Continent's Scourge*. Translated by Arthur Goldhammer. Foreword by Todd Meyers.

Jonathan Strauss, *Human Remains: Medicine, Death, and Desire in Nineteenth-Century Paris*.

Georges Canguilhem, *Writings on Medicine*. Translated and with an Introduction by Stefanos Geroulanos and Todd Meyers.

François Delaporte, *Figures of Medicine: Blood, Face Transplants, Parasites*. Translated by Nils F. Schott. Foreword by Christopher Lawrence.

Juan Manuel Garrido, *On Time, Being, and Hunger: Challenging the Traditional Way of Thinking Life*.

Pamela Reynolds, *War in Worcester: Youth and the Apartheid State*.

Vanessa Lemm and Miguel Vatter, eds., *The Government of Life: Foucault, Biopolitics, and Neoliberalism*.

Henning Schmidgen, *The Helmholtz Curves: Tracing Lost Time*. Translated by Nils F. Schott.

Henning Schmidgen, *Bruno Latour in Pieces: An Intellectual Biography*. Translated by Gloria Custance.

Veena Das, *Affliction: Health, Disease, Poverty*.

Kathleen Frederickson, *The Ploy of Instinct: Victorian Sciences of Nature and Sexuality in Liberal Governance*.

Roma Chatterji (ed.), *Wording the World: Veena Das and Scenes of Inheritance*.

Jean-Luc Nancy and Aurélien Barrau, *What's These Worlds Coming To?* Translated by Travis Holloway and Flor Méchain. Foreword by David Pettigrew.

Anthony Stavrianakis, Gaymon Bennett, and Lyle Fearnley, eds., *Science, Reason, Modernity: Readings for an Anthropology of the Contemporary.*

Richard Baxstrom and Todd Meyers, *Realizing the Witch: Science, Cinema, and the Mastery of the Invisible.*

Hervé Guibert, *Cytomegalovirus: A Hospitalization Diary.* Introduction by David Caron, Afterword by Todd Meyers, Translated by Clara Orban.

Leif Weatherby, *Transplanting the Metaphysical Organ: German Romanticism between Leibniz and Marx.*